Springer Series on
ATOMIC, OPTICAL, AND PLASMA PHYSICS 67

Springer Series on

ATOMIC, OPTICAL, AND PLASMA PHYSICS

The Springer Series on Atomic, Optical, and Plasma Physics covers in a comprehensive manner theory and experiment in the entire field of atoms and molecules and their interaction with electromagnetic radiation. Books in the series provide a rich source of new ideas and techniques with wide applications in fields such as chemistry, materials science, astrophysics, surface science, plasma technology, advanced optics, aeronomy, and engineering. Laser physics is a particular connecting theme that has provided much of the continuing impetus for new developments in the field. The purpose of the series is to cover the gap between standard undergraduate textbooks and the research literature with emphasis on the fundamental ideas, methods, techniques, and results in the field.

Please view available titles in *Springer Series on Atomic, Optical, and Plasma Physics* on series homepage http://www.springer.com/series/411

N. Chandra
R. Ghosh

Quantum Entanglement in Electron Optics

Generation, Characterization, and Applications

With 23 Figures

Springer

N. Chandra
Indian Institute of Technology
Department of Physics and Meteorology
Kharagpur 721302, West Bengal
India
ncphy@phy.iitkgp.ernet.in

R. Ghosh
c/o B.N. Panda
1/A–46(H.I.G.), Lingraj Vihar
Pokhariput 751020, Orissa
India
rama_g69768@yahoo.co.in

Password: 978-3-642-24069-0

Springer Series on Atomic, Optical, and Plasma Physics ISSN 1615-5653
ISBN 978-3-642-24069-0 ISBN 978-3-642-24070-6 (eBook)
DOI 10.1007/978-3-642-24070-6
Springer Heidelberg New York Dordrecht London

Library of Congress Control Number: 2012944986

Printed on acid-free paper

Springer is part of Springer Science+Business Media (www.springer.com)

"An expert is a person
who has made every possible mistake
at least once."

Albert Einstein
14 March 1879 - 18 April 1955

This book is dedicated to all those patient people who taught us how to read, write, think, and do scientific research.

Foreword

Quantum entanglement is one of the most intriguing phenomena in nature. In the words of Albert Einstein, it is a "spooky long distance interaction". In essence, it is the effect of the instantaneous interaction between two quantum objects when their properties are described by quantum variables. The possible eigenvalues of these variables are for a single object completely random. Their actual values become determined by a corresponding measurement of the particular variable only. For the case of an entangled pair of two quantum objects, however, the eigenvalue of a particular variable of the second object will be completely determined by the result of the measurement performed on the first object, although the values associated with the second object are completely random, if its properties are measured independently, which means not in coincidence with the measurement of the first one. This is the result of the joint wave function of an entangled pair of quantum objects possessing joint quantum variables. Moreover, even the selection of the particular quantum variable by the measurement on the first object determines the variable and their eigenvalues of the second object. This is even in contradiction with the laws of quantum mechanics themselves, because the selected variable of the "spooky long distance interaction" does not exist in quantum mechanics - or more precisely - according to Heisenberg's uncertainty principle. This is the reason that Albert Einstein argued in the famous Einstein, Podolsky and Rosen (EPR) paper that the variable must have an existence before its measurement which is determined by so-called hidden variables. This is the basis of the longstanding discussion between the exponents of "pure" quantum mechanics and the exponents of the so-called local realistic theories with hidden variables. The outcome of this dispute is known. All measurements probing Bell's inequality, a quantitative way of distinguishing between the two theories, have shown that quantum mechanics is right. One of the major problems in probing the validity of Bell's inequality was the variable to be chosen. EPR suggested in their Gedankenexperiment that the position in ordinary space be taken. This is, however, even nowadays still not feasible. Bohm suggested the spin of the electron, but due to the fact that the spin of the electron cannot be probed by a Stern-Gerlach experiment, this suggestion has also never been realized. Instead, the polarization properties of the photon, which are equivalent to the electron spin properties and hence are also called photon spin, gave rise to a breakthrough in the Bell-type experiments. At the beginning of the 1980s, Alain Aspect was the first to show the validity of Bell's inequality. Later on, Hans Kleinpoppen, Anton Zeilinger and many others followed. Nowadays, the spooky long distance interaction has been proven for distances of hundreds of kilometres, and experiments extending this interaction distance to thousands of kilometres are underway.

All these epoch-making experiments were done with photons entangled with respect to their pure polarization properties. However, there are other variables in nature, which may be entangled, e.g. the spin of a photoelectron and the polarization of a subsequent fluorescence photon. Such experiments have never been performed because they are extremely difficult. Not even the necessary theoretical framework has been developed.

The establishment of this theoretical framework is the subject of the book by Naresh Chandra and Rama Gosh. They derive concrete formulae for the interpretation of such experiments in a very general fashion. In addition, they provide formulae for specific examples, which may help the experimentalists to analyse their results. This is a pioneering work in the field of quantum entanglement of properties other than polarization variables and will become the standard work for all future theoretical and experimental work in this field.

Berlin-Dahlem Uwe Becker
Fritz-Haber-Institut der Max-Planck-Gesellschaft

Preface

Nobel Laureate astrophysicist, the late Professor Subramanyam Chandrasekhar, was once asked by the journalists what he thought was the greatest triumph of the human race in the twentieth century—building of a super-computer or putting a man on moon? With the given options, Professor Chandrasekhar's answer was obvious because it would not had, probably, been possible for man to put his feet on moon without first building super-computers. But, a super-computer is just one of the several gifts [e.g., televisions, personal computers, mobile phones, *i* phones, internet, world wide web (www), etc.] of the information technology to the present human civilization. These user-friendly important products of the information technology are the visible faces of the classical information science and use many devices (e.g., lasers, transistors, electronic chips, magnetic discs, integrated circuits, etc.) based on the principles discovered using quantum theory. Birth and subsequent development of quantum physics is, probably, the most outstanding achievement of the collective efforts of the human race in the last century. Without quantum physics, neither the super-computers could had, probably, been built nor man could had gone to the moon.

But, quantum theory is just only one of the subjects in modern physics with its mother being, probably, the atomic physics. Efforts to theoretically explain the line spectrum of a hydrogen atom—following the Nobel Prize winning proposal of a nuclear model of atom by Rutherford[1]—and the desire to understand the dynamics, behaviour, properties, etc., of electrons inside atoms/molecules heavier than hydrogen, contributed to the initial birth and the significant subsequent growth of quantum physics.

Since the 1970s of the last century, a new branch in information science is being developed. This is a counterpart of the classical information and has come to be known as quantum information. An important ingredient of this new science of information is the presence of nonlocal correlation—known as entanglement—

[1]E. Rutherford, "The Scattering of α and β Particles by Matter and the Structure of the Atom," Philos. Mag. **21**, 669 (1911).

between two or more particles which may even be separated by large distances and do not necessarily have any interactions among themselves or with any other external system, but each of which possesses at least two simultaneously accessible degrees of freedom. Entanglement, like many others (e.g.,wave-particle duality of light and matter, uncertainty principle), is a purely quantum phenomenon which defies all classical intuitions. Both theoretical and experimental studies performed hitherto in quantum information science have used entanglement primarily based on photon and/or atom optics.

It, indeed, is very surprising that electron optics, which continues to play important roles in classical information, has so far been seldom considered in quantum information. It has already been suggested that an electron, on account of its one-half spin angular momentum which, according to space quantization in quantum mechanics, can have only two orientations in space, has all the necessary properties—and, in some respects, is, probably, better than any other two-level system (e.g., photon, atom, etc)—to form a basic unit in quantum information. Specifically, use of electron optics can be quite convenient and useful for storing information for a long time, as well as in quantum communication and quantum networking wherein entangled basic units of quantum information are needed at separately addressable distant locations.

Studies in electron optics are, usually, considered to be in the realms of atomic and molecular physics. But, this monograph is a not an appendage to a long list of excellent books, monographs, etc., already available on this well-known subject; neither, it is an addition to the rapidly expanding literature on quantum information. It, on the other hand, is a small and rudimentary effort to investigate whether studies in quantum information can more conveniently and fruitfully be performed using atomic and molecular physics, and/or is it possible to gain, using the tools developed in the former, new insight about some of the processes capable of taking place in an atom or a molecule following the absorption of a single photon? In other words, this monograph is, probably, the first attempt to combine—in a simple way—the studies in atomic and molecular physics with those in quantum information, or vice versa.

Chapter 1 mentions some of the processes wherein absorption of a single photon by an atom or a molecule can liberate one/two/three of its bound electrons, or causes emission of one/two electrons plus a photon. Although the absorbing target can be in any of its gaseous, liquid, or solid forms; however, herein it is taken to be either an atom or a rotating linear molecule in gas phase. A reader, unfamiliar with the field of quantum information science, is briefly introduced in Chap. 2 to the later's those basic concepts whose understanding is essential not only to follow the discussions given in the rest of this monograph, but also to take up, if interested, higher studies in this important new branch of science and, hence, be able to understand its ever growing literature published elsewhere (e.g., research journals, review articles, books, etc.). Chapter 3 formulates the necessary theoretical framework for calculating the transition amplitudes and density matrices required for studying—using the tools developed in quantum information science—properties of a state representing a system of all those particles which are emitted during any of the processes mentioned in the introductory Chap. 1 of this monograph.

The remaining eight chapters and five appendices in this book are divided in three parts. The Part I contains Chaps. 4–8. The first three (ie, 4th, 5th, and 6th) chapters in this part describe entanglement properties of the spin states of two of those particles participating in the first three [specified by (1.1)–(1.3)] of the eight processes mentioned in Chap. 1, taking place in an atomic target when the electrons in it experience only the Coulomb interaction. The remaining two (ie, 7th and 8th) chapters in the Part I, describe non-local correlation between particles generated in the presence of Coulomb plus spin-orbit interactions inside an atom during the processes (1.1)–(1.4).

The quantum informatic properties of electrons liberated from a rotating linear molecule, following the absorption of a single photon, are investigated, on the other hand, in Chaps. 9 and 10—the Part II of this book. These two respective chapters contain discussions of properties of electrons emitted during the processes (1.2) and (1.3) taking place in a rotating linear molecule, without and with spin-dependent interactions.

The Chap. 11, and the five appendices which follow it, constitute the Part III of this book. This last chapter—in addition to containing the conclusions of Chaps. 4–10, as well as comparing the properties of Coulombic entanglement of a spin state of two electrons generated in the process (1.2) or (1.3) taking place in an atom or a rotating linear molecule—explores the prospects of the present and future multidisciplinary studies which combine atomic and molecular physics with quantum information science. Chapter 11 suggests also some of the processes whose combination with those investigated herein can help generate entanglement between more than two flying particles—at least one of whom is an electron. The five Appendices A–E contain materials, including computer programmes, relevant to the present and suggested future studies.

The investigations discussed in the present monograph were mainly carried out during the two research projects supported by the Council of Scientific & Industrial Research (CSIR), New Delhi, India, and the financial help provided by the CSIR to Dr. Minakshi Chakraborty, Shri Susil Parida, and to one (RG) of us in different forms (e.g., Junior Research Fellowship, Senior Research Fellowship, Research Associateship) at different times during their work at the Indian Institute of Technology, Kharagpur, India. Susil has helped us also with his knowledge of Linux and LaTex during the preparation of the manuscript for, as well as in checking the proofs of, this monograph.

We are specifically indebted to Professor Dr. U. Becker at the Fritz-Haber-Institut der Max-Planck-Gesellschaft, Berlin, Germany, for his interest in this work and for suggesting as well as encouraging us to publish it in the "Springer Series on Atomic, Optical, and Plasma Physics."

Help and support provided by Dr. habil. C.E. Ascheron, Ms. E. Sauer, and Ms. A. Duhm, Springer Science & Business Media, Heidelberg, Germany, have been very conducive in bringing this effort to a successful end.

Kharagpur

Naresh Chandra
Rama Ghosh

Contents

Chapter 1
Introduction and Preliminaries

The year 2005, under the auspices of the United Nations Organization (UNO), was celebrated as the International (or World) Year of Physics. It was as much in the memory of that *Annus Mirabilis* (Miracle Year) of 1905 in which Albert Einstein published his five famous papers [1–5] more than a century ago, as to honor their author himself by the whole world together. We now know the profound effects those five papers [1–5] have had not only on the development of science, in general, and physics, in particular; but also on the evolution, through these developments, of the human civilization we are enjoying today. One [1] of the five papers which Einstein wrote in the *Annus Mirabilis*, provided an explanation of the experimentally observed photoelectric effect [6, 7]. The subsequent confirmation in an experiment on the scattering of X-rays, performed by Compton [8], of the assumptions made by Einstein in his paper [1] on photoelectric effect [6, 7], following Planck's [9] successful theoretical explanation of black-body radiation by assuming that the energy in an electromagnetic wave is emitted in the form of discrete packets (now called photons), gave rise to the concept of wave-particle duality of light (see, e.g., [10]). This notion led to another equally profound concept of wave-particle duality of matter (see, e.g., [10]). These and other such theoretical developments, sometimes followed, and at other times preceded, by experiments in the very early part of the twentieth century, led to the formulation of the now well-known quantum theory.

Einstein, however, did not agree with the interpretation of the quantum theory applied to a composite system consisting of two or more separated subsystems. In 1935, that is 30 years after the *Annus Mirabilis*, Einstein, with his colleagues B. Podolsky and N. Rosen, published a paper [11] titled "Can quantum-mechanical description of physical reality be considered complete?" Publication of this paper [11] started purely philosophical discussions related to the completeness, or otherwise, of the quantum theory. These philosophical discussions continued for more than 30 years. Then a breakthrough came when Bell [12] developed an inequality to quantitatively answer the questions raised by Einstein–Podolsky–Rosen (EPR) [11] about the completeness of quantum theory.

N. Chandra and R. Ghosh, *Quantum Entanglement in Electron Optics*, Springer Series on Atomic, Optical, and Plasma Physics 67, DOI 10.1007/978-3-642-24070-6_1,

A second breakthrough came with the suggestions that the same property of the states of a composite system, on the basis of which EPR [11] questioned the completeness of quantum theory, can be technologically used to encrypt [13] and teleport [14] information, as well as to build computers [15, 16] which can perform certain tasks much faster than the presently available 'classical' computers.

A model that employs laws of quantum physics for the transmission (communication), processing (computation), and storing of the information has come to be known as quantum information [17]. Similar to the classical bit in classical information, the basic unit of quantum information is the quantum bit or qubit[1] [18]. Availability of two or more qubits with entanglement (i.e., nonlocal correlation) is an essential ingredient for any quantum information related studies. These qubits can be of any kind of particles possessing two independent and simultaneously accessible states. Some of the well-known examples of such qubits are [17]: an electron or any other spin-$\frac{1}{2}$ particle; a two-level atom/ion; a photon with negative (left) and positive (right) helicities (circular polarizations), or with horizontal and vertical linear polarizations; states of two photons entangled with respect to their phase and momentum [19], or energy and momentum [20], *etc.*

Hitherto, most of the work done in the field of quantum information has mainly been based on the use of entangled states of two or more photonic qubits generated in various ways (e.g., parametric down conversion, cascade transitions in atoms from a suitable excited state to the ground state, simultaneous emission from a metastable atomic state in a true second-order decay process, *etc.*). There are several excellent books and/or original articles describing processes presently being used for generating two or more entangled photons: See, for example, ([19–27], *etc.*) for parametric down conversion; ([28, 29], *etc.*) for cascade emission; ([30], *etc.*) for simultaneous emission in second-order decay, *etc.*

Photons are, probably, the best possible carriers of information—nothing can travel faster than an electromagnetic wave—and easily maneuverable with familiar optical elements (e.g., lenses, mirrors, gratings, *etc.*). However, photons are not suitable for storing information for a long period of time. Moreover, any detection of a photon always leads to its destruction, unless one is prepared to perform quantum non-demolition type extremely difficult experiments (see, e.g., [27, 31, 32], *etc.*).

However, both an atom and a positive/negative ion are the particles which are not only stable (unless, the later happens to capture/lose an electron), but also capable of storing information for as long a time as one wants. On account of its electric neutrality, an atom, unlike an ion, cannot be directly detected. Detection of an atom requires use of a fluorescence or ionization type process which leads to a change in its original state. Although, motion of an ion is easily maneuverable compared to that of an atom; however, it is difficult to make either of them to travel fast on account of their large masses.

An electron, on the other hand, is never destroyed unless it happens to interact with a positron; moreover, until it is captured by an atom or a positive ion, an

[1]See, pages 15–22 in Chap. 2 for a detailed description of a qubit.

electron can never lose its identity. Both, bound and free electrons are available in abundance in nature. In addition to being readily observable with the help of an electron spectrometer without affecting any of its properties (e.g., momentum, energy, spin polarization, *etc.*), electronic motions can be easily maneuvered with a suitable combination of electric and magnetic fields. Being the lightest particle next to photon, electrons are capable of travelling with speeds comparable to that of light. Electrons have since long been used to carry information in the form of electric signals and/or for storing it for a long time. Furthermore, it has already been proposed [33] that electron spin can be used in quantum computation [17, 34]. Consequently, creation of a solid-state quantum computer, based on spin qubits, has already been suggested [35, 36] in spintronics [37].[2]

For using any particles (e.g., photons, atoms, electrons, *etc.*) in quantum communication and quantum networking, one requires [38], on the other hand, two or more entangled, but mobile qubits, which are separately addressable because of their being at different locations. In order to have on-chip quantum communication, a number of theoretical proposals [38–42] for electron entanglers have hitherto been made in spintronics. For example, extraction of a Cooper pair from a superconductor [39, 40], or of the singlet ground state of a quantum dot with an even number of electrons [42], are some of the methods which have already been investigated in spintronics [37].

The methods suggested and/or used hitherto for generating quantum entanglement between two or more qubits can be divided in two broad classes:

(*a*) Using internal processes: Here, entanglement is produced due to processes taking place inside an atom or a molecule and, hence, does not depend, in any way whatsoever, on the conditions and/or circumstances existing out side these systems, e.g., entanglement among photons produced in parametric down conversion ([19–27], *etc.*), cascade decay ([28, 29], *etc.*), second-order decay ([30], *etc.*), *etc.* Consequently, now the entanglement is not coupled to the environment external to the system and, hence, is least prone to the harmful effects [17] of decoherence[3] and dissipation.[4] Such flying entangled particles, receding away from their common source, are most suitable for distant quantum communications [17, 34, 43] (e.g., teleportation, cryptography, *etc.*), quantum networking (e.g., "$1 \longrightarrow N$" quantum telecloning [44], i.e., quantum teleportation [14] with one sender and N receivers), distributed quantum computation [17, 45], *etc.*

(*b*) Using external conditions: This is a kind of "brute force" method of entangling totally independent particles kept at a single place, or at more than one places. One, then, creates external conditions (e.g., applied static and/or dynamic fields, number of photons/phonons in a cavity, *etc.*) required for entangling

[2]It is active manipulation of the spin degree of freedom of electrons in a solid-state environment.

[3]For its brief description, see Appendix C on pages 269–271.

[4]For its brief description, see Appendix C on pages 269–271.

these particles. The well-known examples of creating non-separable states of particles using brute force methods are the entanglement generated among atoms inside a quantum cavity, ions in a trap, *etc*. For entangling atomic qubits placed in separate cavities at different locations, use of bipartite continuous variable states [46, 47] and interference of polarized photons leaking out of optical cavities containing individual atoms [48] are among the several methods (e.g., [49–52]) which have hitherto been suggested. Entanglement generated using brute force methods is, naturally, very much dependent upon the external environment and, hence, very likely to be affected with a change in the external conditions. It is, therefore, very fragile and prone to decoherence and dissipation. Particles entangled using such brute force methods are, probably, least suitable for distant quantum communication. Moreover, recent both theoretical [53] and experimental [54,55] studies have shown that entanglement between atomic qubits trapped in separate cavities is lost in finite time under the influence of pure vacuum noise. Thus, the entanglement generated in [46–52], or using other similar methods, is very short-lived.

There are several photon-induced processes in atomic and molecular physics which are capable of producing one or more qubits in the form of photons, electrons, and ions. These processes may take place during the interaction of electromagnetic radiation with atoms and molecules in their gaseous, liquid, or solid phase. In this monograph, we have investigated and characterized the properties of quantum entanglement between two qubits produced in some of these processes.[5] These qubits can either be of the same or of different kinds of particles, but at least one of them is certainly an electron.

In the present study, whereas light has been treated classically, an atom or a molecule is described by quantum mechanical methods. Therefore, the essential theoretical techniques used in this semiclassical study are non-relativistic quantum mechanics ([10, 58–60], *etc*.), density matrix ([60–62], *etc*.) and Racah algebra ([63–68], *etc*.), in addition to those needed for theoretical characterization and quantification of entanglement.

Several things are known to happen when a single photon of appropriate energy is absorbed by an atom or a molecule. One of such possibilities is the emission of one or more particles. Well known and interesting among such processes are:

(1) Emission of a single particle

(a) One-step single photoionization (1-SPI) (γ_r, e_p)

$$\gamma_r + \mathfrak{T} \longrightarrow \mathfrak{T}^{1+} + e_p. \tag{1.1}$$

(2) Emission of two particles

[5] Quantum informatic properties of three electrons produced in the photon-induced processes [(1.6),(1.7)] have also been investigated [56, 57], but have not been discussed in this monograph.

(a) One-step double photoionization (1-DPI) $[\gamma_r, (e_p + e_{p'})]$

$$\gamma_r + \mathfrak{T} \longrightarrow \mathfrak{T}^{2+} + e_p + e_{p'}. \tag{1.2}$$

(b) Two-step double photoionization (2-DPI) $[\gamma_r, (e_p, e_a)]$
Inner-shell photoionization

$$\gamma_r + \mathfrak{T} \longrightarrow \mathfrak{T}^{1+^*} + e_p, \tag{1.3a}$$

followed by spontaneous, single nonradiative decay

$$\mathfrak{T}^{1+^*} \longrightarrow \mathfrak{T}^{2+} + e_a. \tag{1.3b}$$

(c) Photoionization plus fluorescence $[\gamma_r, (e_p, \gamma_d)]$
Inner-shell photoionization

$$\gamma_r + \mathfrak{T} \longrightarrow \mathfrak{T}^{1+^*} + e_p, \tag{1.4a}$$

followed by spontaneous, single, radiative decay

$$\mathfrak{T}^{1+^*} \longrightarrow \mathfrak{T}^{1+} + \gamma_d. \tag{1.4b}$$

(3) Emission of three particles

(a) One-step triple photoionization (1-TPI) $[\gamma_r, (e_p + e_{p'} + e_{p''})]$

$$\gamma_r + \mathfrak{T} \longrightarrow \mathfrak{T}^{3+} + e_p + e_{p'} + e_{p''}. \tag{1.5}$$

(b) Two-step triple photoionization (2-TPI) $[\gamma_r, (e_p, e_a + e_{a'})]$
Inner-shell photoionization

$$\gamma_r + \mathfrak{T} \longrightarrow \mathfrak{T}^{1+^*} + e_p, \tag{1.6a}$$

followed by spontaneous, simultaneous, double nonradiative decay

$$\mathfrak{T}^{1+^*} \longrightarrow \mathfrak{T}^{3+} + e_a + e_{a'}. \tag{1.6b}$$

(c) Three-step triple photoionization (3-TPI) $[\gamma_r, (e_p, e_{a_1}, e_{a_2})]$

Inner-shell photoionization

$$\gamma_r + \mathfrak{T} \longrightarrow \mathfrak{T}^{1+^*} + e_p, \tag{1.7a}$$

followed by spontaneous, cascade, double nonradiative, decay

$$\mathfrak{T}^{1+^*} \longrightarrow \mathfrak{T}^{2+^*} + e_{a_1}, \tag{1.7b}$$

and

$$\mathfrak{T}^{2+^*} \longrightarrow \mathfrak{T}^{3+} + e_{a_2}. \tag{1.7c}$$

(d) Three-step double photoionization plus fluorescence [γ_r, (e_p, e_a, γ_d)]
Inner-shell photoionization

$$\gamma_r + \mathfrak{T} \longrightarrow \mathfrak{T}^{1+^*} + e_p, \tag{1.8a}$$

followed by spontaneous non-radiative decay

$$\mathfrak{T}^{1+^*} \longrightarrow \mathfrak{T}^{2+^*} + e_a, \tag{1.8b}$$

and spontaneous radiative decay

$$\mathfrak{T}^{2+^*} \longrightarrow \mathfrak{T}^{2+} + \gamma_d. \tag{1.8c}$$

We will see in this monograph, among other things, that the quantum informatics properties of the states of the same kind and of same number of particles very much depend upon the above-mentioned process due to which their generation has taken place, while other things (e.g., interactions, states of $\mathfrak{T}$, *etc.*) are identical.

Unless stated otherwise, in each of the processes (1.1)–(1.8) and in the rest of this monograph, γ_r always represents an incident photon of frequency $\nu_{\rm r}$—and, hence, of energy $E_{\rm r} = h\nu_{\rm r}$ (with the subscript "r" standing for the radiation)—absorbed by $\mathfrak{T}$ leading to the later's photoionization. Processes (1.1)–(1.8) can take place in any atom or a molecule. However, the target $\mathfrak{T}$, considered herein, is always an unpolarized atom or an unpolarized rotating linear molecule belonging to either of the $\mathrm{C}_{\infty\mathrm{v}}$ or $\mathrm{D}_{\infty\mathrm{h}}$ point groups [69, 70]. Whenever, one or more electrons bound to a target become free due to the energy directly supplied by an incident electromagnetic wave, the freed electrons are called photoelectrons [71, 72]. Thus, e_p is the only photoelectron in each of (1.1), (1.3), (1.4), and (1.6)–(1.8). But, there are two (e_p, $e_{p'}$) and three (e_p, $e_{p'}$, $e_{p''}$) photoelectrons in the processes (1.2) and (1.5), respectively. Photoelectron (s) may come out from any shell present in the electronic configuration of an atomic or a molecular target $\mathfrak{T}$. However, electrons

e_p, (e_p, $e_{p'}$), and (e_p, $e_{p'}$, $e_{p''}$) have left outer-most shells of their respective parents after the incident photon γ_r was absorbed by $\mathfrak{T}$ in (1.1), (1.2), and in (1.5). Consequently, each of $\mathfrak{T}^{1+}$, $\mathfrak{T}^{2+}$, and $\mathfrak{T}^{3+}$—formed in the respective one-step processes (1.1), (1.2), and (1.5)—is in its ground electronic state and is called a photoion (i.e., singly charged positive ion), photo-dication (i.e., doubly charged positive ion), and photo-trication (i.e., triply charged positive ion).

Although, positive ion $\mathfrak{T}^{1+}$ in the two-step process (1.4) and the dication $\mathfrak{T}^{2+}$ in the [two-step, three-step] processes [(1.3), (1.8)] too are in their ground states; however, neither of these is formed in the same step in which photon γ_r was absorbed, i.e., in the first step of each of the three processes. Further, $\mathfrak{T}^{1+^*}$ in each of (1.3), (1.4), and (1.6)–(1.8) is the excited photoion as it has a vacancy in one of its inner-shells due to the departure of the photoelectron e_p from the parent target $\mathfrak{T}$. It loses its extra energy non-radiatively [73, 74] in the ejection of either a single secondary electron [e_a, e_{a_1}, e_a] in [(1.3b), (1.7b), (1.8b)], or two secondary electrons (e_a, $e_{a'}$) in (1.6b); whereas, in (1.4b), a radiative decay of $\mathfrak{T}^{1+^*}$ takes place in the emission of the photon γ_d. Processes (1.7) and (1.8), on the other hand, show that the excited dication $\mathfrak{T}^{2+^*}$ with two vacancies may lose its extra energy due either to the departure of an electron e_{a_2} in (1.7c), or emission of a photon γ_d in (1.8c). Lastly, the trication $\mathfrak{T}^{3+}$ is in its ground state in each of the two- and three-step processes (1.6) and (1.7).

For an incident photon γ_r possessing energy E_r required for one or the other of the processes (1.1)–(1.8) to take place, it is usually sufficient [10, 60] to treat the ionization step in each of these eight processes in the electric dipole ($E1$) approximation [72]. Herein, a state of this photon is represented by $\left|\left|\ell_r\right| = 1, m_r\right\rangle$, where $\left|\ell_r\right| = 1$ is its angular momentum in the $E1$ approximation [10, 60, 72] and m_r specifies its polarization. For a linearly polarized (LP) γ_r, $m_r = 0$; whereas, $m_r = -1$ and $m_r = +1$ represent left circularly polarized (LCP) and right circularly polarized (RCP) photons, respectively. Unpolarized (UP) incident radiation, on the other hand, is taken to be an even mixture of the LCP and RCP light. In this book, unless stated otherwise, polarization of γ_r also specifies the polar axis of the space-fixed (SF) [also called laboratory-fixed (LF)] frame of reference $OXYZ$ shown in Fig. 1.1. The origin "O" of this SF/LF frame is always at the nucleus (center of mass) of the atomic (molecular) target $\mathfrak{T}$; its polar (i.e., OZ)-axis is along the direction of incidence for circularly polarized or unpolarized γ_r. If, on the other hand, the ionizing radiation in (1.1)–(1.8) happens to be LP, then the direction of its electric field vector defines the OZ-axis of the LF/SF frame. These and other properties of the coordinate system $OXYZ$ are shown in Fig. 1.1.

It has already been mentioned elsewhere in this chapter that each of the species ($\mathfrak{T}$, $\mathfrak{T}^{1+}$, $\mathfrak{T}^{2+}$, $\mathfrak{T}^{3+}$), participating in one or the other of the processes (1.1)–(1.8), is in its ground state. Unless stated otherwise, let us represent these ground states by ($|0\rangle$, $|1^+\rangle$, $|2^+\rangle$, $|3^+\rangle$) and their energies by (E_0, E_{1^+}, E_{2^+}, E_{3^+}). Further, $|1^{+^*}\rangle$ and $E_{1^{+^*}}$ are, respectively, the state and energy of the excited photoion $\mathfrak{T}^{1+^*}$ present in each of [(1.3), (1.4), (1.6)–(1.8)]. Similarly, the excited dication $\mathfrak{T}^{2+^*}$ in [(1.7), (1.8)] is in the state $|2^{+^*}\rangle$ with energy $E_{2^{+^*}}$.

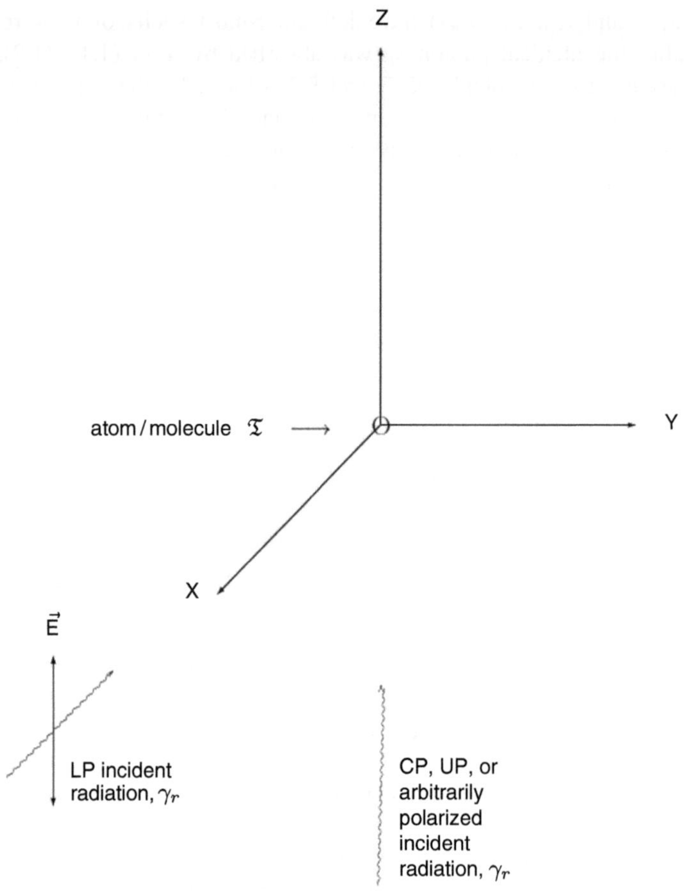

Fig. 1.1 Space-fixed (SF) [also called laboratory-fixed (LF)] frame of reference OXYZ. Its origin "O" is at the nucleus (center of mass) of the atomic (molecular) target $\mathfrak{T}$. As explained in the text, the direction of its polar OZ-axis, shown in this figure, is to be determined by the polarization of the incident radiation γ_r which ionizes $\mathfrak{T}$ in each of the processes (1.1)–(1.8)

The processes (1.1)–(1.8) provide opportunities to investigate entanglement among various particles which may or may not be of the same kind. Nonlocal correlation among particles of different varieties has come to be known as hybrid entanglement. For example, one always studies in (1.1) hybrid entanglement between the photoion $\mathfrak{T}^{1+}$ and the photoelectron e_p. But, in the remaining seven processes entanglement either only among the outgoing particles [e.g., (e_p, $e_{p'}$) in (1.2), (e_p, e_a) in (1.3), (e_p, γ_d) in (1.4), *etc.*], or the emitted particles plus the residual ion [e.g., (e_p, $e_{p'}$, $\mathfrak{T}^{2+}$) in (1.2), (e_p, e_a, $\mathfrak{T}^{2+}$) in (1.3), (e_p, γ_d, $\mathfrak{T}^{1+}$) in (1.4), *etc.*] may be the goal of a particular study. Nevertheless, presence or absence of entanglement in all these cases completely depends upon processes taking place inside the target $\mathfrak{T}$ and/or its ions. Hence, the entanglement generated in the processes (1.1)–(1.8) is of type (a) discussed earlier on page 3 in this chapter.

The next chapter, *ie*, Chap. 2, briefly describes the basic concepts and some of the, hitherto known, most important applications of quantum information science. Therein, we first explain what is information and how do we human beings deal with it. This discussion helps us in explaining about those things which make information classical or quantum. One is then naturally led to introduce the entangled states which are the building blocks of quantum information. It is then followed by a brief description of the methods, currently in vogue, for characterizing and quantifying entanglement of the states of two particles. The discussion in Chap. 2 as well as in the rest of this monograph is limited to the bipartite entanglement of discrete variables.

A proper study of any of the processes (1.1)–(1.8) using the tools developed in quantum information science requires that one must necessarily have for it a density operator and the corresponding density matrix. In order to calculate this density matrix, it is imperative to know both the magnitudes and the phases of the amplitudes describing the various transitions taking place in the particular process one is interested in. Chapter 3 develops a theoretical framework for calculating, for any of the above-mentioned eight processes, the required density operators and the corresponding density matrices using the transition amplitudes obtained therein.

The two dominant forces [10, 60] experienced by an electron inside an atom or a molecule are Coulombic and spin-dependent. The Coulomb force arises due to interactions among the electric charges carried by their nuclei and electrons. Thus, mutual interactions among the charges of all the constituents of an atom or a molecule are the sources of Coulomb force in any target $\mathfrak{T}$. Motions of these charges inside an atom or a molecule generate magnetic fields. The interaction of the magnetic moment associated with the spin angular momentum of an electron with this internal magnetic field gives rise to a spin-dependent force—usually known as [10, 60] as spin-orbit interaction (SOI) in an atom. An exact study of any process taking place in a target $\mathfrak{T}$ requires that both the Coulomb plus spin-dependent interactions be taken into account. But for light atoms and molecules, with small or intermediate values of the atomic number Z, spin-dependent interaction is much weaker than the Coulomb forces experienced by an electron in $\mathfrak{T}$.[6] On the other hand, for a target $\mathfrak{T}$ with large value of Z, spin-dependent interactions too, in addition to the Coulomb interaction, needs to be taken into account. Chapter 3 explains the procedures for calculating transition amplitudes so that one may have a density matrix for a desired process taking place in an atomic or a molecular target $\mathfrak{T}$ without or with spin-dependent interactions.

In Part I, the target $\mathfrak{T}$ in the processes (1.1)–(1.8) always represents an atom. Therein, Chaps. 4–6 are concerned with the entanglement properties, generated in each of the processes (1.1)–(1.3), taking place in an atom in the absence of SOI. Thus, these three chapters describe entanglement which may arise due only to the

[6] See footnote (8) on page 58.

Coulomb forces experienced by electrons in an atom. The non-local correlation, discussed in Chaps. 4–6 has, therefore, been called as "Coulombic entanglement.[7]" Chapter 4 describes, probably, the simplest possible, yet completely realistic, example, of hybrid Coulombic entanglement between the spin angular momenta of the photoion $\mathfrak{T}^{1+}$ and photoelectron e_p generated in the process (1.1); whereas, the Chaps. [5, 6] discuss Coulombic nonlocal correlations between the spins of electrons forming pairs $[(e_p, e_{p'}), (e_p, e_a)]$ which are flying away in the processes [(1.2), (1.3)].

Entanglement generated between two flying qubits in the presence of Coulomb plus spin-orbit interactions (C+SOIs) in an atom is explained in the remaining two chapters of Part I. Here, Chap. 7 discusses entanglement between the spins of photoelectron–photoion $(e_p,\ \mathfrak{T}^{1+})$, two photoelectrons $(e_p,\ e_{p'})$, and those of a photoelectron–Auger electron pair $(e_p,\ e_a)$ generated in the processes (1.1), (1.2), and (1.3), respectively. The last Chap. 8 of Part I describes, on the other hand, entanglement between spin of photoelectron e_p and polarization of photon γ_d sequentially emitted in the process (1.4). It is shown in these last two chapters of Part I that non-separability of the bipartite state for each of the $(e_p,\ \mathfrak{T}^{1+})$, $(e_p,\ e_{p'})$, $(e_p,\ e_a)$, and $(e_p,\ \gamma_d)$ pairs very much depends upon the total angular momentum $\boldsymbol{J}\ (= \boldsymbol{L} + \boldsymbol{S})$ which determines a fine-structure level [10, 60] when both C+SOIs are taken into account in an atom. The non-local correlation studied in the presence of both Coulomb plus spin-orbit interactions in Chaps. 7 and 8 has, therefore, been called as "fine-structure entanglement."

Electronic entanglement generated during the processes (1.2) and (1.3) taking place when $\mathfrak{T}$ happens to be either [69, 70] of the $C_{\infty v}$ and $D_{\infty h}$ molecular target is the subject of the respective Chaps. 9 and 10 present in Part II of this monograph. In each of these two chapters, properties of both the Coulombic entanglement and of the fine-structure entanglement have been investigated. For a proper description of molecular symmetries in this work, the nuclear rotation has fully been taken into account by using parity adapted states [69, 75].

The last Chap. 11, constituting—along with the five Appendices A–E—Part III, contains not only the conclusions of the investigations reported in this monograph, but discusses also the directions in which such multidisciplinary studies combining atomic and molecular physics with quantum information can be useful to both of these important branches of pure as well as applied sciences.

The first appendix, ie, Appendix A, describes, in brief, those properties of density operators and density matrices which are both relevant and useful to the investigations reported herein. The Appendix B, ie, the second appendix, shows as to how one can perform the studies reported in the Chaps. 3–10 using state multipoles (also called statistical tensors) [61, 68] when the ionizing radiation γ_r in processes (1.1)–(1.8) is either of an arbitrary (described by four Stokes parameters [61, 76]) or of a definite (represented by $|1,\ m_r\rangle$) polarization. The third appendix, Appendix C,

[7]It is shown in Chap. 8 that there cannot exist any Coulombic entanglement between spin of photoelectron e_p and polarization of photon γ_d emitted sequentially in the process (1.4).

very briefly and qualitatively explains meanings of decoherence and dissipation, in general, and in the context of quantum information, in particular, including their harmful effects on entanglement. The last two appendices, i.e., Appendices D and E, contain brief write-ups of the computer programmes written in the FORTRAN 90 language for calculating the results described in the present monograph. It is obvious from its various chapters that the calculations of the required density matrices very heavily depend upon an accurate knowledge of the numerical values of 3-j, 6-j, and 9-j symbols [63–68]. Appendix D explains computer programmes for calculating these three different kinds of n-j symbols in terms of the powers of prime numbers so that their numerical values are exact. This appendix contains also a test run for this programme as well as its output. The last appendix, i.e., Appendix E, is about the computer code used in Chap. 7 for the calculation of the density matrix needed for studying the fine-structure entanglement between the spins of photoelectron e_p and Auger electron e_a, sequentially generated in the 2-DPI process (1.3). It is a very generic programme which could readily be adapted to other studies of fine-structure entanglement performed in Chaps. 8 and 10. The complete listings of these two FORTRAN 90 codes can be downloaded from http://extra.springer.com [with the pass phrase (i.e., 978-3-642-24069-0) supplied by the Springer]. The downloaded material will contain, among other things, also test runs and their outputs for the programmes described in the Appendices D and E.

Although, the results presented herein are primarily for the quantum entanglement properties of the bipartite states of electronic qubits, electronic plus photonic qubits, or an electronic qubit and an ionic qudit, formed in one or the other of the first four of the eight photon-induced processes mentioned on pages 4–6 in the present introductory Chap. 1; the theoretical frame work described in the following two chapters, namely Chaps. 2 and 3, is, however, very general which can readily be applied, and/or adapted, to the study of such properties of three or more particles generated in the remaining (1.5)–(1.8), or other similar, processes. For example, Coulombic spin–entanglement among the three outgoing electrons in the processes [(1.6), (1.7)] have already been investigated in detail in References [56, 57] by properly adapting the framework developed in the next two chapters of this monograph.

Chapter 2
Quantum Information: Basic Relevant Concepts and Applications

2.1 Information

The evolution, progress, and survival of human civilization depend equally on creation of knowledge and on the dissemination of the created knowledge to the contemporary and future generations.

2.1.1 A Physical Entity

According to the Oxford dictionary [77], the meaning of information is "facts or knowledge provided or learned." How is knowledge learned? It can be done in several ways. For example: Through thoughts in a human mind, events (e.g., meeting of two persons, accidents, *etc.*), experiments performed in a laboratory, *etc.* All these, and other possible, means of learning knowledge require physical—live or neutral—objects, e.g., human beings, equipments, *etc*. The generated knowledge can be stored (i.e., imprinted) on a physical support (e.g., human mind, leaf of a tree, rocky wall of a cave, sheet of a paper, a magneto-optical disc, *etc.*). How is knowledge transmitted? By mechanical waves, electromagnetic waves, electric signals, writing articles and/or books, *etc*. Processing (i.e., analysis) of information also requires physical objects.

Thus generation, storing, dissemination, transmission, processing, *etc*., of information require physical objects in the form of particles (i.e., matter) and waves. Both of these are the entities whose properties are studied using laws of physics. Moreover, information cannot be transmitted faster than the speed of light in vacuum. Thus, information has a physical nature. Indeed, information itself is a physical entity. It was first realized long ago by Landauer [78–80]. This, in other words, means that every aspect of information, beginning from its generation to processing, is governed by the laws of physics.

N. Chandra and R. Ghosh, *Quantum Entanglement in Electron Optics*, Springer Series on Atomic, Optical, and Plasma Physics 67, DOI 10.1007/978-3-642-24070-6_2,

2.1.2 Classical and Quantum Information

The laws of physics so far known to the human societies can be divided into two broad classes namely, the classical laws and the quantal laws. Although, quantum physics, between these two theories of Nature hitherto known to the mankind, provides a more precise description of the atomic and subatomic world; results obtained using it are, however, in complete agreement with those given by the classical physics wherever this later theory becomes applicable. The theoretical predictions based on quantum physics have experimentally been verified to as high degree of precision as more than ten orders of magnitude.

The mode of information which uses classical laws from its generation to processing is known as classical information. Similarly, the mode of information based on quantum mechanical laws for its generation, coding , processing, *etc*., is known as quantum information.

2.1.3 Discritization of Information

How do we human beings deal with information? We learn and/or provide knowledge and facts by observations, demonstrations, as well as by using words and/or digits to convey it to others. In most of the languages used in this world, each word is a string of certain alphabets. The total number of alphabets, their pronunciations, ways of writing may differ from one language to another, but it is fixed in each language. Total number of digits, on the other hand, is same (i.e., from 0 to 9) in all languages, although their pronunciations and style of writing may yet be different again in different languages.

Each word in the English language, for example, is written in terms of its twenty six Roman alphabets. Similarly, in order to write a word in the Sanskrit language, one uses Devanagari alphabets whose ways of writing, pronunciation, as well as the total number is quite different from those of Roman alphabets. Numbers in all languages, on the other hand, are expressed in terms of ten digits; but, their ways of writing and pronouncing differ from one language to another.

These alphabets or digits can be called as bits. Each bit of any language can be read or replicated as many times as one wants without any difficulties. Any text or numeral, in any language, can be expressed in terms of its bits and, hence, can be called an irreducible package of bits. We, human beings, have thus discretized information in terms of irreducible packages consisting of bits. The same principle of discritization of information in the form of irreducible packages of a finite number of bits is applied to both classical and quantum information.

2.1.4 *Bit*

2.1.4.1 Classical Bit (cbit)

The elementary unit of classical information is classical bit (or cbit). It is a classical system with only two states "0" and "1" implying false and true, or no and yes, *etc.* Any text can be coded into a string of cbits. Some examples of a cbit are an electrical switch ($0 \equiv$ off/on, $1 \equiv$ on/off), a coin ($0 \equiv$ head/tail, $1 \equiv$ tail/head), *etc.* Each cbit can be stored physically. In a classical computer, each cbit is registered as a charged state of a capacitor.

The classical information too depends for its storage, transmission, and processing on equipments designed using quantum mechanics. Nevertheless, it does not change the basic nature of a cbit. That is, it has two distinguishable and stable macroscopic states. These states can be read as many times as one wants without affecting, in any ways whatsoever, their values or any of their other properties. In addition, the states of a cbit can be cloned or replicated any number of times without any difficulties. Moreover, a string of two cbits can exist in any of the 2^2 states 00, 01, 10, 11 at a time. These are the binary system representations of the decimal system of integers 0 to 3 with the condition that each binary representation is to consist of no fewer than two cbits. Thus a string of n cbits can exist in any of the 2^n (from 000...00 to 111...11) states with each of them containing no less than n cbits.

2.1.4.2 Quantum Bit (Qubit)

In quantum information, the analogue of classical bit is the quantum bit. Schumacher [18] called it, in analogy to cbit, as qubit. The concept of a qubit in quantum mechanics can be explained in many ways. It is a two-state quantum system. A spin-$\frac{1}{2}$ particle, a photon polarization, an atomic system with two relevant states, *etc*., are some of the most familiar examples of qubits.

Let us say that the two possible states of a qubit are $|0\rangle$ and $|1\rangle$. The difference between a cbit and a qubit is that, unlike a cbit, a qubit can be in an infinitely many coherent linear superpositions of its states $|0\rangle$ and $|1\rangle$. That is

$$|\psi\rangle = C_0|0\rangle + C_1|1\rangle, \tag{2.1a}$$

where C_0 and C_1 are complex numbers with the condition

$$|C_0|^2 + |C_1|^2 = 1. \tag{2.1b}$$

A measurement of $|\psi\rangle$ will give the state $|0\rangle$ with probability $|C_0|^2$ or $|1\rangle$ with probability $|C_1|^2$, leaving $|\psi\rangle \longrightarrow |\psi_0\rangle \equiv |0\rangle$, or $|\psi\rangle \longrightarrow |\psi_1\rangle \equiv |1\rangle$ state, respectively. Thus, the process of measurement in quantum mechanics has disturbed the system, which never happens in classical mechanics. In quantum mechanics, measurements without disturbance are never possible. The original state $|\psi\rangle$ of (2.1) is said to have been reduced, or collapsed, by the measurement into the new state $|\psi_0\rangle$ or $|\psi_1\rangle$ after the measurement. These new states are the eigenstates of the system under consideration.

As each of C_0 and C_1 in (2.1a) can take an infinite number of values, subject to the condition (2.1b), a qubit can be in any of the infinitely many states (2.1); whereas, a cbit can take only one of the two values "0" or "1" at a time. Consequently, the information contained in a qubit is infinite in comparison to the information in a cbit. However, the result of measurement on a qubit is only $|0\rangle$ or $|1\rangle$, probabilistically.

The state (2.1) of a qubit can also be looked upon as a vector in two-dimensional (2-D) complex vector space. The $|0\rangle$ and $|1\rangle$ form an orthonormal basis for this 2-D vector space and are known as computational basis states. Let us parametrize the complex coefficients in (2.1) as $C_i = r_i \ \mathrm{e}^{\mathrm{i}\,\phi_i}$ with $i = 0$ and 1. Then

$$|\psi\rangle = \mathrm{e}^{\mathrm{i}\,\phi_0} \left(r_0|0\rangle + r_1 \ \mathrm{e}^{\mathrm{i}\,(\phi_1-\phi_0)}|1\rangle\right). \tag{2.2a}$$

One can always ignore the global phase factor $\mathrm{e}^{\mathrm{i}\,\phi_0}$. Then imposing the normalization condition (2.1b), one can write

$$|\psi\rangle = \cos\frac{\theta}{2}\,|0\rangle + \mathrm{e}^{\mathrm{i}\phi} \sin\frac{\theta}{2}\,|1\rangle. \tag{2.2b}$$

Here, r_0 and r_1 have been parametrized by the spherical angles $(\theta, \phi = \phi_1 - \phi_0)$ with $0 \leq \theta \leq \pi$ and $0 \leq \phi \leq 2\pi$. The angles θ and ϕ define a point on a unit, 3-D sphere S^2, called the Bloch sphere, as shown in Fig. 2.1, such that the north and the south poles are taken to be the states $|0\rangle$ and $|1\rangle$, respectively. [In (2.2b), the polar angle θ has been divided by two so that the north and south poles of the Bloch sphere in Fig. 2.1 correspond, respectively, to the states $|0\rangle$ and $|1\rangle$ of a qubit.]

The Bloch sphere provides a simple and useful tool for visualizing the state of a single qubit. The state (2.2) of a single qubit can be parametrized as a point on this sphere. This construction provides nice representations also of the cbits as specific points on this sphere. The cbit "0" (i.e., the qubit state $|0\rangle$) marks the north pole; whereas, the cbit "1" (i.e., the qubit state $|1\rangle$) corresponds to the south pole. Any other point on this sphere represents a non-trivial linear superposition (2.2) of the basis states $|0\rangle$ and $|1\rangle$. The angle

$$\frac{1}{2}\theta = \tan^{-1}\left(\frac{r_1}{r_0}\right) = \tan^{-1}\frac{|C_1|}{|C_0|} \tag{2.2c}$$

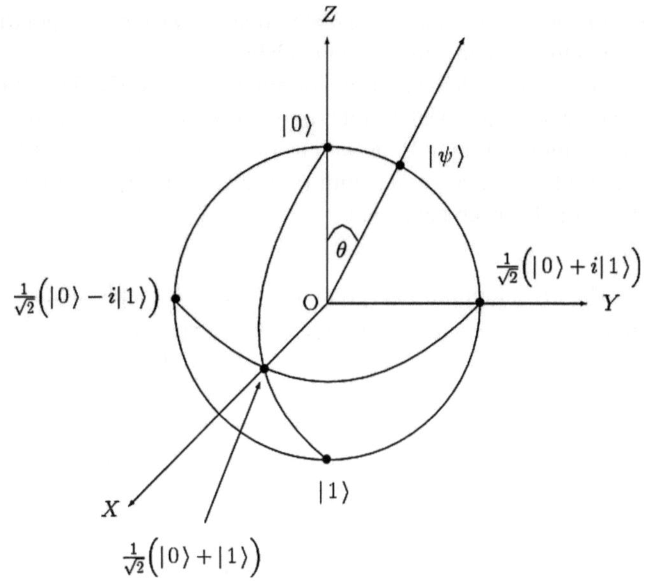

Fig. 2.1 A Bloch sphere wherein state (2.2) of a qubit is shown as a point on its surface. The north (for $\theta = 0$) and the south (for $\theta = \pi$) poles correspond to the pure state $|0\rangle$ and $|1\rangle$, respectively, of this qubit. (First published in [110].)

is related to the proportion of the state $|1\rangle$ to the state $|0\rangle$ in the composition of the state $|\psi\rangle$ in (2.1) and (2.2), while the angle ϕ determines the relative phase of these two basis states.

Examples of Qubits

Photonic Qubit The most popular and very successfully used representation of a qubit is, probably, the electric field vector $\boldsymbol{E}$ (see, e.g., Fig. 1.1) of an electromagnetic wave. The orientation of $\boldsymbol{E}$ determines, in turn, the polarization of a photon. A photon has, therefore, formed—hitherto perhaps the most widely used—two-state system (i.e., a qubit) in quantum information. A photonic qubit can have [17] either linear or circular polarization. The former of these can be in the form of H/V (i.e., horizontal/vertical represented by $|\rightarrow\rangle/|\uparrow\rangle$), or D/A (i.e., diagonal/antidiagonal represented by $|\nearrow\rangle/|\searrow\rangle$, or $|\nwarrow\rangle/|\swarrow\rangle$); whereas, the later is R/L (i.e., right/left circular polarization represented by $|\curvearrowright\rangle/|\curvearrowleft\rangle$). The basis vectors (i.e., H and V, or D and A, or R and L) in each of these three sets are orthogonal. In addition, these representations (i.e., H/V, D/A, R/L) form three linearly independent basis sets. Polarization qubits are very simple to encode and decode using simple optical devices (e.g., mirrors, lenses, half-wave plates, quarter wave plates, polarizing beam splitters, *etc.*) readily available in any optical laboratory.

In addition to the entanglement of two orthogonal states of polarization of light, two photons possessing nonlocal correlation with respect to time and energy [81, 82], position and momentum [83], as well as the spatial modes of electromagnetic field carrying orbital angular momentum (see, e.g., [84, 85] and other references given in [86]) have also been generated.

Spin-$\frac{1}{2}$ Qubit The most familiar interpretation of (2.1) for a physicist or a chemist, for that matter, is that of a state of an object whose spin angular momentum is $\frac{1}{2}$ (e.g., an electron). Then $|0\rangle \equiv |\uparrow_z\rangle$ and $|1\rangle \equiv |\downarrow^z\rangle$ are the spin-up and spin-down states along a particular axis, such as the OZ-axis in Fig. 2.1 for the Bloch sphere. Then two complex numbers (C_0, C_1), or the real spherical angles (θ, ϕ), used to characterize the state $|\psi\rangle$ of a qubit, describe the orientation of the spin in 3-D space in an arbitrary direction. For example, state

$$|\uparrow_x\rangle = \begin{cases} \frac{1}{\sqrt{2}}(|0\rangle + |1\rangle), \\ i.e., \\ \frac{1}{\sqrt{2}}(|\uparrow_z\rangle + |\downarrow^z\rangle) \end{cases} \tag{2.3a}$$

with spin-up along the OX-axis, corresponds to $(\theta = \pi/2,\ \phi = 0)$ in (2.2b). Similarly, the state

$$|\downarrow^x\rangle = \frac{1}{\sqrt{2}}(|\uparrow_z\rangle - |\downarrow^z\rangle), \tag{2.3b}$$

orthonormal to (2.3a), represents a qubit whose spin is down along the OX-axis obtained on taking $(\theta = \pi/2, \phi = \pi)$ in (2.2b). If one measures the spin of a qubit along OZ-axis for either of the states (2.3), one will obtain $|\uparrow_z\rangle$ or $|\downarrow^z\rangle$ with equal probability of 1/2.

Let us now consider the combination

$$\frac{1}{\sqrt{2}}(|\uparrow_x\rangle + |\downarrow^x\rangle). \tag{2.4}$$

A measurement along the OX-axis of the spin of a particle in this state will yield either of $|\uparrow_x\rangle$ and $|\downarrow^x\rangle$ with equal probability of $\frac{1}{2}$. But, what will be the result if the spin of a particle in state (2.4) is measured along the OZ-axis? State (2.4) is an algebraic sum of (2.3) which is equivalent to $|\uparrow_z\rangle$. Therefore, when one performs a measurement on (2.4) along the OZ-axis, the result will always be $|\uparrow_z\rangle$, never $|\downarrow^z\rangle$. Thus, for qubits, probabilities can add in unexpected ways. This phenomenon is known as "quantum interference" and is an important feature of quantum information. This formal equivalence with a spin-$\frac{1}{2}$ object applies to any two-level quantum system.

If one denotes the respective basis kets $|0\rangle$ and $|1\rangle$ by the vectors

$$|0\rangle = |\uparrow_z\rangle \equiv \begin{pmatrix}1\\0\end{pmatrix} \quad \text{and} \quad |1\rangle = |\downarrow^z\rangle \equiv \begin{pmatrix}0\\1\end{pmatrix}, \tag{2.5}$$

then, one knows from the elementary quantum mechanics (see, e.g., [10, 60]), that the operators for measuring along the OX, OY, and OZ axes the spin angular momentum of a spin-$\frac{1}{2}$ particle are the well-known Pauli spin matrices

$$\sigma_x = \begin{pmatrix}0 & 1\\ 1 & 0\end{pmatrix}, \; \sigma_y = \begin{pmatrix}0 & -\mathrm{i}\\ \mathrm{i} & 0\end{pmatrix}, \; \sigma_z = \begin{pmatrix}1 & 0\\ 0 & -1\end{pmatrix}. \tag{2.6}$$

For measuring spin angular momentum along any other unit vector, say

$$\hat{\boldsymbol{u}} = a\,\hat{\mathbf{e}}_x + b\,\hat{\mathbf{e}}_y + c\,\hat{\mathbf{e}}_z, \tag{2.7a}$$

the corresponding operator is

$$\sigma_u = a\,\sigma_x + b\,\sigma_y + c\,\sigma_z = \hat{\boldsymbol{u}}\cdot\boldsymbol{\sigma}. \tag{2.7b}$$

Here, in (2.7), $|a|^2 + |b|^2 + |c|^2 = 1$ and $(\hat{\mathbf{e}}_x, \hat{\mathbf{e}}_y, \hat{\mathbf{e}}_z)$ are the unit vectors along the (OX, OY, OZ) axes of the SF in Fig. 1.1, respectively.

In the context of a general qubit where states are represented by $|0\rangle$ and $|1\rangle$, Pauli operators (2.6) can be written as

$$\left.\begin{aligned} \sigma_x &= |1\rangle\langle 0| + |0\rangle\langle 1|, \\ \sigma_y &= \mathrm{i}(|1\rangle\langle 0| - |0\rangle\langle 1|), \\ \sigma_z &= |0\rangle\langle 0| - |1\rangle\langle 1|. \end{aligned}\right\} \tag{2.8}$$

Operator σ_x turns a $|0\rangle$ state into a $|1\rangle$ state, and vice versa. It is, therefore, called [17] a bit flip operator. On the other hand, σ_z switches between $(C_0|0\rangle + C_1|1\rangle)$ and $(C_0|0\rangle - C_1|1\rangle)$ states. It is, therefore, a phase flip operator [17]. However, all the operators of the form (2.7) and (2.8) have eigenvalues ± 1 for the spin to be parallel and antiparallel to their corresponding axis.

Thus, a qubit has a much richer structure than a cbit. In addition to being capable to be in either of the states $|0\rangle$ and $|1\rangle$ at a time, just as a cbit, a qubit can be in a continuum of the coherent superposition (2.1) [or, (2.2b)] of these two states specified by two parameters (C_0, C_1) [or, (θ, ϕ)]. However, only a single bit of information can be read from a single qubit. When the angular momentum of a spin-$\frac{1}{2}$ particle is measured along an axis, the only possible results are parallel or antiparallel to the chosen axis. After the measurements, the state is reduced to one of the eigenstates of the measured observable. And, from the result of the measurement, one knows what it is changed to. No more information about the original state can be retrieved. Consequently, it is not possible [87] to prepare a copy of an unknown state in quantum mechanics. Thus, unlike a cbit, one cannot replicate, or clone, a state (2.1)/(2.2) of a qubit.

2.1.4.3 Multiple Qubits

Let us consider a system of two qubits, A and B with their states $(|0\rangle_A, |1\rangle_A)$ and $(|0\rangle_B, |1\rangle_B)$, respectively. Then the computational basis states for this two-qubit system are $|00\rangle \equiv |0\rangle_B\ |0\rangle_A$, $|01\rangle \equiv |0\rangle_B\ |1\rangle_A$, $|10\rangle \equiv |1\rangle_B\ |0\rangle_A$, and $|11\rangle \equiv |1\rangle_B\ |1\rangle_A$. A pair of qubits can exist also in a linear superposition of these four bases states. Thus, a most general two-qubit state vector is

$$|\psi\rangle = C_{00}|00\rangle + C_{01}|01\rangle + C_{10}|10\rangle + C_{11}|11\rangle \tag{2.9a}$$

with the normalization condition

$$|C_{00}|^2 + |C_{01}|^2 + |C_{10}|^2 + |C_{11}|^2 = 1. \tag{2.9b}$$

The digit pairs (00, 01, 10, 11), appearing in the computational basis states in (2.9a), are the binary system representations of the digits (0, 1, 2, 3) used in the decimal system. Thus (2.9a) can also be written as

$$|\psi\rangle = C_0|0\rangle + C_1|1\rangle + C_2|2\rangle + C_3|3\rangle. \tag{2.10}$$

Here, we have labeled the respective basis states $(|00\rangle, |01\rangle, |10\rangle, |11\rangle)$ and their amplitudes $(C_{00}, C_{01}, C_{10}, C_{11})$, both appearing in (2.9), by the bases set $|n\rangle$ and the expansion coefficients C_n with $n = 0$–3.

In these equations, $|C_{j\,i}|^2$ is the probability that a simultaneous measurement on the qubits B and A in the two-qubit state (2.9a) will yield states $|j\rangle_B$ and $|i\rangle_A$, respectively. Let us perform a measurement only on the qubit A of this two-qubit system, without worrying about the qubit B. In that case, $|C_{00}|^2 + |C_{10}|^2$ and $|C_{01}|^2 + |C_{11}|^2$ are the probabilities of finding A in its states $|0\rangle_A$ and $|1\rangle_A$, respectively. After these two respective measurements, the two-qubit state (2.9a) reduces, or collapses, to the new states

$$|\psi\rangle \Longrightarrow |\psi'\rangle_A = \frac{C_{00}|0\rangle_B + C_{10}|1\rangle_B}{\sqrt{|C_{00}|^2 + |C_{10}|^2}}\,|0\rangle_A \tag{2.11a}$$

and

$$|\psi\rangle \Longrightarrow |\psi''\rangle_A = \frac{C_{01}|0\rangle_B + C_{11}|1\rangle_B}{\sqrt{|C_{01}|^2 + |C_{11}|^2}}\,|1\rangle_A, \tag{2.11b}$$

respectively. One can write similar expressions if the measurements are made on the qubit B in place of A.

More generally, let us consider a system of N qubits. The computational bases of this system are

$$\begin{aligned}
&|0\rangle_N\,|0\rangle_{N-1}\,\ldots\ldots\ldots\ldots\,|0\rangle_2\,|0\rangle_1 \equiv |00\ldots\ldots\ldots.00\rangle_1\\
&|0\rangle_N\,|0\rangle_{N-1}\,\ldots\ldots\ldots\ldots\,|0\rangle_2\,|1\rangle_1 \equiv |00\ldots\ldots\ldots.01\rangle_2\\
&\qquad\vdots\\
&|0\rangle_N\,|1\rangle_{N-1}\,\ldots\ldots\ldots\ldots\,|1\rangle_2\,|1\rangle_1 \equiv |01\ldots\ldots\ldots.11\rangle_{2^N-1}\\
&|1\rangle_N\,|1\rangle_{N-1}\,\ldots\ldots\ldots\ldots\,|1\rangle_2\,|1\rangle_1 \equiv |11\ldots\ldots\ldots.11\rangle_{2^N}
\end{aligned} \tag{2.12}$$

Hence, in analogy to (2.10), the most general N-qubit state is

$$|\psi\rangle = \sum_{n=0}^{2^N-1} C_n\,|n\rangle \tag{2.13a}$$

with

$$\sum_{n-0}^{2^N-1} |C_n|^2 = 1, \tag{2.13b}$$

and

$$\left.\begin{aligned}
|0\rangle &\equiv |00\ldots\ldots\ldots.00\rangle_1,\\
|1\rangle &\equiv |00\ldots\ldots\ldots.01\rangle_2,\\
&\;\;\vdots\\
|2^N-2\rangle &\equiv |01\ldots\ldots\ldots.11\rangle_{2^N-1}\\
|2^N-1\rangle &\equiv |11\ldots\ldots\ldots.11\rangle_{2^N}.
\end{aligned}\right\} \tag{2.13c}$$

Similar to (2.10), the integer n in (2.13) represents a digit in the decimal system. The binary system representation of each n consists of N zeros (0s) and/or ones (1s). In (2.13a), each basis state $|n\rangle$ represents one of the products (2.13c) of the individual states of N qubits of our system; $|C_n|^2$ in (2.13b) is the probability that when this N-qubit system is in the state represented by $|\psi\rangle$ in (2.13a), a simultaneous measurement on all the N qubits will have the same outcome as the system were simply in the state $|n\rangle$.

Further, let us perform a simultaneous measurement only on, say, first, second, and the Nth qubits of our N-qubit system, while the remaining $N-3$ qubits of this system remains unobserved. If the system was initially in the state (2.13a) and the outcomes of the measurement on these three qubits are $|1\rangle_1$, $|0\rangle_2$, and $|1\rangle_N$, respectively, then this system, after such a measurement, will collapse to the state

$$|\psi\rangle \Longrightarrow |\psi'\rangle \otimes [|1\rangle_N\,|0\rangle_2\,|1\rangle_1]\,, \tag{2.14a}$$

where the state of the remaining $(N-3)$ qubits is given by

$$|\psi'\rangle = \left(\sum_{m=0}^{2^{n-3}-1} |\gamma_m|^2\right)^{-\frac{1}{2}} \sum_{m=0}^{2^{n-3}-1} \gamma_m\,|m\rangle. \tag{2.14b}$$

In the last expression (2.14b), the binary system representation of each m is composed of all those $N-3$ qubits in the original system on which no measurements are made (i.e., other than the observed first, second, and the Nth qubits).

2.1.4.4 Qudits

It is possible to extend two-level qubits to three-level qutrit, or to d (>3)-level qudits. This provides an extension of the binary quantum logic to a three- or d-dimensional Hilbert space. Using qudits, for example, one can reduce the number N_2 of qubits needed to implement a protocol in quantum information by a factor of $|\log_2 d|$, for the Hilbert space of N_d qudits contains the space of N_2 qubits as $d^{N_d} > 2^{N_2}$. Chapter 4, in this monograph, describes ionic qudits and investigates properties of their entangled states formed with electronic qubits. Both, these qudits as well as their bipartite states with electrons can be readily produced in a laboratory.

2.2 Entanglement

It is a concept embedded in one of the fundamental postulates, namely the superposition principle (see, e.g., [58, 59]), of quantum mechanics.

2.2.1 Definition

A quantum system may have one or more parts. The one with more than one parts is called a composite quantum system and its parts are called subsystems. A composite quantum system with two parts is called bipartite, the one with three parts is known as tripartite; whereas, a multipartite quantum system consists of more than three subsystems. Let us consider, for an easy understanding, a bipartite quantum system composed of the parts U and V. These two subsystems are defined in d_u- and d_v-dimensional Hilbert spaces H_u and H_v, respectively. Then, according to the superposition principle [58, 59], the tensor product [60] space $H_u \otimes H_v$ of dimensions $d_u\, d_v$ is associated with the composite system $U \oplus V$. It is, sometimes, possible to decompose a state, say, $|\psi\rangle$ of this bipartite system into a tensor product of the states of its two subsystems, i.e.,

$$|\psi\rangle = \Big\{|u\rangle\Big\} \otimes \Big\{|v\rangle\Big\} \tag{2.15a}$$

with

$$\Big\{|u\rangle\Big\} \equiv \Big(|u_1\rangle, |u_2\rangle, \ldots\ldots\ldots, |u_{d_u}\rangle\Big) \tag{2.15b}$$

and

$$\Big\{|v\rangle\Big\} \equiv \Big(|v_1\rangle, |v_2\rangle, \ldots\ldots\ldots, |v_{d_v}\rangle\Big) \tag{2.15c}$$

describing states of the subsystems U and V, respectively. It, however, is not always possible to write the state of a composite system in the form of (2.15a).

Let us assume, for simplicity, that each of the parts U and V of the composite system $U \oplus V$ is defined in a two-dimensional Hilbert space, i.e., $d_u, d_v = 2$. For this bipartite system, consider the state

$$|\psi\rangle = C_1\,|u_1\rangle \otimes |v_1\rangle + C_2\,|u_2\rangle \otimes |v_2\rangle \equiv C_1\,|u_1\,v_1\rangle + C_2\,|u_2\,v_2\rangle \tag{2.16a}$$

with

$$C_1, C_2 \neq 0 \qquad \text{and} \qquad |C_1|^2 + |C_2|^2 = 1. \tag{2.16b}$$

This state cannot be written as $\big(\alpha_1\,|u_1\rangle + \alpha_2\,|u_2\rangle\big) \otimes \big(\beta_1\,|v_1\rangle + \beta_2\,|v_2\rangle\big)$ for any complex numbers $\alpha_1, \alpha_2, \beta_1, \beta_2$. Two such subsystems are called *entangled* [88]. Bipartite states which can be decomposed in the form of (2.15a) are known, on the other hand, to be product (or separable, or un-entangled) states. The two most famous, and also familiar, examples of bipartite entangled states are the Einstein–Rosen–Podolsky–Bohm (EPRB) states [11, 89] of two spin-$\frac{1}{2}$ particles. Namely,

$$|\psi^{+}\rangle \equiv |S = 1, M = 0\rangle = \begin{cases} \frac{1}{\sqrt{2}}\Big(|0\rangle_v \otimes |1\rangle_u + |1\rangle_v \otimes |0\rangle_u\Big), \\ \qquad\qquad i.e., \\ \frac{1}{\sqrt{2}}\Big(|01\rangle + |10\rangle\Big) \end{cases} \tag{2.17a}$$

and

$$|\psi^{-}\rangle \equiv |S = 0, M = 0\rangle = \begin{cases} \frac{1}{\sqrt{2}}\Big(|0\rangle_v \otimes |1\rangle_u - |1\rangle_v \otimes |0\rangle_u\Big), \\ \qquad\qquad i.e., \\ \frac{1}{\sqrt{2}}\Big(|01\rangle - |10\rangle\Big) \end{cases} \tag{2.17b}$$

for the total angular momenta $S \equiv \left|\mathbf{S}\right| = 1$ and 0, respectively, with their projections $M = 0$ about a given axis (e.g., OZ-axis in Fig. 1.1). Further in [(2.17a), (2.17b)], the respective kets $|0\rangle$ and $|1\rangle$ represent spin-up (i.e., $|\uparrow\rangle$) and spin-down (i.e., $|\downarrow\rangle$) states of a spin-$\frac{1}{2}$ qubit along the same direction for which $M = 0$. States [(2.17a), (2.17b)] are two of the four Bell states [17, 43]. Moreover, as any reader familiar with quantum mechanics (see, e.g., [10, 58–60], *etc.*) will know, (2.17a) and (2.17b) represent, respectively, one of the triplet and the (single) singlet states (each with $M = 0$) of a system of two spin-$\frac{1}{2}$ particles.

The remaining two Bell states for a bipartite system of qubits are [17, 43]

$$|\phi^{\pm}\rangle = \frac{1}{\sqrt{2}}\Big(|00\rangle \pm |11\rangle\Big). \tag{2.17c}$$

States (2.17) can represent two entangled photonic qubits with H/V, D/A, or R/L polarizations as well [17, 24, 43]. Parts U and V forming the composite system $U \oplus V$ which can be described by any of the four Bell states (2.17) are called [17, 43] also the EPRB [11] pairs of subsystems.

Each of the other two kets

$$\begin{aligned} |S = 1, M = +1\rangle &= |00\rangle, \\ |S = 1, M = -1\rangle &= |11\rangle \end{aligned} \tag{2.18}$$

of the triplet-group of states [whose third member is (2.17a)] of two spin-$\frac{1}{2}$ particles is simply written in the form of (2.15a). Each of (2.18) is, therefore, a product or separable state neither of which represents two entangled particles.

Suppose, one makes measurements of the spin component of one of the two qubits in any of the four Bell states given in (2.17). Obviously, there is a equal probability of $\frac{1}{2}$ for observing $|0\,1\rangle$ or $|1\,0\rangle$ in [(2.17a), (2.17b)], or either of $|0\,0\rangle$ and $|1\,1\rangle$ in (2.17c), states of the composite system. This means that if one of the qubits in the bipartite states [(2.17a), (2.17b)] is known to have spin $|\uparrow\rangle$, the other must necessarily have its spin $|\downarrow\rangle$, or vice versa; whereas in the states (2.17c), both particles have either spins $|\uparrow\uparrow\rangle$ or $|\downarrow\downarrow\rangle$. When the spin component of the first qubit is shown to have $|\uparrow\rangle$ in the either of the states $|\psi^{\pm}\rangle$, the measurement apparatus has picked up the product $|0\,1\rangle$ of the composite system. A subsequent measurement of the second qubit must ascertain that its spin state is given by $|\downarrow\rangle$. A similar interpretation is applicable to the state $|\phi^{\pm}\rangle$ where both particles in the composite system are either in $|0\,0\rangle$ or in $|1\,1\rangle$ state.

Thus, an outcome of the measurement of the spin of the second qubit in any of the four Bell states (2.17) of the composite system of two qubits depends upon the result of the measurement of the state of the first qubit. Here, two qubits can be kilometers apart with no possibility of any kind of communication or of mutual interaction between them. It is as though the second qubit knows which spin component of the first qubit is being measured. It is for this reason that the individual states of two qubits are said to be entangled. But this interpretation cannot be applied to the product states (2.18) for the simple reasons that here states of both qubits are known to be same even before the measurements; there is no need to adjust for the second particle its spin component according to the result of the measurement on the first particle.

Let us generalize these concepts to a system of N particles. According to the superposition principle [58, 59] in quantum mechanics, a most general state of this system is given by

$$\Psi = \sum_{i>1} \phi_i(1)\,\psi_i(2)\,\chi_i(3)................\xi_i(N). \tag{2.19}$$

Here, $\phi_i(1)$ is the ith state of the particle 1, and so on. Each term (i.e., $\phi_i(1)\psi_i(2)\chi_i(3).......\xi_i(\text{N})$) in the sum on the right-hand side of the above equation is said to represent a product state of the form (2.15a) of our N-particle system (i.e., it is simply a product of the state of each of the N individual particles the system is made of). Thus, $\phi_i(1)\psi_i(2)\chi_i(3).......\xi_i(\text{N})$ is the ith product state in Ψ. This product state is said to be separable because it is obtained simply by multiplying the state of individual particles.

The state of a system of more than one particle having nonlocal correlation among themselves can never be represented by a single product state. Thus entanglement is that property of Ψ in (2.19) which prohibits it to be represented by a single product state. Here, the sum over i must have more than one product term (i.e., $i > 1$) for Ψ to represent an entangled state of N particles. Moreover, for N particles represented in (2.19) to be entangled, it should not be possible by means of any transformations, whatsoever, to re-write Ψ in a way that it contains only a single product state. It is obvious that for entanglement to be present, the number of particles N in Ψ in (2.19) must be greater than one. However, the number of product terms (i.e., value of i) needs not be equal to N.

The state vector $\phi_i(1)$ in (2.19) are defined on a Hilbert space of dimensions d_1. In order to obtain entanglement, $d_1 \geq 2$. This statement concerns only those degrees of freedom which are being entangled. Clearly a quantum system will not be entangled in all of its properties. Consequently, a particle with $d_1 = 2$ (i.e., a spin-$\frac{1}{2}$ particle, a photon with negative and positive helicities, a 2-level atom, *etc.*) is a qubit. Likewise, a qutrit is a 3-D (i.e., $d_1 = 3$) quantum system (e.g., a 3-level atom, *etc.*). In general, a d-D (i.e., $d_1 = d$) quantum system is called a qudit.

2.2.2 The EPR Paradox

Einstein [11], along with his colleagues Podolsky and Rosen, was first to recognize in 1935 this kind of correlation between two geographically distant parts of a composite quantum system. The word entanglement for such correlation was, however, first coined by Schrödinger [88]. EPR [11] never used this terminology and they discussed it, unlike in [(2.17), (2.19)], for two continuous variables. It was reformulated for discrete variables and interpreted by Bohm [89] for the singlet state of two spin-$\frac{1}{2}$ particles. Schrödinger's work [88] was, however, inspired by the EPR paper [11].

EPR argued that entanglement violated local realism (also called Einstein's locality [11]). Local realism is the idea that objects have definite properties, whether or not they are measured. Measurements of these properties are not affected by the events taking sufficiently far away. Entanglement violates local realism on two accounts [11]: (1) According to quantum mechanics, state of a system is never definitely known when it is measured. (2) Although, two subsystems are far apart, outcome of the measurement of, say, the second subsystem depends upon the results of measurements of the first subsystem. Schrödinger [88] tried to explain this paradox in the following way: "The whole is in a definite state, the parts taken individually are not."

EPR, on the other hand, thought that as the quantum mechanics disallows definite values prior to their measurements, the quantum theory, from a classical perspective, must be incomplete. According to them [11], there should be some variables which should explain the nonlocal correlation in an entangled state of a composite quantum system. These variables came to be called as 'hidden variables' as they are not so far specified by the quantum mechanics.

However, entanglement does not violate Einstein's relativity. The only way a person holding the second qubit of an EPR pair can come to know about the state of his/her subsystem (without making any measurements on it) is by receiving a classical communication (e.g., a phone call) about the result of the measurements performed by the other person on the first qubit (i.e., his/her subsystem) held by him/her. A classical communication cannot propagate faster than the speed of light.

2.2.3 *Bell Inequality*

For about three decades, following the articles published by Einstein et al. [11] and by Schrödinger [88] in 1935, the debate about entanglement and EPR paradox was philosophical in nature discussing, among other things, the incompleteness—or otherwise—of quantum mechanics. Proceeding on the assumption of incompleteness of quantum theory and taking hidden variables into account, Bell [12] derived correlation inequalities that can be violated in quantum mechanics, but have to be satisfied within every model that obeys local realism (i.e., Einstein's locality [11]) and is complete. Such a model is known as hidden-variable model [90, 91]. In this model, measurement is actually fundamentally deterministic, but appears to be probabilistic because some degrees of freedom are not precisely known.

Various forms of Bell's inequality [12] have been reviewed by Clauser and Shimony [92] and by Peres [93]. However, experimental tests (see, for example, [90, 94–97]) of Einstein's locality principle [11] are usually based on a form of a Bell's inequality [12] which was derived by Clauser et al. [92, 98]. It applies to a situation in which two parties A and B have one particle each upon which they can make measurements of two different physical properties of the particle in their respective possession. The party A can measure one of the either observables Ω_a and $\Omega_{a'}$; while, B can measure properties Ω_b and $\Omega_{b'}$. Suppose,[1] following Preskill [99], that the observables Ω_a, $\Omega_{a'}$, Ω_b, $\Omega_{b'}$ take values from ± 1 and are functions of hidden variables.

If the hidden variable model is to be applicable, one can then assign values from ± 1 to all the four observables, even though it is impossible to simultaneously measure both of Ω_a, and $\Omega_{a'}$, or both of Ω_b, and $\Omega_{b'}$. Accordingly, it follows that either $(\Omega_a + \Omega_{a'}) = 0$, in which case $(\Omega_a - \Omega_{a'}) = \pm 2$, otherwise $(\Omega_a - \Omega_{a'}) = 0$ with $(\Omega_a + \Omega_{a'}) = \pm 2$. Therefore

$$\Omega \equiv (\Omega_a + \Omega_{a'})\,\Omega_b + (\Omega_a - \Omega_{a'})\,\Omega_{b'} = \pm\,2. \tag{2.20a}$$

Consequently,

$$\langle |\Omega| \rangle = 2. \tag{2.20b}$$

But

$$\big|\langle \Omega \rangle\big| \;\leq\; \langle |\Omega| \rangle. \tag{2.20c}$$

Then, it follows from (2.20)

[1] The remaining discussion in the present Sect. 2.2.3 has been adapted from [99, 100].

$$\left|\langle \Omega_a\, \Omega_b \rangle \;+\; \langle \Omega_{a'}\, \Omega_b \rangle \;+\; \langle \Omega_a\, \Omega_{b'} \rangle \;-\; \langle \Omega_{a'}\, \Omega_{b'} \rangle\right| \leq 2. \qquad (2.21)$$

This result [92, 98] is called Clauser–Horne–Shimony–Holt (CHSH) inequality. It is applicable for any random variables Ω_a, $\Omega_{a'}$, Ω_b, $\Omega_{b'}$ taking values $\pm\,1$ that are governed by a joint probability distribution.

Let us now calculate the value of the inequality (2.21) using quantum mechanics. For simplicity, we assume that measurements are being performed on two spin-$\frac{1}{2}$ qubits. Then, Ω_a, $\Omega_{a'}$, Ω_b, $\Omega_{b'}$ are obviously the Hermitian operators related to the Pauli matrices defined in (2.6). One now takes, analogously to Eqs. (2.7),

$$\Omega_a = \boldsymbol{\sigma}^A \cdot \hat{\mathbf{u}}_A, \qquad \Omega_{a'} = \boldsymbol{\sigma}^A \cdot \hat{\mathbf{u}}_{A'}, \qquad (2.22a)$$

where

$$\boldsymbol{\sigma}^A \;=\; \hat{\mathbf{e}}_x\, \sigma_x^A \;+\; \hat{\mathbf{e}}_y\, \sigma_y^A \;+\; \hat{\mathbf{e}}_z\, \sigma_z^A \qquad (2.22b)$$

with $\hat{\mathbf{u}}_A$ and $\hat{\mathbf{u}}_{A'}$ being unit vectors, for the qubit in the possession of the party A. Similarly, for the qubit with the party B, one has

$$\Omega_b = \boldsymbol{\sigma}^B \cdot \hat{\mathbf{u}}_B, \qquad \Omega_{b'} = \boldsymbol{\sigma}^B \cdot \hat{\mathbf{u}}_{B'}, \qquad (2.23a)$$

with

$$\boldsymbol{\sigma}^B \;=\; \hat{\mathbf{e}}_x\, \sigma_x^B \;+\; \hat{\mathbf{e}}_y\, \sigma_y^B \;+\; \hat{\mathbf{e}}_z\, \sigma_z^B \qquad (2.23b)$$

Each of the four observables in (2.22a) and (2.23a) has eigenvalues $\pm\,1$. Hence, the results of measurements of these observables shall always be $\pm\,1$, as well.

Let us assume that two spin-$\frac{1}{2}$ qubits, one each possessed by the party A and party B, are in the singlet state (2.17b). Then, one readily finds

$$\langle \Omega_a\, \Omega_b \rangle \;=\; \langle \psi^- |\, \left(\boldsymbol{\sigma}^A \cdot \hat{\mathbf{u}}_A\right) \left(\boldsymbol{\sigma}^B \cdot \hat{\mathbf{u}}_B\right) | \psi^- \rangle \;=\; -\,\hat{\mathbf{u}}_A \cdot \hat{\mathbf{u}}_B\,. \qquad (2.24)$$

The left-hand side of the inequality (2.21), consequently, becomes

$$\begin{aligned} &\left|\hat{\mathbf{u}}_A \cdot \hat{\mathbf{u}}_B + \hat{\mathbf{u}}_{A'} \cdot \hat{\mathbf{u}}_B + \hat{\mathbf{u}}_A \cdot \hat{\mathbf{u}}_{B'} - \hat{\mathbf{u}}_{A'} \cdot \hat{\mathbf{u}}_{B'}\right| \\ &= \left|\cos\phi_{AB} + \cos\phi_{A'B} + \cos\phi_{AB'} - \cos\phi_{A'B'}\right|, \end{aligned} \qquad (2.25)$$

where $\phi_{A\,B} \;=\; \cos^{-1}\left(\hat{\mathbf{u}}_A \cdot \hat{\mathbf{u}}_B\right)$, and so on. Let us consider [100], for simplicity, geometries wherein all the four unit vectors ($\hat{\mathbf{u}}_A$, $\hat{\mathbf{u}}_B$, $\hat{\mathbf{u}}_{A'}$, $\hat{\mathbf{u}}_{B'}$) are in a single plane in Fig. 2.2. On taking ($\phi_{A\,B} \;=\; \phi_{A'\,B} \;=\; \phi_{A\,B'} \;=\; \pi/4$, $\phi_{A'\,B'} \;=\; 3\pi/4$) or

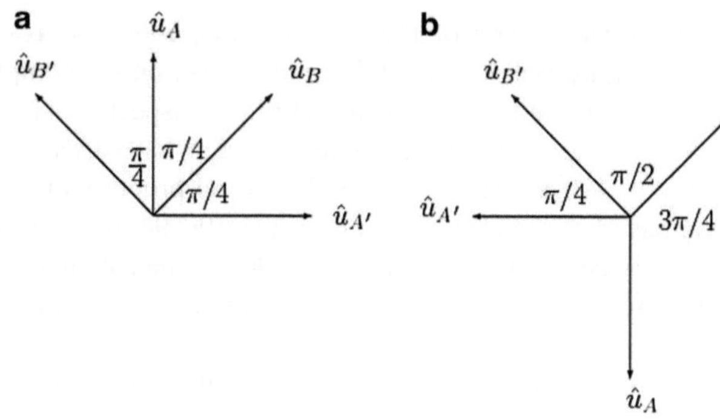

Fig. 2.2 Two [100] of the possible one-planar configurations of the unit vectors $\hat{\mathbf{u}}_A, \hat{\mathbf{u}}_{A'}, \hat{\mathbf{u}}_B, \hat{\mathbf{u}}_{B'}$ defined in (2.22a) and (2.23a) for which maximal violation of the inequality (2.21) takes place. (First published in [100].)

($\phi_{AB} = \phi_{A'B} = 3\pi/4, \phi_{AB'} = 5\pi/4, \phi_{A'B'} = \pi/4$), shown as (a) or (b) in Fig. 2.2, respectively. In view of the relationship (2.25), the CHSH [92, 98] inequality (2.21) in either of these two cases becomes

$$4\frac{1}{\sqrt{2}} = 2\sqrt{2} > 2. \tag{2.26}$$

This is a clear violation of (2.21) by the quantum mechanical predictions. Any configuration sufficiently near to those two shown in Figs. 2.2 will violate the inequality (2.21). But, the two configurations in this figure are those for which Bell's inequality (2.21) [92, 98] is maximally violated.

There are indeed infinitely many configurations of directions $\hat{\mathbf{u}}_A, \hat{\mathbf{u}}_B, \hat{\mathbf{u}}_{A'}$, and $\hat{\mathbf{u}}_{B'}$ for which the quantum mechanical predictions do not satisfy the inequality (2.21).

This shows that quantum mechanics is not consistent with the EPR's idea of local realism or Einstein's locality [11]. This, in other words, means that either quantum mechanics must be considered intrinsically nonlocal, or we must accept a theory whose measured quantities do not corresponds to the elements of reality. This also means that there are no hidden variables [11] for the quantum mechanics to take into account. Hence, quantum theory is complete.

Entangled states, like those given in (2.17), can be generated in a laboratory and the expectation values required in the inequality (2.21) can directly be measured. Thus, Bell's inequality, in a form derived by Clauser et al. [92, 98], provides a direct experimental test for the completeness of quantum mechanics which can be performed in a laboratory, rather than being merely a philosophical device.

Almost every experiment (see, e.g., [90, 94–97], *etc.*), hitherto performed to test Bell's inequality, has yielded results that violate (2.21) and, consequently, agrees with the predictions of quantum mechanics. Due to the possibilities [90, 101] of the *communication loophole*[2] and of the *detection loophole*,[3] the results of the early experiments violating Bell's inequalities and, hence, on the completeness of quantum mechanics, could not be regarded as definitive. Aspect [96] has discussed *communication loophole* and recent experiments which have tried to overcome this loophole (see, e.g., [90, 97, 102]). Although, several proposals for closing the *detection loophole* have been made (see, for example, [90, 94, 101], *etc.*), however, the experiment recently performed by Rowe et al. [94] on the entangled $^{9}Be^{1+}$ ions is probably the first to do so [90]. In fact, implementing a *loophole*-free test of Bell inequality violation has technologically proved to be a very challenging task. Nevertheless, no experiment has hitherto contradicted quantum mechanics.

2.3 Entanglement in Quantum Information Science

It is obvious from the discussions in Sect. 2.2 that measurements on the entangled states of spatially separated subsystems allow physicists to test fundamental notions about the nature of the physical world in general, quantum theory in particular. The two kinds of systems whose entangled states have hitherto been investigated [103] are those in which one can study properties of individual particles, or wherein collective measurements are possible. The later type includes, for example, cold clouds of 10^7 atoms (e.g., [104]), optical lattices consisting of 10^5 two-level atoms (e.g., [105]), *etc*. These [104, 105], and other (e.g., [106, 107], *etc*.), collective measurements have shown [103] that multiparticle entanglement is capable of influencing macroscopic thermodynamical properties (e.g., magnetic susceptibility, heat capacity, *etc*) of solids.

The quantum information, on the other hand, provides a new way of thinking about the entanglement as a resource [108] to carry out tasks in data processing and communication which would otherwise be very hard, or even impossible, to perform. It includes entangled states of two, or more, particles which are at different locations and can individually be addressed. Here, one can study properties of individual particles constituting a composite system. Some of the, hither to known, most important applications of entanglement in quantum information are briefly described in the following Sects. 2.3.1–2.3.5 herein. Although, the following discussion specifically refers to entangled qubits, however, it can readily be generalized to nonseparable states of qudits as well.

---

[2]It (see, e.g., [90, 96, 97, 101, 102], *etc*) is related to the spatial separation of the detectors used by the parties A and B in their experiment to test Bell's inequality (2.21).

[3]This loophole (see, [90, 94, 101] and references therein) refers to the problem arising due to low total detection efficiency in an experiment performed to test the inequality (2.21). One, therefore, assumes in such experiments that the detected pairs of particles is a fair representation of the emitted pairs.

2.3.1 *Quantum Cryptography*

One of earliest discoveries in quantum information science was that the quantum mechanics can be used to do key distribution in such a way that the securities of the two parties A and B cannot be comprised by any third party C. This procedure is known as quantum cryptography or quantum key distribution. The basic idea in quantum cryptography is to exploit the quantum mechanical principle that a system cannot be observed without being disturbed [58, 59].

The first protocol to use the entanglement between two qubits for cryptographic purpose was developed by Ekert [13]. In this protocol, Bell's inequality [12] safeguards the confidentiality in the transmission of pairs of entangled qubits. Six months after Ekert's work [13], Bennett et al. [109] presented a very simple scheme that still uses entangled bipartite states of qubits, but does not invoke Bell's theorem [12] for safe guarding confidentiality in the transmission. In this later scheme [17, 109, 110], two parties A and B, share a supply of known Bell state, say, $|\psi^-\rangle$ defined in (2.17b). Each of the parties A and B makes measurements on the qubits of each entangled pair in either of the eigen bases $\{|0\rangle, |1\rangle\}$ and $\{(|0\rangle + |1\rangle)/\sqrt{2}, (|0\rangle - |1\rangle)/\sqrt{2}\}$ of σ_z and of σ_x, respectively. Here, σ_x and σ_z are two of the three Cartesian components of the Pauli spin matrix $\boldsymbol{\sigma}$ defined in (2.6) [or, (2.8)]. For each pair, A and B chose the bases at random. They keep the record of the outcome of each measurement. Then, A and B make public the bases that they used for each measurement. After they have discarded all outcomes where their bases did not match, A and B then compare a sub-set of remaining data publicly. If the results correlate as expected, A and B are then assured that no eavesdropper has intercepted, or tempered with, their entangled states. The remaining data is then used as a reset key.

2.3.2 *Dense Coding*

Bennett and Weisner [111] showed that in quantum information, one can send to a distant location classical information encoded in two cbits (which may have any of the four values 00, 01, 10, 11) with the help of only a single qubit. It is possible provided [17, 110, 111] each of the physically separated sender and receiver of information holds one qubit which are entangled. On account of the nonlocal correlation, any operations (corresponding to the values 00, 01, 10, 11 of two cbits) performed by the sender on his/her qubit, also affect that qubit which is possessed by the person who will be receiving the classical information. Consequently, after the single qubit, dispatched by the sender, has reached its destination, the receiver performs appropriate operations on the two entangled qubits—now both of them under his/her possession—for decoding the information meant to be sent to him/her by the other person, no matter how far the two observers are from each other. Dense coding is not possible unless the qubit possessed by one observer is entangled to the qubit with the other observer, irrespective of the distance between the two.

2.3.3 *Quantum Teleportation*

The no-cloning theorem [87] tells us that an unknown quantum state cannot be copied. This theorem is based on one of the fundamental postulates of quantum mechanics mentioned in the previous subsection of this chapter, namely, quantum mechanical measurements always disturbs a system (see, e.g., [58, 59]). Therefore, the very process used to learn about an unknown state, before it can be cloned, changes it and, hence, the validity of the no-cloning theorem for an unknown state. It means that in order to transfer an unknown quantum state, say, $|\psi_0\rangle = \alpha_0 |0\rangle + \beta_0 |1\rangle$ from one place to another, one has to transport it physically. But, Bennett et al. [14, 110] developed a protocol by which the unknown quantum state $|\psi_0\rangle$ can be transmitted to a remote location without physically carrying it, provided one is willing to destroy the original state. This procedure is known as quantum teleportation [14, 110] and is possible because of the presence of entanglement between two remote qubits.

Unlike many other applications in quantum information science, which utilize entanglement to improve upon what is possible in classical physics, or with reasoning based on classical physics, quantum teleportation achieves a purely quantum mechanical task which cannot be performed using classical physics. Measurements on an unknown state reveals minimum information about it and at the same time alters it. No scheme, other than teleportation, based on measurements and classical communications, will allow faithful transfer of an unknown state from one location to another, without physically transporting it.

Quantum teleportation emphasizes the interchangeability of different resources in quantum mechanics, showing that one shared EPR pair together with two classical bits of communication is a resource equal to one qubit of communication at least.

2.3.4 *Quantum Computation*

It has long been known that simulation of quantum systems is difficult as the complexity of their description grows exponentially with the system size. Deutsch [112] postulated that a computer based on the laws of quantum mechanics would be so powerful that it could simulate any physical system efficiently. Shor's algorithm [113] for efficient prime factorization of a large number demonstrated the power of quantum computation. Entanglement plays an important role in quantum computation. Jozsa and Linden [114] showed that entangled states must be present in a quantum computer at some stage of an algorithm for it to exhibit exponential speed-up compared to classical computers. Second, information must be encoded in entangled states if quantum computers are going to be able to operate in the presence of noise (Shor [113]). Finally, paradigms for a quantum computer have been proposed, based solely on measurements made on certain entangled states which acts as a resource for quantum computation [15, 115, 116].

2.3.5 *Other Applications of Entanglement*

There are many other applications of entanglement in quantum information. For example, separated parties can use entangled states to improve the performance in certain competitive games [117], in distributed quantum computing [45, 118], distributed quantum networks [44], *etc*. Entangled states can allow more precise measurements in, for example, frequency standards [119], lithography [120], etc., that would, otherwise, not be possible.

2.4 Criteria and Measures of Entanglement

For many important reasons, some of which have already been discussed in the last two sections of this chapter, the quantum entanglement has now become a resource for quantum communication and information processing [108]. Consequently, entanglement has been the subject of much study in recent years. The main theme of the present volume too is the generation and characterization of entanglement in, primarily, bipartite states of flying qubits which are at different locations and separately addressable. One or more of these entangled particles is necessarily an electron. In accordance with the other resources like energy, information, *etc*., it is very much desirable to have a theory of entanglement giving specific rules about as to how it can, or cannot, be manipulated and quantified. Due to the intense efforts made during the last two–three decades, such a theory has indeed emerged, at least for bipartite states of qubits. The first step in the development of such a theory is to detect its (i.e., entanglement's) presence and to quantify it. In this section, we summarize, in brief, the criteria which make one to decide whether a given state is separable or entangled. The criteria discussed herein can always be applied to bipartite states of qubits described by a known density matrix (DM), say, ρ. Entanglement measures, on the other hand, quantify the amount of entanglement and, hence, attempt to provide a number to explain the degree of entanglement of a given state.

2.4.1 *Criteria of Entanglement*

According to one of the fundamental postulates [58, 59] of quantum theory, the complete information about a system is contained in its wavefunction. But, such a description of a system in quantum mechanics by a single wavefunction is possible if it (i.e., the system) happens to be in one of its pure states. This, in turn, implies that all states of a given system are not necessarily pure; states which are not pure are called mixed states. A mixed state of any system can never be represented by a single wavefunction. Studies of the properties of a mixed state requires applications of the methods of density operator and/or density matrix (see, e.g., [60–62], *etc.*). A pure state, on the other hand, can equally be described by both a single wavefunction, or by a density operator and/or matrix. Consequently, the criteria developed for the nonseparability of a pure state of a system composed of two or more subsystems are not necessarily applicable to a mixed state of the same system; whereas, those suitable for mixed states can always be used to study entanglement properties of pure states as well.

Appendix A contains a brief discussion of a pure state, mixed state, and of the relevant properties of a density operator and of a density matrix representing such states. A pure state $|\psi_{i\,j}\rangle$ of two subsystems (say, A and B) is said to be separable if it can be written in the form of (A.20) [i.e., (20) in Appendix A], otherwise it is entangled. The well-known examples of pure separable states are, among others, two triplet components (2.18) of a composite system of two spin-$\frac{1}{2}$ particles. The four Bell states (2.17), on the other hand, are the well-known examples of pure entangled states. A mixed state is called separable if it can be prepared by two parties, A and B, in a classical way, i.e., by agreeing over phone on the local preparations of their respective states. A density matrix that has been created in this way can only contain classical correlations. Mathematically, a mixed state ρ of a bipartite system is called separable iff it can be written as [121]

$$\rho = \sum_i p_i\, |a_i\rangle\langle a_i| \otimes |b_i\rangle\langle b_i| \; = \sum_i p_i\, |\psi_{i\,i}\rangle\langle\psi_{i\,i}|, \tag{2.27}$$

[e.g., (2.39a) and (A.4)], otherwise it is entangled. Equation (2.27) is a bipartite specialization of a more general separable multipartite state (2.39a). The second term on the right-hand side of the above equation has been written using the definition (A.20). As defined in Appendix A, the coefficients p_i are the probabilities satisfying the condition (A.2). In general, $\langle a_i|\, a_j\rangle,\ \langle b_i|\, b_j\rangle \neq\ \delta_{i\,j}$. An example of a mixed, separable bipartite state of qubits that contains classical correlation is $\rho = (|0\,0\rangle\langle 0\,0| + |1\,1\rangle\langle 1\,1|)/2$.

The well-known state [121]

$$\rho_W = p\,\sigma_1 + (1 - p)\,\sigma_0, \tag{2.28a}$$

on the other hand, is an example of a mixed entangled state of two qubits. Here,

$$\sigma_0 = \frac{1}{4}\,\mathbb{I}_4 \tag{2.28b}$$

and

$$\sigma_1 = |\mathrm{Bell}\rangle\langle\mathrm{Bell}|, \tag{2.28c}$$

with $\mathbb{I}_4$ in (2.28b) an unit matrix of dimensions (4×4) and the ket $|\text{Bell}\rangle$ in (2.28c) representing one of the four Bell states given in (2.17). The diagonal matrix σ_0 describes two qubits to be in any of the pure separable states $|00\rangle$, $|01\rangle$, $|10\rangle$, $|11\rangle$ with an equal probability of $\frac{1}{4}$ and is, therefore, called a maximally chaotic [22], or white noise [122], state. But σ_1 is, obviously, a pure entangled state because each of (2.17) is also pure as well as nonseparable.

The presence of both separable and nonseparable states in (2.28a) means that ρ_W is, in general, necessarily neither a pure nor an entangled state. Its degree of mixedness as well as nonseparability depends upon the amount $(1 - p)$ of the maximally chaotic state σ_0 present in (2.28a). The bipartite states of the kind of (2.28a) for two qubits were first introduced in quantum information by Werner [121] with p appropriately called the mixing parameter. Obviously, for $p = 1$, ρ_W represents the pure entangled state (2.28c)—i.e., one of the four Bell states (2.17); whereas, for $p = 0$, (2.28a) reduces to the maximally chaotic state (2.28b). It is discussed on page 128 in Chap. 5 that the density matrix (2.28a) describes, on the other hand, a mixed entangled state of two qubits if

$$\frac{1}{3} < p \leq 1. \tag{2.28d}$$

Otherwise, a bipartite state represented by (2.28a) is always separable. Although, the mixing parameter p in (2.28a) can, in principle, take any values between $-1/3$ to $+1$ (see discussion on page 128); however, for ρ_W to be qualified to be a Werner [121] state, one must always have

$$0 \leq p \leq 1. \tag{2.28e}$$

Decomposition (2.27) of a mixed separable state is not unique. An example [123], reproduced in (A.19), gives some of the possible representations of a $(2 \otimes 2)$ maximally chaotic state [22, 122]. Consequently, finding a decomposition, as in (2.27), for a given state ρ, or proving that it does not exist, is a nontrivial—sometimes formidable—task which has hitherto been satisfactorily performed [124, 125] only for the states of two qubits or one qubit and one qutrit. Therefore, the naive looking definition (2.27) of separability is not "user-friendly," and one requires criteria which are easier to apply for testing separability, or otherwise, of a given state.

For the reasons discussed in the preceding paragraphs herein, one can classify the criterion for deciding whether a given density operator/matrix represents an entangled, or a separable, state according to their applicability to pure or mixed states.[4] For an unambiguous conclusion on the presence/absence of entanglement

[4] A criterion applicable to test nonseparability of a mixed states can always be used for pure states as well, but not vice-versa.

in a given state, the condition to be used must be necessary as well as sufficient. Such conditions have been discussed and reviewed in detail at various places (see, e.g., [24,126,127], *etc*.). Following are some of the criteria often used to test whether a given state of two qubits is entangled or not.

2.4.1.1 Pure States

Bell Inequality

Bell's inequality (2.21) was originally formulated [92, 98] to quantify Einstein's locality principle [11] and, hence, to experimentally [90] test the hidden variable theories. After every experiment [90, 94–97], performed hitherto, was found to violate these inequalities, they were then began to be used as a criterion to test whether a given bipartite state is entangled or not. It was, however, discovered by several workers (see, e.g., [121, 124, 125, 128–130]) that, although a product state always satisfies Bell's inequalities, their nonviolation does not necessarily mean that the state is separable. This situation arises when the two subsystems form a nonpure, i.e., a mixed state. Hence, one of the earliest known criterion for the existence of the entanglement is the violation of Bell's inequalities. But, this criterion forms a necessary and sufficient condition only for pure bipartite states of qubits.

The operator form of the inequality (2.21) for a system of two qubits, A and B, in a pure state represented by the density operator ρ^{AB} is [131, 132]

$$\begin{aligned}\Big|\mathrm{Tr}\Big[\rho^{AB}\Big\{\Big(\boldsymbol{\sigma}^{A}\cdot\hat{\mathbf{u}}_{A}\Big)\otimes\Big(\boldsymbol{\sigma}^{B}\cdot\hat{\mathbf{u}}_{B}\Big)\Big\}\Big]&+\\ \mathrm{Tr}\Big[\rho^{AB}\Big\{\Big(\boldsymbol{\sigma}^{A}\cdot\hat{\mathbf{u}}_{A'}\Big)\otimes\Big(\boldsymbol{\sigma}^{B}\cdot\hat{\mathbf{u}}_{B}\Big)\Big\}\Big]&+\\ \mathrm{Tr}\Big[\rho^{AB}\Big\{\Big(\boldsymbol{\sigma}^{A}\cdot\hat{\mathbf{u}}_{A}\Big)\otimes\Big(\boldsymbol{\sigma}^{B}\cdot\hat{\mathbf{u}}_{B'}\Big)\Big\}\Big]&-\\ \mathrm{Tr}\Big[\rho^{AB}\Big\{\Big(\boldsymbol{\sigma}^{A}\cdot\hat{\mathbf{u}}_{A'}\Big)\otimes\Big(\boldsymbol{\sigma}^{B}\cdot\hat{\mathbf{u}}_{B'}\Big)\Big\}\Big]\Big|&\leq 2\end{aligned} \tag{2.29}$$

Here, Tr stands for trace defined in (A.8); the other quantities present in this operator-inequality have their meanings identical to those implied by (2.22) and (2.23). Further discussions on Bell inequalities, both for bipartite and multipartite systems, can be found, for example, in [24, 133].

Schmidt Decomposition

It is a very simple, necessary and sufficient test for the separability. It is, however, applicable only to pure bipartite states.

Let us consider the bipartite system specified in Sect. A.3 of Appendix A. Then, an arbitrary state in the Hilbert space of this system can be expanded as[5]

$$|\psi\rangle = \sum_{i\,j} C_{i\,j}\,|a_i\rangle\,|b_j\rangle = \sum_i |a_i\rangle\,|\mathfrak{b}_i\rangle, \tag{2.30a}$$

where

$$|\mathfrak{b}_i\rangle \equiv \sum_j C_{i\,j}|\,b_j\rangle. \tag{2.30b}$$

Although, $|a_i\rangle$ and $|b_j\rangle$ form orthonormal basis in the respective spaces of their subsystems A and B, however, $|\mathfrak{b}_i\rangle$ need not be mutually orthogonal or normalized.

If the composite system $A \otimes B \equiv A\,B$ is assumed to be in a pure state, then according to (A.13), its density operator is given by

$$\rho^{A\,B} = |\psi\rangle\langle\psi|. \tag{2.31}$$

In view of (A.25),

$$\rho^A \equiv \mathrm{Tr}_B\left(\rho^{A\,B}\right) = \mathrm{Tr}_B\left(|\psi\rangle\langle\psi|\right) \tag{2.32}$$

is the reduced density operator for the component A of the bipartite system $A \otimes B$. On substituting relation (2.30a), density operator (2.32) becomes

$$\begin{aligned}
\rho^A &= \mathrm{Tr}_B\left[\left(\sum_i |a_i\rangle\,|\mathfrak{b}_i\rangle\right)\left(\sum_{i'} \langle a_{i'}|\,\langle\mathfrak{b}_{i'}|\right)\right] \\
&= \sum_{i\,i'} \left[\mathrm{Tr}_B\left(|\mathfrak{b}_i\rangle\langle\mathfrak{b}_{i'}|\right)\right]\left(|a_i\rangle\langle a_{i'}|\right) \\
&= \sum_{i\,i'}\left(\sum_k \langle b_k|\,\mathfrak{b}_i\rangle\,\langle\mathfrak{b}_{i'}|\,b_k\rangle\right)\left(|a_i\rangle\langle a_{i'}|\right) \\
&= \sum_{i\,i'}\left(\sum_k \langle\mathfrak{b}_{i'}|\,b_k\rangle\langle b_k|\,\mathfrak{b}_i\rangle\right)\left(|a_i\rangle\langle a_{i'}|\right) \\
&= \sum_{i\,i'} \langle\mathfrak{b}_{i'}|\,\mathfrak{b}_i\rangle\left(|a_i\rangle\langle a_{i'}|\right),
\end{aligned} \tag{2.33}$$

where $|\mathfrak{b}_k\rangle$ are already defined in (2.30b). If $|a_i\rangle$ are the bases in which ρ^A is diagonal with eigenvalues p_i, then

$$\rho^A = \sum_i p_i\,|\,a_i\rangle\langle a_i\,|. \tag{2.34}$$

---

[5]The following demonstration is based upon that given in [134].

On comparing (2.33) and (2.34), one finds

$$\langle \mathfrak{b}_{i'} | \mathfrak{b}_i \rangle = p_i \, \delta_{i\,i'} \tag{2.35}$$

Thus, the $|\mathfrak{b}_i\rangle$, defined in (2.30b), are also orthogonal. In order to normalize these new kets, we write them as follows:

$$\begin{aligned} |b_i'\rangle &= p_i^{-\frac{1}{2}} \, |\mathfrak{b}_i\rangle \\ &= p_i^{-\frac{1}{2}} \sum_j C_{i\,j} | b_j \rangle. \end{aligned} \tag{2.36}$$

Here, one may assume $p_i \neq 0$, because one will have only those p_i's appearing in (2.36) which make a nonzero contribution to (2.34). Now (2.30a) becomes

$$|\psi\rangle = \sum_{i=1}^{n_s} \sqrt{p_i} \, |a_i\rangle |b_i'\rangle \tag{2.37}$$

in terms of the particular orthonormal basis of A and B. Any bipartite pure state can be expressed in the form of (2.37) wherein the bases will depend upon the pure state being expanded. Kets $|a_i\rangle$ and $|b_i'\rangle$, present in (2.37), are called Schmidt bases for the respective subsystems A and B. The number n_s of nonzero values present in (2.37), where $n_s \leq \text{Min}(d_A, d_B)$, is called the Schmidt number for the state $|\psi\rangle$. A given pure state is separable if its Schmidt number is one [134].

Although, Schmidt decomposition is not unique for a given pure state of a bipartite system, it is, nevertheless, different for different pure states of the same system. The Schmidt decomposition cannot, in general, be extended to a quantum system composed of more than two subsystems, even though it is in a pure state.

2.4.1.2 Mixed States

For mixed states, the situation about the criterion for classifying an entangled state is less simple. Several necessary and/or sufficient conditions for the separability of the state of a system consisting of more than one particle have been developed. These have aptly been reviewed in several articles published by various workers (see, e.g., [24, 126, 127, 132, 135], *etc.*). In the following, we briefly describe only those separability conditions which are user friendly and, hence, in vogue. As these are the criteria obeyed by a separable density matrix, their violation naturally means that the state represented by the corresponding density operator is entangled.

Peres–Horodecki Criterion (Positive Partial Transpose)

As the name suggests, the partial transpose (PT) of the density matrix of a multipartite system is obtained by transposing it with respect to the indices of only one of its subsystems. Consider a multipartite system which is composed of the parts $A_1, A_2, A_3, \ldots, A_N, \ldots$. The elements of this multipartite system's density matrix that is partially transposed with respect to, say, its subsystem A_N are given by [124]

$$\left(\rho^{\mathrm{T}_N}\right)_{i_1 i_2 \ldots i_{N-1} i_N i_{N+1} \ldots;\, j_1 j_2 \ldots j_{N-1} j_N j_{N+1} \ldots} = \rho_{i_1 i_2 \ldots; i_{N-1} j_N i_{N+1} \ldots; j_1 j_2 \ldots j_{N-1} i_N j_{N+1} \ldots}. \tag{2.38}$$

Here we have defined, for brevity, $\rho^{A_1 A_2 \ldots A_N \ldots} \equiv \rho$, with $(i_1, j_1), (i_2, j_2), \ldots, (i_N, j_N), \ldots$ subscripts referring to the subsystems $A_1, A_2, \ldots, A_N, \ldots$, respectively. Let, ρ_{sep} be a separable state of this multipartite system. then, according to (2.27), it can be decomposed as

$$\rho_{\mathrm{sep}} = \sum_k p_k\, |a_k^{(1)}\rangle\langle a_k^{(1)}| \otimes |a_k^{(2)}\rangle\langle a_k^{(2)}| \otimes \ldots \otimes |a_k^{(N-1)}\rangle\langle a_k^{(N-1)}| \otimes |a_k^{(N)}\rangle\langle a_k^{(N)}| \otimes |a_k^{(N+1)}\rangle\langle a_k^{(N+1)}| \otimes \ldots \tag{2.39a}$$

Here, $\{|a_k^{(1)}\rangle\}$ (with $k = 1, 2, \ldots, d_1$), $\{|a_k^{(2)}\rangle\}$ (with $k = 1, 2, \ldots, d_2$), *etc.*, are the complete orthonormal basis sets of sizes d_1, d_2, *etc.*, for the parts A_1, A_2, *etc.*, constituting our multipartite system. On defining $\rho_k^{A_1} \equiv |a_k^{(1)}\rangle\langle a_k^{(1)}|$, *etc.*, expression (2.39a) can equivalently be written as

$$\rho_{\mathrm{sep}} = \sum_k p_k\, \rho_k^{A_1} \otimes \rho_k^{A_2} \otimes \ldots \otimes \rho_k^{A_N} \otimes \ldots \tag{2.39b}$$

If the sums in (2.39) contain only a single (i.e., $k = 1$) product term, then ρ_{sep} is a pure, otherwise a mixed, separable state.

In view of the definition (2.38), the PT of (2.39) with respect to the subsystem, say, A_N is given by

$$\rho_{\mathrm{sep}}^{\mathrm{T}_N} = \sum_k p_k\, \rho_k^{A_1} \otimes \rho_k^{A_2} \otimes \ldots \otimes \left(\rho_k^{A_N}\right)^{\mathrm{T}_N} \otimes \ldots \tag{2.40a}$$

Since $\left(\rho_k^{A_N}\right)^{\mathrm{T}_N}$ is again a valid DM for the subsystem A_N and identical to its (i.e., A_N's) DM $\left(|a_k^{(N)}\rangle\langle a_k^{(N)}|\right)$, one immediately finds that, similar to ρ_{sep} in (2.39), its PT (2.40a) must also be positive (i.e., $\rho_{\mathrm{sep}}^{\mathrm{T}_N} \geq 0$)[6] for the former to represent a

[6] An operator, say, Ω is said to be positive if each of its eigenvalues is real and non-negative. It is denoted as $\Omega \geq 0$. The expectation value of a positive operator for any state is also positive. If each of the eigenvalues of this operator is strictly greater than zero, then Ω is said to be positive definite. See Appendix A for more details.

physically acceptable state of our separable multipartite system $A_1 \otimes A_2 \otimes \ldots \otimes A_N \otimes \ldots$ described by the state (2.39). The same holds if the operation of partial transposition on the state (2.39) is performed on a part other than A_N of this multipartite system. Thus, PT of a separable state ρ with respect to any of its component systems is also positive [124].[7] Such multipartite states are called fully separable [136, 137] and, hence, have the property

$$\rho_{\text{sep}}^{\mathrm{T}_1} \geq 0,\ \rho_{\text{sep}}^{\mathrm{T}_2} \geq 0,\ \ldots,\ \rho_{\text{sep}}^{\mathrm{T}_N} \geq 0,\ \ldots \tag{2.40b}$$

Thus, the positivity of the PT of a given state of more than one particles is a necessary condition for its separability. It was shown by Horodecki et al. [125] that the converse (i.e., if $\rho^{\mathrm{T}_N} \geq 0$, then ρ is separable) is true only for the low-dimensional systems, namely bipartite states of two, say, qubits A and B (i.e., $d_A = d_B = 2$), or of one qubit and one qutrit (i.e., $d_A = 2$, $d_B = 3$, or vice-versa). Hence, in these two cases, the positivity of the partial transpose (i.e., PPT) is a necessary and sufficient condition for the separability of a bipartite state. For higher dimensional bipartite states, it is only necessary, and the existence of entangled states with PPT has been shown [126, 132, 138]. Nonseparable states with dimensions $d_A\, d_B > 6$ possessing PPT are called [126, 132, 138] bound entangled states.[8] Krauss et al. [139] and Horodecki et al. [140] have, in addition, shown that a bipartite state represented by a density matrix possessing PPT and with a rank[9] less than or equal to max (d_A, d_B) is separable as well; on the other hand, a PPT bipartite state has bound entanglement if its rank $\geq$ max (d_A, d_B).

Sanpera et al. [141] have shown that out of the four possible eigenvalues of the PT of a DM of two qubits, only one can be negative if this bipartite system is in a nonseparable state. On the other hand, PT of an entangled state of a composite (bipartite/tripartite/multipartite) system of dimensions > 4 can possess even more than one negative eigenvalues [142]. Sanpera et al. [143] have indeed shown that partial transposition can be interpreted as a time reversal operation. Wolf [144] has studied the properties of the operation of partial transposition in quantum information in detail.

An application of one of the well-known properties [145] of the direct product of matrices to a separable state like (2.39) immediately gives

$$(\rho_{\text{sep}})^{\mathrm{T}} = \sum_k p_k \left(\rho_k^{A_1}\right)^{\mathrm{T}_1} \otimes \left(\rho_k^{A_2}\right)^{\mathrm{T}_2} \otimes \ldots \otimes \left(\rho_k^{A_N}\right)^{\mathrm{T}_N} \otimes \ldots \tag{2.41a}$$

Here, the superscript "T" stands for the operation of full transposition which too is positive operator. Using the property (2.41a), one then immediately has

[7] Although, the form of a partially transposed matrix (e.g., ρ^{T_1}, or ρ^{T_2}, *etc*) depends on the choice of the bases, its eigenvalues are, however, always independent of the chosen bases.

[8] Sect. 4.2.2 on pages 106–112 in Chap. 4 contains a brief description of bound entanglement.

[9] Rank of a matrix is equal to the number of its nonzero eigenvalues [132].

$$\left[\left(\rho^A\right)^{\mathrm{T}_A} \otimes \rho^B\right]^{\mathrm{T}} = \rho^A \otimes (\rho^B)^{\mathrm{T}_B} \tag{2.41b}$$

for the product state $\rho^A \otimes \rho^B$ of the subsystems A and B of any dimensions. The full transposition of a matrix does not change its eigenvalues. For a separable bipartite state, therefore, eigenvalues of $\left(\rho^A\right)^{\mathrm{T}_A} \otimes \rho^B$ must be identical to those of $\rho^A \otimes (\rho^B)^{\mathrm{T}_B}$, while the two states may themselves differ from each other. However, the property (2.41b) may not, in general, hold for nonseparable bipartite states which cannot be written in the form of (2.41a).

Reduction Criteria

It is a necessary condition for the separability of a bipartite state of qudits. According to this test [127, 132, 146], if ρ is the density matrix of a separable bipartite state, then it must satisfy the inequalities

$$\left(\rho^A \otimes \mathbb{I}_B\right) - \rho \geq 0 \tag{2.42a}$$

and

$$\left(\mathbb{I}_A \otimes \rho^B\right) - \rho \geq 0. \tag{2.42b}$$

Here, ρ^A and ρ^B are the reduced matrices for the subsystems A and B calculated according to [(A.24), (A.26), (A.27)] from their composite matrix ρ; each of $\mathbb{I}_A$ and $\mathbb{I}_B$ is a unit matrix with sizes same as that of ρ^A and ρ^B, respectively. The positivity of the left-hand sides of (2.42) is a separability criterion because it corresponds to an application of the positive map [125,127,146] to B's subsystem or to A's subsystem. A positive map [125,132,146] $\Lambda(\sigma) = \mathrm{Tr}(\sigma) \otimes I - \sigma$ applied to one subsystem of a separable state σ preserves the properties of a density matrix. The resulting density matrix, therefore, has to remain positive [125, 132, 146].

2.4.2 *Entanglement Measures*

This section contains a brief discussion of some of the methods mostly used for quantifying entanglement present in a given nonseparable state. Such methods are also called entanglement measures. There are many articles (e.g., [24, 147–151], *etc.*) which have discussed entanglement measures in greater details and provide extensive references to original literature on this important topic. Although, there are a large number of entanglement measures; however, a satisfactory understanding of such measures has so far been obtained for states representing two subsystems together, each with finite or infinite degrees of freedom. It has been shown [142,152–154] that a good entanglement measure must satisfy at least four conditions. Let us denote by $\mathcal{E}(\rho)$ the amount of entanglement quantified in a bipartite state $\rho^{AB} \equiv \rho$ by a given entanglement measure. Here, A and B are two parties each holding one qubit. For this entanglement measure to be good, $\mathcal{E}(\rho)$ must satisfy following conditions (see, for example, [142, 148, 152–154], *etc*)

(i) $0 \leq \mathcal{E}(\rho) \leq 1$ always; $\mathcal{E}(\rho) = 0$ for a separable (i.e., product, or un-entangled)[10] ρ; $\mathcal{E}(\rho) > 0$ for a nonseparable (i.e., nonproduct, or entangled)[11] ρ; $\mathcal{E}(\sigma_1) = 1$, where σ_1 is the Bell state (2.28c).

(ii) $\mathcal{E}(\rho)$ does not change under any local operations (LO).[12]

(iii) $\mathcal{E}(\rho)$ is not increased by local operations combined with classical communications (LOCC),[13] performed on the two parts of the bipartite system.

(iv) The measure must satisfy the condition of monotonicity under discarding information, i.e., $\mathcal{E}\left(\Sigma_i\ p_i\ \rho_i\right) \leq \Sigma_i\ p_i\ \mathcal{E}(\rho_i)$.

The following measures are currently in vogue for calculating the amount of entanglement present in a nonseparable state of two qubits.

2.4.2.1 Pure States

Quantification of entanglement in a bipartite pure state is very simple. The reduced density matrix ρ^A or ρ^B of a sub-system is always pure if the state ρ of the composite system is pure as well as separable. However, if ρ is entangled, then both ρ^A are ρ^B are necessarily mixed even though ρ itself may be pure. Hence, the degree of entanglement of a pure state ρ of a bipartite system is associated with the degree of mixing of its reduced states. It is the von Neumann [155] entropy E_{vN} which is generally used to quantify the degree of mixing of a reduced state ρ^A or ρ^B. Then the entanglement measure for a pure state of a composite system is

$$E_{vN} = -\mathrm{Tr}\left[\rho^A \log_2\left(\rho^A\right)\right] = -\mathrm{Tr}\left[\rho^B \log_2\left(\rho^B\right)\right]. \tag{2.43}$$

---

[10]For ρ to represent a product state, it should be possible to write it in the form of (2.39) [or, equivalently, (A.4)].

[11]A non-separable state ρ cannot be written in the form of (2.39) [or, equivalently, (A.4)].

[12]It consists of unitary transformations or (generalized) measurements, say, Ω_A and Ω_B, performed separately by A and B on their respective qubits of a bipartite state. Such a local transformation/operation is written as $\Omega_A \otimes \Omega_B$.

[13]It is the local operations performed in (ii) which have been correlated by classical communications. That is, the parties A and B communicate to each other the results of their measurements on their respective qubits of a shared bipartite state using presently available telecom technologies based on classical information. For more on LOCC see, for example, [126] and references therein.

Since the entropy of a pure state is always zero, the entropies of both ρ^A and ρ^B of a bipartite pure state ρ of a composite system must be equal. The right-hand side of (2.43) is calculated [43] in the functional calculus with the definition $0 \log_2 0 \equiv 0$. The logarithm to the base 2 is used because each of the subsystems A and B in (2.43) is a qubit. It also means that the unit of entropy in the present case is a "bit" [43].

The von Neumann entropy (2.43) possesses properties to be a suitable measure of entanglement [153] for a pure bipartite state. It, in addition, (i) is additive, i.e., the entanglement of two copies $|\psi\rangle \otimes |\psi\rangle$ of the same state is double of the entanglement of a single copy $|\psi\rangle$ and (ii) has asymptotic continuity. This later property roughly means that small changes in a state leads to small changes in its entropy.

If the reduced density matrix of a pure bipartite state is a multiple of the identity matrix, the composite system is said to be maximally entangled because the entropy E_{vN} is then maximum. For example, each of the four Bell states (2.17) is maximally entangled. That is,

$$\rho^A = \mathrm{Tr}_B\left(|\psi^\pm\rangle\langle\psi^\pm|\right) = \mathrm{Tr}_B\left(|\phi^\pm\rangle\langle\phi^\pm|\right) = \frac{1}{2}\mathbb{I}_2 \tag{2.44a}$$

and

$$\rho^B = \mathrm{Tr}_A\left(|\psi^\pm\rangle\langle\psi^\pm|\right) = \mathrm{Tr}_A\left(|\phi^\pm\rangle\langle\phi^\pm|\right) = \frac{1}{2}\mathbb{I}_2. \tag{2.44b}$$

Here, $\mathbb{I}_2$ is a (2×2) unit matrix.

A maximally entangled pure state of two spin-$\frac{1}{2}$ qubits (say, A and B) physically means [99] that if one measures spin of the particle A along any axis, the result is completely random—one finds spin up with probability $\frac{1}{2}$ and spin down with the same probability of $\frac{1}{2}$.[14] Therefore, if one performs any local measurements on A or B, one requires no information about the preparation of their corresponding states, one instead generates a random qubit. von Neumann entropy (2.44) for a maximally entangled pure state is $\log_2 2$.

The situation is, however, quite different with the case of a single qubit in a pure state. There, one can store a bit by preparing, say, either $|0\rangle \equiv |\uparrow_z\rangle$, or $|1\rangle \equiv |\downarrow^z\rangle$, and one can recover that bit reliably by measuring along the OZ-axis. With two qubits, one ought to be able to store two bits, but in the maximally entangled Bell states (2.17) this information is hidden, i.e., one can not acquire it by measuring A or B.

Maximally entangled two qubit states play central role in quantum information science. For example, since entanglement cannot be increased by LOCC, these are

[14] A maximally entangled mixed state, on the other hand, is the one [142] which, for a given mixedness, achieves the greatest possible entanglement. See, for example, discussion given on pages 128 and 129.

the only states from which any other state can be obtained asymptotically by LOCC alone [156].

2.4.2.2 Mixed Sates

For mixed states, the picture is quite complicated. Several [142] entanglement measures for such states have so far been introduced for quantifying their degree of entanglement. Presently, the most widely used computational measures of entanglement for mixed, bipartite, states of qubits are (a) negativity, (b) concurrence, and (c) entanglement of formation. The following contains a brief description of each of these three measures without their technical derivations which can always be found in the corresponding original references cited herein.

Negativity

According to the Peres–Horodecki [124, 125] requirement (herein discussed on pages 38–41), PPT is a necessary and sufficient condition for identifying a product state represented by a matrix of size ≤ 6. Zyczowski et al. [157] related this condition for the separability with the trace norm[15] $\| \rho \|_1$ of the DM ρ of a given state. Later on, Vidal and Werner [159] showed that for a physically acceptable DM ρ normalized to unit trace, $\| \rho \|_1 = \mathrm{Tr}\rho = 1$ and argued that $\| \mathrm{PT}(\rho) \|_1 = 1 + 2\,\big|$sum of the negative eigenvalues of $\mathrm{PT}(\rho)\big|$ as $\mathrm{PT}(\rho)$ may have negative eigenvalues as well. Here, and in the following, $\mathrm{PT}(\rho)$ always means, unless stated otherwise, partial transpose [124] of the DM ρ with respect to one of its subsystems. The negativity of the state ρ is then defined as [159]

$$\mathcal{N}(\rho) = \left| \sum \left[\text{negative eigenvalues of the PT}(\rho) \right] \right|. \tag{2.45a}$$

The PT of a DM too is Hermitian and, hence, $\| \ \mathrm{PT}(\rho) \ \|_1 = \mathrm{Tr}\,[\mathrm{PT}(\rho)] = 1$ for a separable state ρ normalized to unit trace. Thus, negativity of a DM gives us the amount (or degree) by which a given state ρ of more than one particles violates the Peres–Horodecki [124, 125] condition for the separability. It has also been proved [159] that negativity is a good measure of entanglement. In the case of bipartite states of qubits, wherein the partial transpose of an entangled state can have only one negative eigenvalue [141], negativity (2.45a) can also be written as [147]

---

[15]The trace norm of a matrix M is defined by (see, e.g., Sect. VI 6 in [158]) $\| M \|_1 \equiv \mathrm{Tr}\sqrt{M\,M^\dagger}$. Square roots of the eigenvalues of $M\,M^\dagger$ are called singular values of the matrix M. Remembering that the trace of a matrix is always equal to the sum of its eigenvalues, the trace norm of a matrix is, therefore, the sum of its singular values. If M is a Hermitian matrix then, obviously, $\| M \|_1 = \sum |\text{eigenvalues of } M|$.

$$\mathcal{N}(\rho) = \left|\min\left[0,\ \text{smallest eigenvalue of PT}\,(\rho)\right]\right|. \tag{2.45b}$$

However, some workers (see, e.g., [142]) have defined negativity to be twice of that given in [(2.45a), (2.45b)]. Hence,

$$\mathcal{N}(\rho) = 2\ \max\left[0,\ -\sum\left(\text{negative eigenvalues of PT}\,(\rho)\right)\right] \tag{2.45c}$$

for states of more than one particles with dimension ≥ 4.

Thus, for a Werner [121] state (2.28), $\mathcal{N}(\rho_W) = 0$ for mixing parameter $p < \frac{1}{3}$, $\mathcal{N}(\rho_W) = 1$ for $p = 1$, or $0 < \mathcal{N}(\rho_W) < 1$ for $\frac{1}{3} < p < 1$, i.e., for the separable, maximally entangled, or for a mixed entangled states ρ_W, respectively. Negativity is probably the only measure of entanglement which is easiest to calculate as it requires simple operation of PT followed by diagonalization of a matrix.

Concurrence

This is a very successful and widely used measure for quantifying entanglement of a given bipartite state. Concurrence has been theoretically calculated as well as experimentally measured. (See, e.g., [160] for a review on concurrence.) The concept of concurrence was first introduced in [161, 162] for pure bipartite states of qubits. It was later generalized to mixed bipartite states of qubits by Wootters [163]. Rungta et al. [164] applied this concept to bipartite, pure states of qudits. Mintert et al. [165] have, since then, introduced concurrence for mixed multipartite states of qubits and obtained necessary expressions for its calculation for such systems.

According to Wootters [149, 163], concurrence for a pure/mixed state ρ of two qubits is given by

$$\mathcal{C}(\rho) = \max\left(0,\ \lambda_1 - \lambda_2 - \lambda_3 - \lambda_4\right). \tag{2.46a}$$

Here, λ_i's are the eigenvalues of the matrix

$$R = \left[\sqrt{\rho}\,\tilde{\rho}\,\sqrt{\rho}\right]^{\frac{1}{2}} \tag{2.46b}$$

in decreasing order with

$$\tilde{\rho} = \left(\sigma_y \otimes \sigma_y\right) \rho^* \left(\sigma_y \otimes \sigma_y\right) \tag{2.46c}$$

the spin-flipped state of ρ. In expression (2.46c), σ_y is the y-component of the Pauli spin matrix $\boldsymbol{\sigma}$ given in (2.6) and the elements of ρ^* are complex conjugate of those of the density matrix ρ. Negativity and concurrence have same values [142] for a pure state of two qubits; although, these two measures of entanglement may differ for mixed, bipartite entangled states. A state ρ of two qubits is separable if its concurrence is zero, it is maximally entangled for $\mathcal{C}(\rho) = 1$; otherwise, $0 < \mathcal{C}(\rho) < 1$ if ρ happens to represent a mixed state.

Hitherto, concurrence is, probably, the only measure which has very successfully been used to experimentally quantify entanglement in a quantum state of more than one qubits. For example, Walborn et al. [166, 167] measured concurrence for twin photons entangled in two degrees of freedom, prepared in pure as well as identical states of polarization and momentum. Subsequent to these [166, 167] experiments, and with a view to extend such measurements of entanglement to mixed multipartite states, both lower [168, 169] and upper [170] bounds of concurrence for arbitrary bipartite [168], as well as multipartite [169, 170], states of qubits have been developed. Huang et al. [171] have measured both of such [168, 170] bounds for a mixed state of two photonic qubits.

Entanglement of Formation

It is, probably, the most basic of the measures hitherto developed in quantum information for quantifying entanglement present in a given bipartite state of qubits. In (2.43), entanglement of a pure, bipartite state is defined to be the von Neumann [155] entropy of any of its two members. The entanglement of formation [herein denoted by $\mathcal{E}_F(\rho)$] of a mixed state ρ of two qubits is defined [161], on the other hand, to be the minimum average entanglement of an ensemble of pure states needed to represent ρ. In other words [162], $\mathcal{E}_F(\rho)$ is the minimum number of the singlets (2.17b) required for generating an ensemble of pure states which can represent the mixed state ρ possessing the desired amount of entanglement.

It was originally introduced by Bennett et al. [161]. An explicit formula, applicable to a special class of states of two qubits, was conjectured for $\mathcal{E}_F(\rho)$ in [162]. Finally, Wootters [149, 163] gave following expression:

$$\begin{aligned}\mathcal{E}_F(\rho_f) = &-\frac{1+\sqrt{1-\mathcal{C}(\rho_f)^2}}{2}\log_2\left(\frac{1+\sqrt{1-\mathcal{C}(\rho_f)^2}}{2}\right)\\ &-\frac{1-\sqrt{1-\mathcal{C}(\rho_f)^2}}{2}\log_2\left(\frac{1-\sqrt{1-\mathcal{C}(\rho_f)^2}}{2}\right)\end{aligned} \tag{2.47}$$

for calculating entanglement of formation for any arbitrary bipartite state of qubits. Thus, according to (2.47), entanglement of formation is an increasing function of the concurrence $\mathcal{C}(\rho)$ already defined in (2.46).

Chapter 3
Theory

3.1 Density Operators

It is obvious from the discussion given in Chap. 2 that for a proper study of the entanglement properties of a system of more than one particle, one needs to know the density matrix of the quantum state one is interested in. For an ab-initio calculation of such a density matrix, the corresponding density operator is always required. This chapter presents useful methodologies for calculating the density operators and density matrices for the processes (1.1)–(1.8).

In general, the density operator of a state ρ of a system generated in a process, which takes place in the initial state ρ' of the same system, is given by [61]

$$\rho = F_{\text{op}}\, \rho' \left[F_{\text{op}}\right]^{\dagger}. \tag{3.1}$$

Here, F_{op} is the operator which performs the desired conversion of the system from its initial state ρ' to the final state ρ. The density operators ρ and ρ' in (3.1) are independent of the dynamical models one may like to use in a study. In (3.1) and elsewhere in this book, superscript $\dagger$ always means, unless stated otherwise, Hermitian conjugate [58–61].

3.1.1 Photoabsorption

Absorption of a single incident photon γ_r by a target $\mathfrak{T}$ may result in several possibilities. The density operator for such an absorption process is given by (see, for example, [61, 62, 172, 173])[1]

[1]For meanings of the symbols $\mathcal{K}_{\text{p}}$ and F_p—present in (3.2) and elsewhere in this monograph—see (3.6) and the related discussion.

N. Chandra and R. Ghosh, *Quantum Entanglement in Electron Optics*, Springer Series on Atomic, Optical, and Plasma Physics 67, DOI 10.1007/978-3-642-24070-6_3,

$$\rho_{\mathrm{p}} = \mathcal{K}_{\mathrm{p}}\, F_{\mathrm{p}}\, \rho_i\, F_{\mathrm{p}}^{\dagger}. \tag{3.2}$$

In none of the processes (1.1)–(1.8), there is any interaction between γ_r and $\mathfrak{T}$ before the absorption of this incident photon by the target takes place. Hence,

$$\rho_i = \rho_0 \otimes \rho_{\mathrm{r}} \tag{3.3a}$$

in (3.2) is the direct product of the density operators

$$\rho_0 = |0\rangle\langle 0| \tag{3.3b}$$

of the target $\mathfrak{T}$ (assuming it absorbs γ_r in its state $|0\rangle$) and

$$\rho_{\mathrm{r}} = |1m_{\mathrm{r}}\rangle\langle 1m_{\mathrm{r}}| \tag{3.3c}$$

of the absorbed radiation γ_r (in the $E1$ approximation).[2] Thus, ρ_i in (3.2) represents the uncorrelated (polarized photon + polarized target)-system before γ_r is absorbed by $\mathfrak{T}$.

If, on the other hand, the target $\mathfrak{T}$ is initially unpolarized, then all its states degenerate to $|0\rangle$ are equally accessible [10, 61, 68] in an experiment. The density operator (3.3b) then needs to be averaged over all such degenerate states. Consequently, one writes [61]

$$\rho_0 = \frac{1}{n_{d_0}} \sum_{d_0} |0\rangle\langle 0| \tag{3.3d}$$

for the density operator of a target in its unpolarized state $|0\rangle$. Here, d_0 represents all those quantum numbers needed to specify states of $\mathfrak{T}$ degenerate to $|0\rangle$, and the total number of these states is n_{d_0}.

Thus, on defining

$$|0;\, 1m_r\rangle \equiv |0\rangle|1m_r\rangle, \tag{3.4a}$$

one has

$$\rho_i = |0;\, 1m_r\rangle\langle 0;\, 1m_r| \tag{3.4b}$$

---

[2]Equation (3.3c) is appropriate for describing a γ_r in a definite state of polarization (i.e., LP, RCP, LCP, or UP). Density matrix for an arbitrarily polarized photon is given in the Appendix B on pages 265–268.

for the density operator of an uncorrelated system of polarized target $\mathfrak{T}$ in its state $|0\rangle$ plus an incident photon γ_r in the $E1$ approximation in the state $|1m_r\rangle$ of a definite polarization.

However, if the state $|0\rangle$ of the target $\mathfrak{T}$ happens to be unpolarized described by the density operator (3.3d), then

$$\rho_i = \frac{1}{n_{d_0}} \sum_{d_0} |0;\ 1m_r\rangle\langle 0;\ 1m_r| \tag{3.4c}$$

represents uncorrelated system of (polarized photon + unpolarized target).

Hence, the substitution of (3.4b) in (3.2) will give us the density operator

$$\rho_p \longrightarrow \rho^{(1)} = \mathcal{K}_p F_p \left(|0;\ 1m_r\rangle\langle 0;\ 1m_r|\right) F_p^\dagger \tag{3.5a}$$

for any 1-step process taking place due to the absorption, in $E1$ approximation, of a polarized photon γ_r by a polarized target $\mathfrak{T}$ in its state $|0\rangle$.

If, on the other hand, the target $\mathfrak{T}$ is unpolarized, then

$$\rho_p \longrightarrow \rho^{(1)} = \frac{\mathcal{K}_p}{n_{d_0}} \sum_{d_0} F_p \left(|0;\ 1m_r\rangle\langle 0;\ 1m_r|\right) F_p^\dagger \tag{3.5b}$$

is the appropriate density operator to be used in place of (3.5a). This last expression is obtained on combining (3.2) and (3.4c).

In each of the processes (1.1)–(1.8), considered in this monograph, it is always the unpolarized target $\mathfrak{T}$ in its state $|0\rangle$ which absorbs—in $E1$ approximation—a single photon γ_r in its state $|1m_r\rangle$ of definite polarization.

Further, in (3.1), (3.5), and hereafter in this monograph, unless stated otherwise, F_p is the appropriate photoabsorption operator with its associated quantity $\mathcal{K}_p$. For example, as is the case with the processes (1.1)–(1.8), if the absorption of γ_r happens to eject an electron, say, e_p from $\mathfrak{T}$, then

$$F_{\mathrm{p}} = \sqrt{\frac{m_{\mathrm{e}}}{\hbar^2}}\ \mathcal{F} \tag{3.6a}$$

is the required photoionization operator with m_{e} being the reduced mass of the photoelectron e_p and (see, for example, [72, 174])

$$\mathcal{F} = \sum_{i=1}^{n_{\mathrm{e}}} \hat{\boldsymbol{\xi}}_{m_{\mathrm{r}}} \cdot \boldsymbol{r}_i \tag{3.6b}$$

in the $E1$ length approximation, or

$$\mathcal{F} = \sum_{i=1}^{n_e} \hat{\xi}_{m_r} \cdot \nabla_i \tag{3.6c}$$

in the $E1$ velocity approximation. In (3.6), n_e is the total number of electrons in $\mathfrak{T}$ [i.e., the atomic number Z of $\mathfrak{T}$ if it happens to represent an atom in (1.1)–(1.8)] before its photoionization takes place; $\hat{\xi}_{m_r}$ (with $m_r = 0, +1, -1$) is the spherical unit vector [65] specifying polarization[3] of the ionizing radiation γ_r in the SF frame $OXYZ$ shown in Fig. 1.1; $\boldsymbol{r}_i$, and $-\mathrm{i}\,\hbar\,\nabla_i$ are the respective position and linear momentum operators—both in the SF—of the i-th electron in $\mathfrak{T}$.

Furthermore, the quantity $\mathcal{K}_p$—present in the density operator (3.2) and elsewhere in this monograph—contains energy of the absorbed photon and some fundamental constants (*e.g.*, [72], charge and mass of e_p, fine-structure constant α, Bohr radius a_0, etc.). It is, thus, totally independent of all the spatial and spin variables of the particles participating in the processes (1.1)–(1.8). Its explicit form [10, 72, 174] very much depends upon the kind of the energy normalization used for the continuum orbital [i.e., $R^-_{\ell_p}(r_p; k_p)$ in (3.22a) on page 61] representing the photoelectron e_p. This explicit form of $\mathcal{K}_p$, however, will not be needed in the present study because a density matrix to be used by us has to be normalized always to unit trace.

3.1.2 Spontaneous Emission

Auger [175] discovered that if an atomic or molecular target happens to have a vacancy in one of its inner-shells, it can spontaneously lose its extra energy by emitting either an electron (say, e_a), or photon (say, γ_d). The e_a emitted in Fig. 3.1 in an inner-shell vacancy's non-radiative decay has come to be known [73, 74, 176] as Auger electron; whereas, photon emitted in Fig. 3.2 in the radiative decay of an inner-shell vacancy is usually called [175] fluorescence (or X-ray). Auger [73, 74, 175, 176] further showed that a vacancy in the outer (inner)-shell of a heavy (light) element usually decays non-radioactively; but, preferably a radiative decay of an inner-core excited state of a heavy element takes place. Thus, both the non-radiative and radiative emissions (1.3) and (1.4), respectively, are 2-step processes wherein the first step consists of creation—due to the absorption of γ_r—of an inner-shell vacancy whose spontaneous decay constitutes the second step. Auger effect [73, 74, 175, 176] has played an important role both in basic and in applied sciences since its discovery.

[3] $\hat{\xi}_0 = \hat{e}_z$: linear polarization along the OZ-axis;
$\hat{\xi}_{+1} = -\frac{1}{\sqrt{2}}(\hat{e}_x + \mathrm{i}\hat{e}_y)$: right circular polarization;
$\hat{\xi}_{-1} = \frac{1}{\sqrt{2}}(\hat{e}_x - \mathrm{i}\,\hat{e}_y)$: left circular polarization.

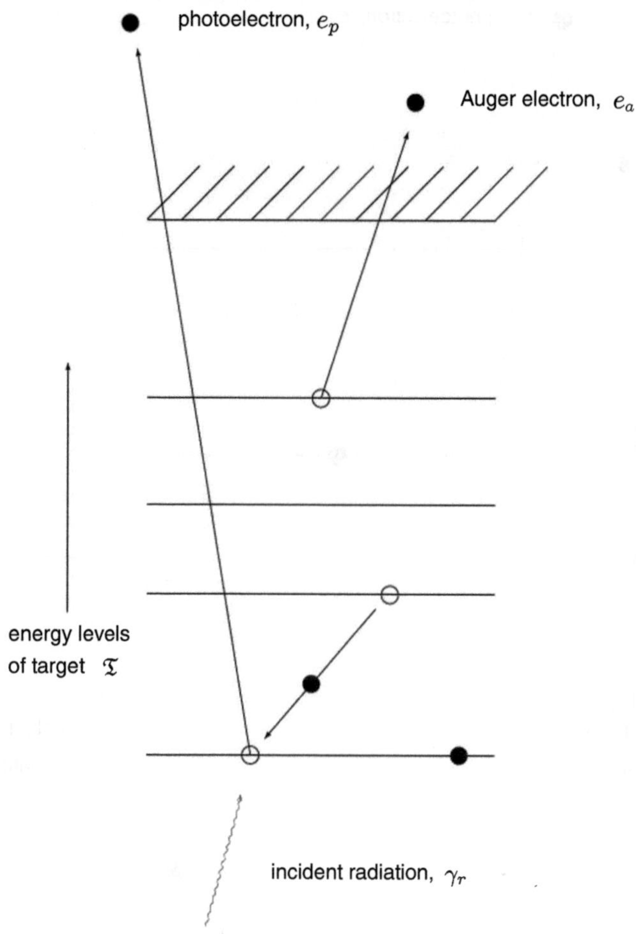

Fig. 3.1 Non-radiative decay (1.3) of an inner-shell vacancy (created in photoionization) taking place in an atom or a molecule due to rearrangement of electrons caused by inter-electronic Coulomb interaction (3.8). (First published in [183].)

Density operator for spontaneous emission from state ρ' of an atom or a molecule with an inner-shell vacancy is given by (see, for example, [177])

$$\rho_{\rm s} = \mathcal{K}_{\rm s}\, F_{\rm s}\, \rho'\, F_{\rm s}^{\dagger}. \tag{3.7}$$

Here, $F_{\rm s}$ is the quantum mechanical form of the interaction responsible for the spontaneous decay of the core excited state ρ' of an atomic or molecular target. The constant $\mathcal{K}_{\rm s}$, associated with the operator $F_{\rm s}$, is explained in the following remaining part of the present section.

The emission of an Auger electron e_a in the decay process (1.3) is due to the inter-electronic Coulomb interaction giving rise to electrons' rearrangement

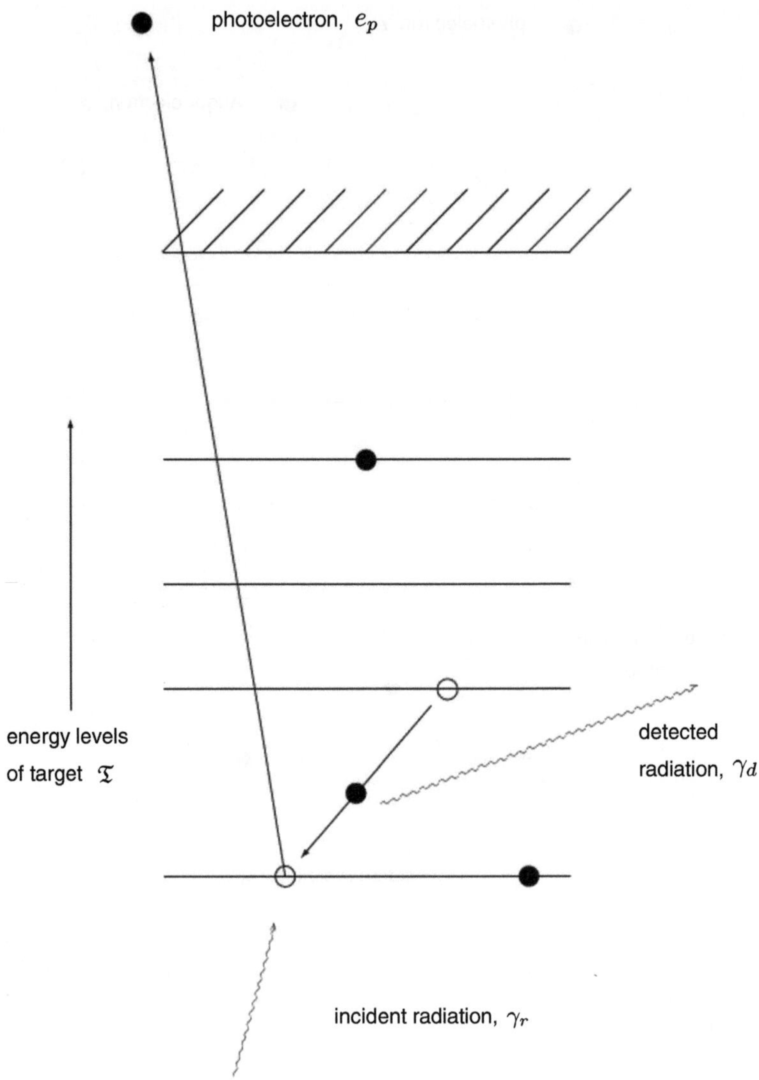

Fig. 3.2 Radiative decay (1.4) of an inner-shell vacancy (created in photoionization) taking place in an atom or a molecule. The quantum mechanical operator which performs this process in the $E1$ approximation is (3.9)

within the target possessing the inner-shell vacancy (see, Fig. 3.1). Hence, (see, for example, [177])

$$F_s \longrightarrow F_a = \frac{1}{4\pi\epsilon_0} \sum_{i \neq j}^{n'_e} |\boldsymbol{r}_i - \boldsymbol{r}_j|^{-1} \tag{3.8}$$

for the non-radiative decay of the state ρ'. Sum in (3.8) involves all electrons n'_e (with the position vector of the i-th electron given by $\boldsymbol{r}_i$) contained in the core-excited state ρ' of the target.

For the radiative decay of the core-excited state ρ', the appropriate operator F_s to be used in (3.7) is, on the other hand, given by [61, 177, 178]

$$F_s \longrightarrow F_\gamma = (\hat{\boldsymbol{u}}_d)^* \cdot \sum_{i=1}^{n'_e} \boldsymbol{r}_i . \tag{3.9}$$

Here, $\hat{\boldsymbol{u}}_d$ is the polarization vector for the emitted fluorescence γ_d referred to the photo-spectrometer's frame of reference. Expression in (3.9) is applicable to transitions allowed in the $E1$ approximation.

The quantity $\mathcal{K}_s$, present in (3.7), has different forms for two different kinds of spontaneous decays of the core-excited state ρ' (see, for example, [10, 177], *etc*). Similar to $\mathcal{K}_p$ in the photoabsorption (3.2), $\mathcal{K}_s$ too depends—but in different ways—on the energy of the emitted particle (i.e., Auger electron e_p, or fluorescence γ_d) and on some of the fundamental constants. Its exact form is determined, again, from the normalization (with respect to energy and/or otherwise) of the particles involved in a spontaneous decay process. Although its dependence (on energy and fundamental constants) is not same for the radiative and non-radiative decays, however, $\mathcal{K}_s$ in neither case contains any spatial or spin variables of any particle which may be participating in either of the spontaneous emissions (1.3) and (1.4). Consequently, contribution of $\mathcal{K}_s$ in (3.7), similar to that of $\mathcal{K}_p$ in (3.2), is eliminated due to the use of a normalized density matrix needed in a study relevant to quantum information and involving either, or both, of the photoabsorption and spontaneous decay processes.

3.1.3 Photoabsorption Followed by Sequential Spontaneous Emissions

Let us now consider an experiment wherein creation of an inner-shell vacancy due to the absorption of a single photon in a target $\mathfrak{T}$ is immediately followed by the spontaneous decay of the consequent core excited state. The density operator

$$\rho_s \longrightarrow \rho^{(2)} = \mathcal{K}_s\, F_s\, \rho_p\, F_s^\dagger = \mathcal{K}_s\, \mathcal{K}_p \left[F_s F_p\right] \rho_i \left[F_s F_p\right]^\dagger \tag{3.10}$$

for this 2-step sequential process is readily obtained by combining the density operators given in (3.1) and (3.7), i.e., on substituting for ρ_p from (3.2) for the core-excited state ρ' in (3.7). Similarly, if the state $\rho^{(2)}$ happens to subsequently decay [*e.g.*, in the processes (1.7c), (1.8c)], the density operator for the consequent 3-step sequential process is readily obtained to be [56, 57]

$$\rho^{(3)} = \mathcal{K}_s' \, F_s' \, \rho^{(2)} \left[F_s'\right]^\dagger \tag{3.11a}$$

$$= \mathcal{K}_s' \, \mathcal{K}_s \, \mathcal{K}_p \left[F_s' F_s F_p\right] \rho_i \left[F_s' F_s F_p\right]^\dagger. \tag{3.11b}$$

Here, the coefficient K_s' and operator F_s' are for the process of second spontaneous decay [e.g., (1.7c) or (1.8c)] of the original state ρ_i. The meanings of ($\mathcal{K}_s'$, F_s') are identical to those of ($\mathcal{K}_s$, F_s) explained in the Sect. 3.1.2 corresponding to, of course, the appropriate (non-radiative or radiative) decay taking place in the third step of the processes (1.7) or (1.8).

The superscripts (2) and (3) on the density operators on the left hand sides of (3.10) and (3.11) mean that $\rho^{(2)}$ and $\rho^{(3)}$ are the states of our atomic or molecular target formed after the completion of two and three sequential steps involved, for example, in the processes (1.3) [or (1.4), (1.6)] and (1.7) [or (1.8)], respectively.

An inner-shell vacancy in an atomic or molecular target $\mathfrak{T}$ can be caused without or with its ionization. The present monograph will, primarily, be concerned only with those inner-shell vacancies which are created in ionization due to the absorption—in $E1$ approximation—of a single photon with well-defined frequency and polarization by one of the electrons in a core orbital of of an unpolarized target's state $|0\rangle$ (see Fig. 3.1 or 3.2). Thus, e_p comes out from one of the inner-shells of $\mathfrak{T}$ forming the excited photoion $\mathfrak{T}^{1+^*}$. One, therefore, substitutes (3.4c) for ρ_i in the expression (3.10), and obtains

$$\rho^{(2)} = \mathcal{K}_p \, \mathcal{K}_s \left[F_s F_p\right] \left(\frac{1}{n_{d_0}} \sum_{d_0} |0;\, 1m_r\rangle\langle 0;\, 1m_r| \right) \left[F_s F_p\right]^\dagger \tag{3.12}$$

for the density operator for a 2-step process consisting of inner-shell photoionization of $\mathfrak{T}$ and the subsequent spontaneous decay of the core excited photoion $\mathfrak{T}^{1+^*}$. After the emission of electron e_p from one of the inner-shells of $\mathfrak{T}$, the excited photoion $\mathfrak{T}^{1+^*}$ is left in one of its degenerate states $|1^{+^*}\rangle$. On twice introducing the unit operator

$$\sum_{d_{1^{+*}}} |1^{+^*}\rangle\langle 1^{+^*}| \equiv 1 \equiv \sum_{d_{1'^{+*}}} |1'^{+^*}\rangle\langle 1'^{+^*}| \tag{3.13}$$

the density operator (3.12) becomes

$$\langle \mu_p \, \hat{\boldsymbol{u}}_p \, \boldsymbol{k}_p | \rho^{(2)} | \mu_p' \, \hat{\boldsymbol{u}}_p \, \boldsymbol{k}_p \rangle$$

$$= \frac{\mathcal{K}_p \, \mathcal{K}_s}{n_{d_0}} \sum_{d_0 \, d_{1^{+*}} \, d_{1'^{+*}}} F_s |1^{+^*}\rangle \, \langle 1^{+^*};\, \mu_p \, \hat{\boldsymbol{u}}_p \, \boldsymbol{k}_p | F_p \, |0;\, 1m_r\rangle$$

$$\times \, \langle 0;\, 1m_r | F_p^\dagger \, |1'^{+^*};\, \mu_p' \, \hat{\boldsymbol{u}}_p \, \boldsymbol{k}_p\rangle \langle 1'^{+^*} | F_s^\dagger. \tag{3.14}$$

In this expression, and hereafter, $d_{1^{+*}}$ (or $d_{1'^{+*}}$) stands for all the degenerate states $|1^{+^*}\rangle$ (or $|1'^{+^*}\rangle$) of $\mathfrak{T}^{1+^*}$ possessing energy $E_{1^{+*}}$; $|\mu_p \hat{\boldsymbol{u}}_p \boldsymbol{k}_p\rangle^-$ (or $|\mu'_p\, \hat{\boldsymbol{u}}_p\, \boldsymbol{k}_p\rangle^-$) denotes a spin–orbital of the photoelectron e_p. In this orbital, unit vector $\hat{\boldsymbol{u}}_p\,(\vartheta_p, \varphi_p)$ specifies e_p's direction of spin polarization; whereas, its direction of propagation is given by the vector $\boldsymbol{k}_p\,(k_p, \hat{\boldsymbol{k}}_p\,(\theta_p, \phi_p))$ for linear momentum. Both, $\hat{\boldsymbol{u}}_p$ and $\boldsymbol{k}_p$ are referred to the SF $OXYZ$ shown in each of the Figs. 4.1 (page 95), 5.1 (page 124), 6.1 (page 137), 8.1 (page 182), *etc*—after e_p has left its parent atom $\mathfrak{T}$ following the absorption of a single photon γ_r in the $E1$ approximation. The magnitude of the propagation vector is related to the photoelectron's mass m_e and kinetic energy ϵ_p by $k_p\ (\equiv |\boldsymbol{k}_p|) = \sqrt{2\, m_e\ \epsilon_p/\hbar^2}$. Further in (3.14), $\mu_{\mathrm{p}},\ \mu'_p = \pm\frac{1}{2}$ are the projections of this electron's spin angular momentum along $\hat{\boldsymbol{u}}_p$. A "–" superscript on the e_p's spin orbital means that it asymptotically satisfies incoming wave boundary conditions [179]. (For brevity, this superscript has not been shown in (3.14).)

The density operator (3.14) can readily be adapted to any of the 2-step processes (1.3), (1.4), or (1.6). It can also be used in (3.11a) for the 3-step processes (1.7) and (1.8). Let us first consider 2-DPI (1.3). After the emission of Auger electron e_a in the non-radiative spontaneous decay, represented in the second step (1.3b) of this two-step process, the excited photoion $\mathfrak{T}^{1+^*}$ is left in the ground state $|2^+\rangle$ of the dication $\mathfrak{T}^{2+}$. A spin orbital of e_a (with appropriate asymptotic boundary conditions [179]) is herein taken to be $|\mu_a \hat{\boldsymbol{u}}_a \boldsymbol{k}_a\rangle^-$. Similar to that of the photoelectron e_p, $\hat{\boldsymbol{u}}_a\,(\vartheta_a, \varphi_a)$ is the direction of space quantization of the spin angular momentum of e_a; whereas, $\boldsymbol{k}_a\ \ (k_a,\ \hat{\boldsymbol{k}}_a\,(\theta_a,\phi_a))$ is the vector for linear momentum of this electron with $k_a\ (\equiv |\boldsymbol{k}_a|) = \sqrt{2\, m_e\, \epsilon_a/\hbar^2}$ corresponding to the kinetic energy ϵ_a. Each of these two vectors $\hat{\boldsymbol{u}}_a$ and $\boldsymbol{k}_a$ is shown in the SF frame in Fig. 6.1 (on page 137). The density matrix for the 2-step process (1.3) consisting of photoionization followed by Auger decay finally becomes

$$
\begin{aligned}
&\langle \mu_p\, \hat{\boldsymbol{u}}_p\, \boldsymbol{k}_p;\ \mu_a\, \hat{\boldsymbol{u}}_a\, \boldsymbol{k}_a |\, \rho^{(2)}\, | \mu'_p\, \hat{\boldsymbol{u}}_p\, \boldsymbol{k}_p;\ \mu'_a\, \hat{\boldsymbol{u}}_a\, \boldsymbol{k}_a\rangle \\
&\quad = \frac{\mathcal{K}_{\mathrm{p}}\, \mathcal{K}_{\mathrm{s}}}{n_{d_0}} \sum_{d_0\, d_{2^+}\, d_{1^{+*}}\, d_{1'^{+*}}} \langle 2^+;\ \mu_a\, \hat{\boldsymbol{u}}_a\, \boldsymbol{k}_a |\, F_{\mathrm{s}}\, |1^{+^*}\rangle \langle 1^{+^*};\ \mu_p\, \hat{\boldsymbol{u}}_p\, \boldsymbol{k}_p |\, F_{\mathrm{p}}\, |0;\ 1m_{\mathrm{r}}\rangle \\
&\quad \times \langle 0;\ 1m_{\mathrm{r}} |\, F_{\mathrm{p}}^{\dagger}\, |1'^{+^*};\ \mu'_p\, \hat{\boldsymbol{u}}_p\, \boldsymbol{k}_p\rangle \langle 1'^{+^*} |\, F_{\mathrm{s}}^{\dagger}\, |2^+;\ \mu'_a\, \hat{\boldsymbol{u}}_a\, \boldsymbol{k}_a\rangle. \qquad (3.15)
\end{aligned}
$$

Here, sum over d_{2^+} indicates that all the degenerate states $|2^+\rangle$ of the unobserved dication $\mathfrak{T}^{2+}$ have been included in (3.15); μ_a in $|\mu_a \hat{\boldsymbol{u}}_a \boldsymbol{k}_a\rangle^-$ and μ'_a in $|\mu'_a\, \hat{\boldsymbol{u}}_a\, \boldsymbol{k}_a{}^-\rangle$ are the projections of the spin angular momentum of the Auger electron e_a along $\hat{\boldsymbol{u}}_a$ and, hence, each can independently be either of $\pm\,\frac{1}{2}$.

Electrons $(e_p,\, e_a)$ in the 2-DPI process (1.3) may have same $[\epsilon_p = \epsilon_a$ (i.e., $k_p = k_a$)] or different $[\epsilon_p \neq \epsilon_a$ (i.e., $k_p \neq k_a$)] energies [subject to satisfying two different conservation conditions in (6.1)—one for each step in (1.3)]—and move in any two directions $\hat{\boldsymbol{k}}_p$ and $\hat{\boldsymbol{k}}_a$. A study of the properties of the entanglement between the spins of these two electrons requires a density matrix, which is diagonal in their energies as well as in their directions of propagation (*i.e.*, diagonal in $\boldsymbol{k}_p$ and $\boldsymbol{k}_a$). But, this density matrix must necessarily be non-diagonal with respect to the components ($\mu_{\rm p}$, $\mu_{\rm a}$) along the quantization directions ($\hat{\boldsymbol{u}}_p$, $\hat{\boldsymbol{u}}_a$) of the spin angular momenta of the (photoelectron e_p, Auger electron e_a).

Thus, expression (3.15) represents a four-dimensional matrix for a 2×2 system, i.e., a state of two qubits. It contains, among other things, complete information on photoelectron e_p and Auger electron e_a ejected sequentially in the 2-step process (1.3). For this matrix to be physically acceptable, it should be: (a) Positive, i.e., none of its four eigenvalues be less than zero [see footnote (6) on page 39]
(b) Hermitian, i.e.,

$$\begin{aligned}&\langle \mu_p\,\hat{\boldsymbol{u}}_p\,\boldsymbol{k}_p;\mu_a\,\hat{\boldsymbol{u}}_a\,\boldsymbol{k}_a|\rho^{(2)}|\mu_p'\,\hat{\boldsymbol{u}}_p\,\boldsymbol{k}_p;\mu_a'\,\hat{\boldsymbol{u}}_a\,\boldsymbol{k}_a\rangle\\&=\langle \mu_p\,\hat{\boldsymbol{u}}_p\,\boldsymbol{k}_p;\mu_a\,\hat{\boldsymbol{u}}_a\,\boldsymbol{k}_a|\rho^{(2)^\dagger}|\mu_p'\,\hat{\boldsymbol{u}}_p\,\boldsymbol{k}_p;\mu_a'\,\hat{\boldsymbol{u}}_a\,\boldsymbol{k}_a\rangle\\&=\langle \mu_p'\,\hat{\boldsymbol{u}}_p\,\boldsymbol{k}_p;\mu_a'\,\hat{\boldsymbol{u}}_a\,\boldsymbol{k}_a|\rho^{(2)}|\mu_p\,\hat{\boldsymbol{u}}_p\,\boldsymbol{k}_p;\mu_a\,\hat{\boldsymbol{u}}_a\,\boldsymbol{k}_a\rangle^*.\end{aligned}\tag{3.16}$$

(c) Each of its four diagonal elements must possesses the property

$$\langle \mu_p\,\hat{\boldsymbol{u}}_p\,\boldsymbol{k}_p;\mu_a\,\hat{\boldsymbol{u}}_a\,\boldsymbol{k}_a|\rho^{(2)}|\mu_p\,\hat{\boldsymbol{u}}_p\,\boldsymbol{k}_p;\mu_a\,\hat{\boldsymbol{u}}_a\,\boldsymbol{k}_a\rangle \geq 0.\tag{3.17}$$

(d) In addition, one will like (3.15) to be normalized to unit trace,[4] i.e.,

$$\begin{aligned}&\mathrm{Tr}\Big[\langle \mu_p\,\hat{\boldsymbol{u}}_p\,\boldsymbol{k}_p;\mu_a\,\hat{\boldsymbol{u}}_a\,\boldsymbol{k}_a|\rho^{(2)}|\mu_p'\,\hat{\boldsymbol{u}}_p\,\boldsymbol{k}_p;\mu_a'\,\hat{\boldsymbol{u}}_a\,\boldsymbol{k}_a\rangle\Big] \equiv\\&\sum_{\mu_{\rm p}=-\frac{1}{2}}^{+\frac{1}{2}}\ \sum_{\mu_{\rm a}=-\frac{1}{2}}^{+\frac{1}{2}} \langle \mu_p\,\hat{\boldsymbol{u}}_p\,\boldsymbol{k}_p;\mu_a\,\hat{\boldsymbol{u}}_a\,\boldsymbol{k}_a|\rho^{(2)}|\mu_p\,\hat{\boldsymbol{u}}_p\,\boldsymbol{k}_p;\mu_a\,\hat{\boldsymbol{u}}_a\,\boldsymbol{k}_a\rangle = 1.\end{aligned}\tag{3.18}$$

The non-diagonal ($\mu_p' \neq \mu_{\rm p}$, $\mu_a' \neq \mu_{\rm a}$) elements in (3.15) represent the coherence effects[5]; whereas the diagonal ($\mu_p' = \mu_{\rm p}$, $\mu_a' = \mu_{\rm a}$) elements describe angular distribution of spin-resolved e_p and e_a. Thus, $\langle \mu_p\,\hat{\boldsymbol{u}}_p\,\boldsymbol{k}_p;\mu_a\,\hat{\boldsymbol{u}}_a\,\boldsymbol{k}_a|\rho^{(2)}|$ $\mu_p\,\hat{\boldsymbol{u}}_p\,\boldsymbol{k}_p;\mu_a\,\hat{\boldsymbol{u}}_a\,\boldsymbol{k}_a\rangle$ in (3.15) gives the probability[6] that photoelectron e_p in (1.3a) is emitted in the direction $\boldsymbol{k}_p$ with its spin component $\mu_{\rm p}$ along $\hat{\boldsymbol{u}}_p$, while the subsequent Auger electron e_a in (1.3b) has propagation vector $\boldsymbol{k}_a$ and spin projection $\mu_{\rm a}$ in the direction $\hat{\boldsymbol{u}}_a$ in Fig. 6.1 (page 137). Thus, diagonal terms in the density matrix (3.15), in addition to characterizing the angular correlation of the

[4]For a definition of trace of a matrix, see, for example, (A.8) on page 254.

[5]A somewhat detailed, but quantitative, discussion of this topic is given in the Appendix C on pages 269–271.

[6]It is for this important reason that these diagonal elements must simultaneously meet the requirements (3.17) and (3.18).

simultaneously observed (e_p, e_a), specify also the orientations in space of the spins of these two sequentially emitted electrons.[7]

Further, if for given values of $(\boldsymbol{k}_p, \boldsymbol{k}_a)$, all the four diagonal elements in (3.15) are equal, then there are no preferred quantization directions $(\hat{\boldsymbol{u}}_p, \hat{\boldsymbol{u}}_a)$ for the orientations of the spins of (e_p, e_a) and, hence, the spins of both of these electrons are randomly oriented in space. This, among other things, means that the angular- and spin-correlation between (e_p, e_a) becomes independent of the orientations of the spins of these two electrons. If, in addition, non-diagonal elements vanish identically, then (3.15) reduces to a diagonal form. A diagonal matrix with equal elements represents a mixed separable (i.e., product or unentangled) state which is called maximally chaotic [22] or white noise [122] state in quantum information science.

3.2 Amplitudes for Atomic Transitions

The bracket $\langle 1^{+^*}, \mu_p \hat{\boldsymbol{u}}_p \boldsymbol{k}_p | F_{\rm p} | 0, 1m_{\rm r} \rangle$ in (3.15) is the transition amplitude for the photoionization step in each of the processes (1.3), (1.4), and (1.6)–(1.8). The Auger decay amplitude—needed for (1.3b), (1.7b), and (1.8b)—is, on the other hand, represented by $\langle 2^+, \mu_a \hat{\boldsymbol{u}}_a \boldsymbol{k}_a | F_{\rm s} | 1^{+^*} \rangle$ in the expression (3.15). Each transition amplitude is, in general, complex containing both magnitudes and phases. These amplitudes can be calculated in a hierarchy of approximations—beginning from semi-empirical phenomenological methods to ab-initio theories. The calculation of these amplitudes from first principles may itself be the goal of a particular theoretical study. Several sophisticated computer programmes have already been developed for this purpose (see, for example, [183]).

The aim of this study is, on the other hand, to investigate the properties of electron optics relevant to quantum information science. For this purpose, one needs an explicit form of a density matrix representing one or the other of the processes (1.1)–(1.8) of interest. The remaining sections in the present Chap. 3, therefore, describe methodologies and develop theoretical frameworks for expressing the transition amplitudes, needed in the description of the required density matrices, in terms of the quantum numbers, among other things, of the states which participate in any of the eight processes listed in the Chap. 1. The procedures described herein are independent of any dynamical models one may like to use in a particular calculation of a required transition amplitude.

While the amplitude for the 1-SPI process (1.1) is readily obtained on replacing in $\langle 1^{+^*}, \mu_p \hat{\boldsymbol{u}}_p \boldsymbol{k}_p | F_{\rm p} | 0, 1m_{\rm r} \rangle$ the state $|1^{+^*}\rangle$ for the excited photoion $\mathfrak{T}^{1+^*}$ by $|1^+\rangle$ of the photoion $\mathfrak{T}^{1+}$ in its ground state; amplitudes needed for some of the other processes [e.g., 1-DPI (1.2), radiative decay (1.4b)] mentioned in Chap. 1, can be calculated with the methods similar to those used in the following

[7]The diagonal elements of a density matrix are called [180–182] population as well. See, for example, Appendix C for further discussion.

with their details described in appropriate chapters of this monograph. Likewise, the procedure used in the present Chap. 3 for calculating the Auger amplitude $\langle 2^+;\, \mu_a\, \hat{\boldsymbol{u}}_a\, k_a |\, F_s\, | 1^{+^*} \rangle$ can readily be generalized for the calculation of the amplitudes needed for two- and three-step processes like (1.6) and [(1.7), (1.8)], respectively.

On the basis of the interactions one takes into account, studies of each of the processes (1.1)–(1.8) in an atomic or molecular target $\mathfrak{T}$ can conveniently be divided into two broad classes. One possibility is to consider only the electrostatic Coulomb forces experienced by constituents of $\mathfrak{T}$. In the second possibility, forces originating from spins of electrons are also taken into account. In the case of atoms, the primary spin-dependent force [10, 60, 184] is SOI; whereas, in a molecule [68, 185], spin–rotation interaction (SRI), in addition to SOI, will also be present. Thus, spin-dependent interactions (SDIs) in this monograph always mean, unless stated otherwise, only the SOI for atomic targets, but both SOI plus SRI if a molecule happens to be the target in a particular study. Moreover, for obvious physical reasons, the quantum numbers needed to specify a state in a linear molecule are totally different from those appropriate for an atomic state. Consequently, studies of the processes (1.1)–(1.8) for a molecule require a mathematical framework which is very different from the one needed when $\mathfrak{T}$ happens to be an atom.

In general [10, 60, 184, 186], the relative magnitudes of the Coulomb and spin–orbit interactions[8] determine as to which of the angular momentum coupling schemes be used for studying any processes taking place in an atom $\mathfrak{T}$. For example, the frequently encountered case of Russell–Saunders (L-S) coupling [10, 60, 184, 186] is most suitable for atoms with small or intermediate values of the atomic number Z when the SOI is much smaller than the Coulomb forces experienced by an electron in an atom.[9] For atoms with large Z, on the other hand, one needs to use j-j coupling [10, 60, 184, 186] as the SOI in such cases dominates the Coulomb interactions.

The case of atoms for which these two interactions are comparable requires that both Coulomb and SOI be treated on the same footing. This situation is known as intermediate coupling [10, 184, 186]. Comparisons of the calculations reported by Lohmann [183] for two of the spin polarization parameters present in an angle- and spin-resolved Auger spectroscopy suggest that there is little difference between

---

[8]For an n_e-electron atom with nuclear charge Ze, the SOI is given by [10, 60, 184, 186]

$$\sum_i^{n_e} \xi(r_i)\, \boldsymbol{\ell}_i \cdot \boldsymbol{s}_i \qquad \text{with} \qquad \xi(r_i) = \frac{\hbar^2}{2m_e^2c^2}\, \frac{Ze^2}{4\pi\epsilon_0}\, \frac{1}{r_i^3}.$$

Here, while r_i is the distance of the i-th electron from the nucleus, $\boldsymbol{\ell}_i$ and $\boldsymbol{s}_i$ are its orbital and spin angular momenta, respectively. Electrons only in an incomplete shell contribute to this sum as the contributions of those in a full shell add to zero [59].

[9]For a hydrogenic atom [i.e., $n_e = 1$ in footnote (8) in the present Chap. 3] with atomic number Z, the ratio of the SOI to the Coulomb interaction [i.e., $\frac{Ze^2}{4\pi\epsilon_0}\frac{1}{r}$] is [58] $\simeq$ $(Z\alpha)^2$, where $\alpha = 1/137$ is the fine structure constant.

the results obtained using L-S or intermediate coupling for most of the transitions in atoms such as Ar, Kr, Xe, *etc*. In the present monograph, we have, therefore, investigated the entanglement properties of a spin-state of two particles emitted from an atomic target $\mathfrak{T}$ only in two coupling schemes, namely the L-S and the j-j couplings. An interested reader can readily apply the methodology suggested herein to study these properties in the intermediate coupling [10,184,186] of angular momenta, if such a need arises.

The following two Sects. 3.2.1 and 3.2.2 contain evaluation of amplitudes without and with SOI, respectively, both for photoionization and Auger decay in an atom. Amplitudes for photoionizing and Auger transitions in a linear molecule, both without and with SDIs, are calculated in the Sect. 3.3.

3.2.1 Transition Amplitudes Without Spin–Orbit Interaction

Neither the $E1$ operators (3.6) nor the spontaneous decay operators [(3.8), (3.9)] depend on any spin variables. Then in the absence of SOI when only the electrostatic Coulomb interaction is present, each of the total orbital and total spin angular momenta is a constant of motion in each of the steps of the processes (1.1)–(1.8). This, in other words, means that these two angular momenta are individually conserved in these processes. In such a situation, the Russell–Saunders (*i.e.*, L-S) coupling [10, 60, 184, 186] becomes naturally applicable.

On the inclusion of the SOI, in addition to the Coulomb interaction, neither the total orbital nor total spin angular momentum is a constant of motion. These two angular momenta are never individually conserved in any of the steps of the processes (1.1)–(1.8) in the simultaneous presence of both of the Coulomb plus spin-orbit interactions (C+SIOs). It, consequently, renders [10, 60, 184, 186] L-S coupling inapplicable in such a physical situation. On the other hand, the total angular momentum—a sum of the total orbital and total spin angular momenta—is now a good quantum number for each step of the processes (1.1)–(1.8). Thus [10, 60, 184, 186], j-j, not L-S, coupling describes the physical situation arising from the simultaneous inclusion of both the Coulomb forces and SOI experienced by electrons in an atomic target $\mathfrak{T}$.

3.2.1.1 Russell–Saunders (i.e., L-S) Coupling

Let ($\boldsymbol{L}_0$, $\boldsymbol{L}_{1^{+*}}$, $\boldsymbol{L}_{2^+}$) and ($\boldsymbol{S}_0$, $\boldsymbol{S}_{1^{+*}}$, $\boldsymbol{S}_{2^+}$) be the total orbital and total spin angular momenta of the (atom $\mathfrak{T}$, excited photoion $\mathfrak{T}^{1+^*}$, dication $\mathfrak{T}^{2+}$) participating in the 2-DPI process (1.3). In the absence of SOI, the conservation conditions for consecutive photoionization (1.3a) and Auger emission (1.3b) are, respectively,

$$\boldsymbol{\ell}_r + \boldsymbol{L}_0 = \boldsymbol{L}_{1^{+*}} + \boldsymbol{\ell}_p \tag{3.19a}$$

and

$$\boldsymbol{L}_{1^{+*}} = \boldsymbol{L}_{2^+} + \boldsymbol{\ell}_a \tag{3.19b}$$

for the orbital angular momenta and

$$\boldsymbol{S}_0 = \boldsymbol{S}_{1^{+*}} + \boldsymbol{s}_p \left[= \left(\tfrac{1}{2}\right)_p \right] \tag{3.20a}$$

and

$$\boldsymbol{S}_{1^{+*}} = \boldsymbol{S}_{2^+} + \boldsymbol{s}_a \left[= \left(\tfrac{1}{2}\right)_a \right] \tag{3.20b}$$

for the spin angular momenta. Here, $[\boldsymbol{\ell}_p,\ \boldsymbol{\ell}_a]$ in [(3.19a), (3.19b)] and $[\boldsymbol{s}_p,\ \boldsymbol{s}_a]$ in [(3.20a), (3.20b)] are the respective orbital and spin angular momenta of $[e_p, e_a]$.

The electronic states of $\mathfrak{T}$, $\mathfrak{T}^{1+^*}$, and $\mathfrak{T}^{2+}$ in the L-S coupling are given by [10, 60] $|0\rangle \equiv |L_0\, S_0\, M_{L_0}\, M_{S_0}\rangle$, $|1^{+^*}\rangle \equiv |L_{1^{+*}}\, S_{1^{+*}}\, M_{L_{1^{+*}}}\, M_{S_{1^{+*}}}\rangle$, and $|2^+\rangle \equiv |L_{2^+}\, S_{2^+}\, M_{L_{2^+}}\, M_{S_{2^+}}\rangle$, respectively. In these states, $(M_{L_0},\ M_{L_{1^{+*}}}, M_{L_{2^+}})$ and $(M_{S_0},\ M_{S_{1^{+*}}},\ M_{S_{2^+}})$ are the projections of $(\boldsymbol{L}_0,\ \boldsymbol{L}_{1^{+*}},\ \boldsymbol{L}_{2^+})$ and of $(\boldsymbol{S}_0,\ \boldsymbol{S}_{1^{+*}},\ \boldsymbol{S}_{2^+})$ along the quantization axis in the SF shown in Figs. 1.1 (page 8), 6.1 (page 137), *etc*. That is, $M_{L_0} = \hat{\boldsymbol{e}}_z \cdot \boldsymbol{L}_0$, *etc*, where $\hat{\boldsymbol{e}}_z$ is an unit vector along the OZ-axis in these figures.

Photoionization Amplitude

The left-hand sides of the conservation conditions [(3.19a), (3.20a)], applicable in the L-S coupling, makes one to write

$$|0,\ 1m_r\rangle \ \equiv\ |L_0 S_0 M_{L_0} S_{M_0},\ 1m_r\rangle = \sum_{L\, M_L} (-1)^{1-L_0-M_L} \times \sqrt{2L+1} \begin{pmatrix} L_0 & 1 & L \\ M_{L_0} & m_r & -M_L \end{pmatrix} |(L_0 1) L\, M_L;\ S_0\, M_{S_0}\rangle. \tag{3.21}$$

for a product state of the $(\gamma_r + \mathfrak{T})$-system. Here, $\begin{pmatrix} \cdot\cdot & \cdot\cdot & \cdot\cdot \\ \cdot\cdot & \cdot\cdot & \cdot\cdot \end{pmatrix}$ is a 3-j symbol [64, 65, 187].

In order to obtain a state of $(\mathfrak{T}^{1+^*} + e_p)$ in the L-S coupling, one needs following partial wave expansion [10]

$$\langle \boldsymbol{r}_p | \tfrac{1}{2}\, \nu_p\, \boldsymbol{k}_p \rangle = |\tfrac{1}{2}\, \nu_p \rangle \sum_{\ell_p m_p} i^{\ell_p}\, \mathrm{e}^{-i\sigma_{\ell_p}}\, R^-_{\ell_p}(r_p; k_p) \left[Y^{m_p}_{\ell_p}(\hat{\boldsymbol{k}}_p) \right]^* Y^{m_p}_{\ell_p}(\hat{\boldsymbol{r}}_p) \quad (3.22a)$$

for the spin–orbital of the photoelectron e_p. Here, $\boldsymbol{r}_p$ is the position vector of e_p, $R^-_{\ell_p}(r_p; k_p)$ is the properly normalized[10] radial function satisfying incoming wave boundary conditions [179] (indicated by the superscript "–"), ν_p is the projection of the spin of this electron along the polar OZ-axis of our SF (e.g., in Fig. 1.1), and σ_{ℓ_p} is the Coulomb phase (see, for example, [72, 174]) for the ℓ_p-th partial wave of e_p. The following rotation

$$\langle \boldsymbol{r}_p |\, \mu_p\, \hat{\boldsymbol{u}}_p\, \boldsymbol{k}_p \rangle = \sum_{\nu_p} \left[\mathcal{D}^{\frac{1}{2}}_{\mu_p\, \nu_p}(\omega_p) \right]^* \langle \boldsymbol{r}_p | \tfrac{1}{2}\, \nu_p\, \boldsymbol{k}_p \rangle \quad (3.22b)$$

gives us the desired spin–orbital (in L-S coupling) for a photoelectron whose spin has a component μ_p along an arbitrary direction $\hat{\boldsymbol{u}}_p$ in, e.g., Figs. 4.1 (page 95), 6.1 (page 137), *etc*. The rotational harmonics [65] $\mathcal{D}$ in the expansion (3.22b) are used for rotating photoelectron's spin quantization axis $\hat{\boldsymbol{e}}_z$ along the direction $\hat{\boldsymbol{u}}_p$—both defined in the SF.

Next, the antisymmetrized product of the state $|1^{+^*}\rangle$ (in L-S coupling) of the excited photoion $\mathfrak{T}^{1+^*}$ and of the spin–orbital (3.22b) of e_p is

$$|1^{+^*}, \mu_p\, \hat{\boldsymbol{u}}_p\, \boldsymbol{k}_p\rangle \equiv \mathcal{A}\Big(|1^{+^*}\rangle\, |\mu_p \hat{\boldsymbol{u}}_p \boldsymbol{k}_p\rangle \Big) = \mathcal{A}\Bigg[|L_{1^{+*}}\, S_{1^{+*}}\, M_{L_{1^{+*}}}\, M_{S_{1^{+*}}}\rangle \times \sum_{\ell_p m_p \nu_p} i^{\ell_p}\, \mathrm{e}^{-i\sigma_{\ell_p}} \left[Y^{m_p}_{\ell_p}(\hat{\boldsymbol{k}}_p) \right]^* \left[\mathcal{D}^{\frac{1}{2}}_{\mu_p\, \nu_p}(\omega_p) \right]^* |\ell_p\, m_p\, \nu_p\rangle \Bigg]. \quad (3.23a)$$

Here, we have defined

$$|\ell_p\, m_p\, \nu_p\rangle \equiv |\tfrac{1}{2}\, \nu_p\rangle\, R^-_{\ell_p}(r_p; k_p)\, Y^{m_p}_{\ell_p}(\hat{\boldsymbol{r}}_p). \quad (3.23b)$$

In (3.23a), operator $\mathcal{A}$ antisymmetrizes the photoelectron e_p with the bound electrons of the excited photoion $\mathfrak{T}^{1+^*}$.

On introducing the desired L-S coupling, according to the prescription shown on the right hand sides of the conservation conditions [(3.19a), (3.20a)], between the angular momenta of the excited photoion $\mathfrak{T}^{1+^*}$ and photoelectron e_p, one obtains [188]

[10] See the discussion given on pages 50 in the last paragraph of Sect. 3.1.1.

$$\mathcal{A}\Big(|L_{1^{+*}}\, S_{1^{+*}}\, M_{L_{1^{+*}}}\, M_{S_{1^{+*}}}\rangle\, |\ell_p\, m_p\, \nu_p\rangle\Big) = \sum_{L\, M_L\, S\, M_S}$$

$$\times (-1)^{\frac{1}{2}+\ell_p-L_{1^{+*}}-S_{1^{+*}}-M_L-M_S} \sqrt{(2L+1)(2S+1)} \begin{pmatrix} L_{1^{+*}} & \ell_p & L \\ M_{L_{1^{+*}}} & m_p & -M_L \end{pmatrix}$$

$$\times \begin{pmatrix} S_{1^{+*}} & \frac{1}{2} & S \\ M_{S_{1^{+*}}} & \nu_p & -M_S \end{pmatrix} |(L_{1^{+*}}\, \ell_p)L\, M_L;\ (S_{1^{+*}}\, \frac{1}{2})S\, M_S\rangle, \tag{3.23c}$$

Substitution of (3.23c) in (3.23a) finally gives us the following expression [188]

$$|1^{+^*};\ \mu_p\, \hat{\boldsymbol{u}}_p\, \boldsymbol{k}_p\rangle = (-1)^{\frac{1}{2}-L_{1^{+*}}-S_{1^{+*}}} \sum_{\ell_p\, m_p\, \nu_p} \sum_{\substack{L\, M_L \\ S\, M_S}} i^{\ell_p}\, e^{-i\sigma_{\ell_p}}\, (-1)^{\ell_p-M_L-M_S}$$

$$\times \sqrt{(2L+1)(2S+1)} \begin{pmatrix} L_{1^{+*}} & \ell_p & L \\ M_{L_{1^{+*}}} & m_p & -M_L \end{pmatrix} \begin{pmatrix} S_{1^{+*}} & \frac{1}{2} & S \\ M_{S_{1^{+*}}} & \nu_p & -M_S \end{pmatrix}$$

$$\times \left[Y_{\ell_p}^{m_p}(\hat{\boldsymbol{k}}_p)\right]^* \left[\mathcal{D}^{\frac{1}{2}}_{\mu_p\, \nu_p}(\omega_p)\right]^* |(L_{1^{+*}}\, \ell_p)L\, M_L;\ \left(S_{1^{+*}}\, \frac{1}{2}\right) S\, M_S\rangle^- \tag{3.24}$$

for a state of the ($\mathfrak{T}^{1+^*}$ + e_p)-system with their orbital and spin angular momenta coupled according to the right-hand sides of the conservation conditions [(3.19a), (3.20a)] in the Russell–Saunders scheme.

Let us now use the kets (3.21) and (3.24) in the amplitude $\langle 1^{+^*}, \mu_p\, \hat{\boldsymbol{u}}_p\, \boldsymbol{k}_p | F_p\, |0, 1 m_r\rangle$. The consequent expression will have, among other things, the Dirac's bracket $\langle (L_{1^{+*}}\, \ell_p)L\, M_L;\ (S_{1^{+*}}\, \frac{1}{2})S\, M_S | F_p\, |(L_0 1)L'\, M_{L'};\ S_0\, M_{S_0}\rangle$. Remembering that the photoionization operator F_p in (3.6) contains only the space coordinates, summation over the spin variables can readily be performed. On using the conservation conditions [(3.19a), (3.20a)], applicable in L-S coupling, one finds

$$\begin{aligned} &\langle (L_{1^{+*}}\, \ell_p)L\, M_L;\ \left(S_{1^{+*}}\, \frac{1}{2}\right) S\, M_S | F_p\, |(L_0 1)L'\, M_{L'};\ S_0\, M_{S_0}\rangle \\ &= \delta_{L\, L'}\, \delta_{M_L\, M_{L'}}\, \delta_{S\, S_0}\, \delta_{M_S\, M_{S_0}}\, \langle (L_{1^{+*}}\, \ell_p)L | F_p\, |(L_0 1)L\rangle. \end{aligned} \tag{3.25}$$

For brevity, the superscript "–" has not explicitly been shown in (3.25) as well as in the following. Amplitude for the $E1$ ionizing transition [e.g., in (1.3a)] finally becomes

$$\mathcal{P}_{0\longrightarrow 1^{+*}} \equiv \langle 1^{+^*};\ \mu_p\, \hat{\boldsymbol{u}}_p\, \boldsymbol{k}_p | F_p\, |0;\ 1\, m_r\rangle = (-1)^{\frac{1}{2}+L_0+L_{1^{+*}}+S_{1^{+*}}+M_{S_0}}$$

$$\times \sqrt{2S_0+1} \sum_{\substack{\ell_p\, m_p\, \nu_p \\ L\, M_L}} (-i)^{\ell_p}\, (-1)^{\ell_p}\, e^{i\sigma_{\ell_p}}\, (2L+1)$$

$$\times \begin{pmatrix} L_{1^{+*}} & \ell_p & L \\ M_{L_{1^{+*}}} & m_p & -M_L \end{pmatrix} \begin{pmatrix} L_0 & 1 & L \\ M_{L_0} & m_r & -M_L \end{pmatrix} \begin{pmatrix} S_{1^{+*}} & \frac{1}{2} & S_0 \\ M_{S_{1^{+*}}} & \nu_p & -M_{S_0} \end{pmatrix}$$

$$\times Y_{\ell_p}^{m_p}(\hat{\boldsymbol{k}}_p)\, \mathcal{D}^{1/2}_{\mu_p \nu_p}(\omega_p)\, \langle (L_{1^{+*}}\, \ell_p)L | F_p | (L_0 1)L \rangle. \tag{3.26}$$

in the L-S coupling suggested by the conservation [(3.19a), (3.20a)] applicable in the absence of SOI when only the Coulomb forces are taken into account. In (3.26) and elsewhere in this monograph, $\mathcal{P}_{0 \longrightarrow 1^{+*}}$ stands for amplitude for the photoionizing transition from the ground state $|0\rangle$ of $\mathfrak{T}$ to the excited state $|1^{+*}\rangle$ of the photoion $\mathfrak{T}^{1+^*}$. Although, the bracket $\langle (L_{1^{+*}}\, \ell_p)L | F_p | (L_0 1)L \rangle$ in (3.26) does not explicitly contain any spin quantum numbers, it does, nevertheless, has an implicit dependence on the multiplicities $[(2S_0+1), (2S_{1^{+*}}+1)]$ of the electronic states $\left[|0\rangle, |1^{+*}\rangle\right]$ of $[\mathfrak{T}, \mathfrak{T}^{1+^*}]$ which determine energies $[E_0,\ E_{1^{+*}}]$ not only of these two atomic species but also that (i.e., ϵ_p) of the photoelectron e_p [see, for example, (6.1a)].

In the calculation of the density matrix (3.15) and other similar expressions occurring elsewhere in this monograph, a knowledge of the following

$$\frac{1}{2S_0+1} \sum_{M_{S_0}} \mathcal{P}_{0 \longrightarrow 1^{+*}} (\mathcal{P}_{0 \longrightarrow 1'^{+*}})^\dagger \equiv$$
$$\frac{1}{2S_0+1} \sum_{M_{S_0}} \langle 1^{+*};\ \mu_p\, \hat{\boldsymbol{u}}_p\, \boldsymbol{k}_p | F_p | 0;\ 1\, m_r \rangle \langle 0;\ 1\, m_r | F_p^\dagger | 1'^{+*};\ \mu'_p\, \hat{\boldsymbol{u}}_p\, \boldsymbol{k}_p \rangle \tag{3.27}$$

sum will also be required. Here, $\mathcal{P}_{0 \longrightarrow 1'^{+*}}$ is photoionization amplitude for the transition $|0\rangle$ $(\equiv |L_0\, S_0\, M_{L_0}\, M_{S_0}\rangle) \longrightarrow |1'^{+*}\rangle$ $(\equiv |L_{1^{+*}}\, S_{1^{+*}}\, M'_{L_{1^{+*}}}\, M'_{S_{1^{+*}}}\rangle)$, with both $|1^{+*}\rangle$ and $|1'^{+*}\rangle$ being the degenerate electronic states for energy $E_{1^{+*}}$ of the excited photoion $\mathfrak{T}^{1+^*}$. After substituting (3.26) and Hermitian conjugate of $\mathcal{P}_{0 \longrightarrow 1'^{+*}}$, the resulting (3.27) will have, among other things, sums over eleven variables. In order to simplify this expression, one may proceed as follows: (a) Add two spherical harmonics, each with argument $\hat{\boldsymbol{k}}_p$, using the identity (4.6.5) [65] {i.e., (4.6.5) from Edmonds [65]}; (b) use addition theorem (4.3.2) [65] for the rotational harmonics with Euler angles ω_p; (c) evaluate triple sum over the product of three 3-j symbols using the identity (6.2.8) [65]. These, and some additional simplifications, reduce (3.27) to the form

$$\frac{1}{2S_0+1} \sum_{M_{S_0}} \mathcal{P}_{0 \longrightarrow 1^{+*}} (\mathcal{P}_{0 \longrightarrow 1'^{+*}})^\dagger = (-1)^{S_0 + M_{S_{1^{+*}}} + \mu'_p} \frac{1}{2\sqrt{\pi}}$$

$$\times \sum_{\substack{\ell_p\, \ell'_p\, L\, L'\, B\, B_1 \\ m_p\, m'_p\, M_L\, M'_L\, \beta\, \beta_1\, \beta'_1}} (-1)^{m'_p} (2B_1+1) \sqrt{2B+1} \begin{pmatrix} \ell_p & \ell'_p & B \\ 0 & 0 & 0 \end{pmatrix}$$

$$
\times \begin{pmatrix} L_0 & 1 & L \\ M_{L_0} & m_r & -M_L \end{pmatrix} \begin{pmatrix} L_0 & 1 & L' \\ M_{L_0} & m_r & -M'_L \end{pmatrix} \begin{pmatrix} \ell_p & \ell'_p & B \\ m_p & -m'_p & \beta \end{pmatrix} \begin{pmatrix} L_{1+^*} & \ell_p & L \\ M_{L_{1+^*}} & m_p & -M_L \end{pmatrix}
$$

$$
\times \begin{pmatrix} L_{1+^*} & \ell'_p & L' \\ M'_{L_{1+^*}} & m'_p & -M'_L \end{pmatrix} \begin{pmatrix} \frac{1}{2} & \frac{1}{2} & B_1 \\ \mu_p & -\mu'_p & \beta_1 \end{pmatrix} \begin{pmatrix} S_{1+^*} & S_{1+^*} & B_1 \\ M'_{S_{1+^*}} & -M_{S_{1+^*}} & \beta'_1 \end{pmatrix}
$$

$$
\times \begin{Bmatrix} \frac{1}{2} & \frac{1}{2} & B_1 \\ S_{1+^*} & S_{1+^*} & S_0 \end{Bmatrix} \left[Y_B^{\beta}(\hat{\boldsymbol{k}}_p) \right]^* \left[\mathcal{D}^{B_1}_{\beta_1\, \beta'_1}(\omega_p) \right]^*
$$

$$
\times \left[(-\mathrm{i})^{\ell_p} \, \mathrm{e}^{\mathrm{i}\sigma_{\ell_p}} \, (-1)^{\ell_p} (2L+1) \sqrt{2\ell_p + 1} \, \langle (L_{1+^*} \, \ell_p) L | F_p | (L_0 \, 1) L \rangle \right]
$$

$$
\times \left[(-\mathrm{i})^{\ell'_p} \, \mathrm{e}^{\mathrm{i}\sigma_{\ell'_p}} \, (-1)^{\ell'_p} (2L'+1) \sqrt{2\ell'_p + 1} \langle (L_{1+^*} \, \ell'_p) L' | F_p | (L_0 \, 1) L' \rangle \right]^* . \tag{3.28}
$$

The Dirac brackets, present in the last two lines of (3.28), are those already defined in (3.25). Herein, and elsewhere in this monograph, $\begin{Bmatrix} \cdots\cdots\cdots \\ \cdots\cdots\cdots \end{Bmatrix}$ is a 6-j symbol [65, 187].

Auger Decay Amplitude

In order to calculate the amplitude $\langle 2;\, \mu_a \, \hat{\boldsymbol{u}}_a \, \boldsymbol{k}_a |\, F_s \, | 1^{+^*} \rangle$ [needed for the processes like (1.3), (1.7), and (1.8)] for the non-radiative spontaneous decay in the absence of SOI, one needs a state for the ($\mathfrak{T}^{2+}$ + e_a)-system with orbital and spin angular momenta coupled according to the right hand sides of the conservation conditions (3.19b) and (3.20b), respectively, in Russell–Saunders scheme. The required state is readily obtained by following a procedure similar to that used in (3.22)–(3.25) for deriving an expression for the state of the ($\mathfrak{T}^{1+^*}$ + e_p)-system. This gives us

$$
\begin{aligned}
|2^+, \mu_a \, \hat{\boldsymbol{u}}_a \, \boldsymbol{k}_a \rangle &\equiv \mathcal{A}\Big(|2^+\rangle | \mu_a \, \hat{\boldsymbol{u}}_a \, \boldsymbol{k}_a \rangle \Big) \\
&= \mathcal{A}\left[|2^+\rangle \sum_{\ell_a m_a \, \nu_a} i^{\ell_a} \, \mathrm{e}^{-\mathrm{i}\sigma_{\ell_a}} \left[Y_{\ell_a}^{m_a}(\hat{\boldsymbol{k}}_a) \right]^* \left[\mathcal{D}^{\frac{1}{2}}_{\mu_a \, \nu_a}(\omega_a) \right]^* | \ell_a \, m_a \, \nu_a \rangle \right] \\
&= (-1)^{\frac{1}{2} - L_{2+} - S_{2+}} \sum_{\substack{\ell_a \, m_a \, \nu_a \\ L \, M_L \, S \, M_S}} i^{\ell_a} \, \mathrm{e}^{-\mathrm{i}\sigma_{\ell_a}} \, (-1)^{\ell_a - M_L - M_S} \\
&\qquad \times \sqrt{(2L+1)(2S+1)} \begin{pmatrix} L_{2+} & \ell_a & L \\ M_{L_{2+}} & m_a & -M_L \end{pmatrix} \begin{pmatrix} S_{2+} & \frac{1}{2} & S \\ M_{S_{2+}} & \nu_a & -M_S \end{pmatrix} \\
&\qquad \times \left[Y_{\ell_a}^{m_a}(\hat{\boldsymbol{k}}_a) \right]^* \left[\mathcal{D}^{\frac{1}{2}}_{\mu_a \, \nu_a}(\omega_a) \right]^* | (L_{2+} \, \ell_a) L \, M_L;\, \left(S_{2+} \, \frac{1}{2} \right) S \, M_S \rangle^-
\end{aligned} \tag{3.29}
$$

with σ_{ℓ_a} being the Coulomb phase [72, 174] for the ℓ_a-th partial wave of the Auger electron e_a. Expression (3.29) has been arrived at by using

$$|\ell_a\, m_a\, \nu_a\rangle = |\frac{1}{2}\, \nu_a\rangle\, R^{-}_{\ell_a}(r_a; k_a)\, Y^{m_a}_{\ell_a}(\hat{\boldsymbol{r}}_a) \tag{3.30a}$$

and

$$\mathcal{A}\Big(|L_{2^+}\, S_{2^+}\, M_{L_{2^+}}\, M_{S_{2^+}}\rangle\, |\ell_a\, m_a\, \nu_a\rangle\Big) = \sum_{L\, M_L\, S\, M_S} (-1)^{\frac{1}{2}+\ell_a-L_{2^+}-S_{2^+}-M_L-M_S}$$
$$\times \sqrt{(2L+1)(2S+1)} \begin{pmatrix} L_{2^+} & \ell_a & L \\ M_{L_{2^+}} & m_a & -M_L \end{pmatrix} \begin{pmatrix} S_{2^+} & \frac{1}{2} & S \\ M_{S_{2^+}} & \nu_a & -M_S \end{pmatrix}$$
$$\times |(L_{2^+}\, \ell_a)L\, M_L;\, \left(S_{2^+}\, \frac{1}{2}\right) S\, M_S\rangle. \tag{3.30b}$$

Here, [(3.30a), (3.30b)] for Auger electron e_a are the counterparts of [(3.23b), (3.23c)] applicable to the photoelectron e_p.

The desired amplitude (herein denoted by $\mathcal{A}_{1^{+*}\longrightarrow 2^+}$) for the emission of an Auger electron in a spontaneous non-radiative transition from state $|1^{+^*}\rangle$ of the excited photoion $\mathfrak{T}^{1+^*}$ to the ground state $|2^+\rangle$ of the dication $\mathfrak{T}^{2+}$ is now readily obtained to be

$$\mathcal{A}_{1^{+*}\longrightarrow 2^+} \equiv \langle 2^+;\, \mu_a\, \hat{\boldsymbol{u}}_a\, \boldsymbol{k}_a | F_a\, | 1^{+^*}\rangle = (-1)^{L_{2^+}+S_{2^+}+M_{L_{1^{+*}}}+M_{S_{1^{+*}}}-\frac{1}{2}}$$
$$\times \sqrt{(2L_{1^{+*}}+1)(2S_{1^{+*}}+1)} \sum_{\ell_a\, m_a\, \nu_a} (-\mathrm{i})^{\ell_a}(-1)^{\ell_a}\, \mathrm{e}^{\mathrm{i}\sigma_{\ell_a}} \begin{pmatrix} L_{2^+} & \ell_a & L_{1^{+*}} \\ M_{L_{2^+}} & m_a & -M_{L_{1^{+*}}} \end{pmatrix}$$
$$\times \begin{pmatrix} S_{2^+} & \frac{1}{2} & S_{1^{+*}} \\ M_{S_{2^+}} & \nu_a & -M_{S_{1^{+*}}} \end{pmatrix} Y^{m_a}_{\ell_a}(\hat{\boldsymbol{k}}_a)\, \mathcal{D}^{1/2}_{\mu_a\,\nu_a}(\omega_a)\, \langle(L_{2^+}\, \ell_a)L_{2^+}|F_a\, |L_{2^+})\rangle. \tag{3.31}$$

Here, angular momentum conservation conditions [(3.19b), (3.20b)] have been used in the following form

$$\langle(L_{2^+}\, \ell_a)L\, M_L;\, \left(S_{2^+}\, \frac{1}{2}\right) S\, M_S|F_a\, |L_{1^{+*}}\, M_{L_{1^{+*}}};\, S_{1^{+*}}\, M_{S_{1^{+*}}}\rangle$$
$$= \delta_{L\, L_{1^{+*}}}\, \delta_{M_L\, M_{L_{1^{+*}}}}\, \delta_{S\, S_{1^{+*}}}\, \delta_{M_S\, M_{S_{1^{+*}}}}\, \langle(L_{1^{+*}}\, \ell_a)L_{2^+}|F_a\, |L_{2^+}\rangle. \tag{3.32}$$

Similar to the bracket $\langle(L_{1^{+*}}\, \ell_p)L|F_p\, |(L_0 1)L\rangle$ present in the photoionization amplitude (3.26) and thereafter, the bracket $\langle(L_{2^+}\, \ell_a)L_{2^+}|F_a\, |L_{2^+})\rangle$ in the Auger amplitude (3.31) and hereafter, too has an implicit dependence on the spin multiplicities $[(2S_{1^{+*}}+1), (2S_{2^+}+1)]$ of the electronic states $[|1^{+^*}\rangle, |2^+\rangle]$ of ions $[\mathfrak{T}^{1+^*}, \mathfrak{T}^{2+}]$ in the L-S coupling [10, 60, 72, 186]. This, in other words, means—among other things—energies $(E_{1^{+*}}, E_{2^+})$ of states $(|1^{+^*}\rangle, |2^+\rangle)$, as well as ϵ_a of the Auger electron e_a [see (6.1b)], will also depend upon the spins $(S_{1^{+*}}, S_{2^+})$ of $(\mathfrak{T}^{1+^*}, \mathfrak{T}^{2+})$.

Let us also calculate the following [188] sum for further use in this monograph

$$
\begin{aligned}
&\sum_{M_{L_{2^+}}\, M_{S_{2^+}}} \mathcal{A}_{1^{+*}\longrightarrow 2^+}\,(\mathcal{A}_{1'^{+*}\longrightarrow 2^+})^\dagger \\
&\equiv \sum_{M_{L_{2^+}}\, M_{S_{2^+}}} \langle 2^+;\, \mu_a\, \hat{\boldsymbol{u}}_a\, \boldsymbol{k}_a | F_{\mathrm{a}} | 1^{+^*}\rangle \langle 1'^{+^*} |\, F_{\mathrm{s}}^\dagger\, | 2^+;\, \mu_a'\, \hat{\boldsymbol{u}}_a\, \boldsymbol{k}_a\rangle \\
&= (-1)^{\mu_a' + L_{2^+} + S_{2^+} + 2M_{S_{1^{+*}}} - M'_{L_{1^{+*}}} - M'_{S_{1^{+*}}}}\, \frac{1}{2\sqrt{\pi}}\, (2L_{1^{+*}}+1) \\
&\times (2S_{1^{+*}}+1) \sum_{\substack{\ell_a\, \ell_a'\, C\, C_1 \\ \gamma\, \gamma_1\, \gamma_1'}} (-1)^{C+C_1}\, (2C_1+1)\sqrt{2C+1} \begin{pmatrix} \ell_{\mathrm{a}} & \ell_a' & C \\ 0 & 0 & 0 \end{pmatrix} \\
&\times \begin{pmatrix} L_{1^{+*}} & L_{1^{+*}} & C \\ M'_{L_{1^{+*}}} & -M_{L_{1^{+*}}} & -\gamma \end{pmatrix} \begin{pmatrix} \frac{1}{2} & \frac{1}{2} & C_1 \\ \mu_{\mathrm{a}} & -\mu_a' & \gamma_1 \end{pmatrix} \begin{pmatrix} S_{1^{+*}} & S_{1^{+*}} & C_1 \\ M'_{S_{1^{+*}}} & -M_{S_{1^{+*}}} & -\gamma_1' \end{pmatrix} \\
&\times \begin{Bmatrix} \ell_{\mathrm{a}} & \ell_a' & C \\ L_{1^{+*}} & L_{1^{+*}} & L_{2^+} \end{Bmatrix} \begin{Bmatrix} \frac{1}{2} & \frac{1}{2} & C_1 \\ S_{1^{+*}} & S_{1^{+*}} & S_{2^+} \end{Bmatrix} \left[Y_C^\gamma(\hat{\boldsymbol{k}}_a)\right]^* \left[\mathcal{D}^{C_1}_{\gamma_1\,\gamma_1'}(\omega_p)\right]^* \\
&\times \left[(-\mathrm{i})^{\ell_{\mathrm{a}}}\, \mathrm{e}^{\mathrm{i}\sigma_{\ell_{\mathrm{a}}}}\, (-1)^{\ell_{\mathrm{a}}} \sqrt{2\ell_{\mathrm{a}}+1}\, \langle (L_{2^+}\, \ell_{\mathrm{a}}) L_{1^{+*}} | F_{\mathrm{a}} | L_{1^{+*}}\rangle \right] \\
&\times \left[(-\mathrm{i})^{\ell_a'}\, \mathrm{e}^{\mathrm{i}\sigma_{\ell_a'}}\, (-1)^{\ell_a'} \sqrt{2\ell_a'+1}\, \langle (L_{2^+}\, \ell_a') L_{1^{+*}} | F_{\mathrm{a}} | L_{1^{+*}}\rangle \right]^* .
\end{aligned}
\tag{3.33}
$$

The procedure followed in simplifying the expression obtained after substituting (3.31) for $\mathcal{A}_{1^{+*}\longrightarrow 2^+}$ and its Hermitian conjugate for $(\mathcal{A}_{1'^{+*}\longrightarrow 2^+})^\dagger$ in the first sum on the right-hand side of (3.33) is very similar to that used in obtaining (3.28).

3.2.2 Transition Amplitudes with Spin–Orbit Interaction

In the presence of the SOI, in addition to the Coulomb interaction, neither the orbital nor the spin angular momenta are individually conserved in an atom [10, 60, 184, 186]—although both of the operators (3.6) and (3.8) for photoionization and for non-radiative spontaneous decay, respectively, are still independent of the spins. It, in other words, means [10, 60, 184, 186] that in such a physical situation the Russell–Saunders scheme of coupling the angular momenta is no longer applicable. Now the total angular momentum (i.e., sum of the orbital and spin angular momenta) is separately conserved in each of the photoionization and Auger emission processes. This implies that the angular momenta should now be coupled according to the j-j coupling scheme [10, 60, 184, 186] for studying a process taking place in an atomic target in the presence of C+SOIs.

3.2.2.1 j-j Coupling

The total angular momenta of the initial target $\mathfrak{T}$, excited photoion $\mathfrak{T}^{1+^*}$, and of the dication $\mathfrak{T}^{2+}$ are given by $\boldsymbol{J}_0 = \boldsymbol{L}_0 + \boldsymbol{S}_0$, $\boldsymbol{J}_{1+^*} = \boldsymbol{L}_{1+^*} + \boldsymbol{S}_{1+^*}$, and $\boldsymbol{J}_{2+} = \boldsymbol{L}_{2+} + \boldsymbol{S}_{2+}$, respectively. The electronic states of the species ($\mathfrak{T}$, $\mathfrak{T}^{1+^*}$, $\mathfrak{T}^{2+}$) in j-j coupling are, respectively, written as [10, 60, 184, 186] ($|0\rangle \equiv |J_0\, M_0\rangle$, $|1^{+^*}\rangle \equiv |J_{1+^*}\, M_{1+^*}\rangle$, $|2^+\rangle \equiv |J_{2+}\, M_{2+}\rangle$). Here, M_0, M_{1+^*}, and M_{2+} (with $M_0 = \boldsymbol{J}_0 \cdot \hat{e}_z$, *etc.*) are the projections of the respective angular momenta $\boldsymbol{J}_0$, $\boldsymbol{J}_{1+^*}$, and $\boldsymbol{J}_{2+}$ along the OZ-axis of the SF in Fig. 1.1. The conditions for the conservation of angular momenta, applicable in the present j-j coupling scheme, are

$$\boldsymbol{\ell}_r + \boldsymbol{J}_0 = \boldsymbol{J}_{1+^*} + \boldsymbol{j}_p \left\{ = \boldsymbol{\ell}_p + \boldsymbol{s}_p \left[= \left(\tfrac{1}{2}\right)_p \right] \right\} \tag{3.34a}$$

for the photoionization (1.3a) and

$$\boldsymbol{J}_{1+^*} = \boldsymbol{J}_{2+} + \boldsymbol{j}_a \left\{ = \boldsymbol{\ell}_a + \boldsymbol{s}_a \left[= \left(\tfrac{1}{2}\right)_a \right] \right\} \tag{3.34b}$$

for the Auger decay (1.3b) in 2-DPI process (1.3). In (3.34), each of $\boldsymbol{j}_p$ and $\boldsymbol{j}_a$ are sum of the orbital and spin angular momenta of photoelectron e_p and Auger electron e_u, respectively, generated in the steps (1.3a) and (1.3b) of 2-DPI.

Following contains derivations of various useful expressions similar to those derived in the previous Sect. 3.2.1 with the important difference that one now needs to work in j-j, rather than in L-S, coupling.

Photoionization Amplitude

According to the left-hand side of the conservation condition (3.34a) for the angular momenta, one has

$$|0, 1m_{\rm r}\rangle \equiv |J_0 M_0, 1m_{\rm r}\rangle = \sum_{J\,M_J} (-1)^{1-J_0-M_J}$$

$$\sqrt{2J+1}\begin{pmatrix} J_0 & 1 & J \\ M_0 & m_{\rm r} & -M_J \end{pmatrix} |(J_0 1) J M_J\rangle \tag{3.35}$$

for a coupled state of the non-interacting $(\mathfrak{T}, \gamma_r)$-system. The spin–orbital, corresponding to (3.22a), of the photoelectron in the present case is

$$\left\langle \boldsymbol{r}_p | \frac{1}{2}\, \nu_p\, \boldsymbol{k}_p \right\rangle = |\frac{1}{2}\, \nu_p\rangle \sum_{\ell_{\rm p}\, m_p} {\rm i}^{\ell_{\rm p}}\, {\rm e}^{-{\rm i}\sigma_{\ell_{\rm p}\,\nu_p}}$$

$$R^-_{\ell_{\rm p}\,\nu_p}(r_p; k_p) \left[Y^{m_p}_{\ell_{\rm p}}(\hat{\boldsymbol{k}}_p) \right]^* Y^{m_p}_{\ell_{\rm p}}(\hat{\boldsymbol{r}}_p). \tag{3.36a}$$

Here, unlike in (3.22a), both the Coulomb phase $\sigma_{\ell_{\rm p}\,\nu_p}$ and the radial function $R^-_{\ell_{\rm p}\,\nu_p}(r_p; k_p)$ depend [189] upon the projection $\nu_p\ (= \boldsymbol{s}_p \cdot \hat{\boldsymbol{e}}_Z)$ of the spin angular momentum $\boldsymbol{s}_p$ of the photoelectron e_p. For expressing this orbital in the j-j coupling [according to the scheme shown within the braces on the right-hand side of (3.34a)], one writes

$${\rm e}^{-{\rm i}\sigma_{\ell_{\rm p}\,\nu_p}} R^-_{\ell_{\rm p}\,\nu_p}(r_p; k_p)\, Y^{m_p}_{\ell_{\rm p}}(\hat{\boldsymbol{r}}_p)\, |\frac{1}{2}\, \nu_p\rangle = \sum_{j_p\, m_{j_p}} (-1)^{s_p - \ell_{\rm p} - m_{j_p}}\, {\rm e}^{-{\rm i}\sigma_{\ell_{\rm p}\, j_p}}$$

$$\times \sqrt{2j_p+1} \begin{pmatrix} \ell_{\rm p} & s_p & j_p \\ m_{\ell_{\rm p}} & \nu_p & -m_{j_p} \end{pmatrix} |j_p\, m_{j_p}\rangle\, R^-_{j_p}(r_p; k_p). \tag{3.36b}$$

Here, $|j_p\, m_{j_p}\rangle$ contains, among other things, the angular coordinates specifying photoelectron's position vector $\boldsymbol{r}_p\ (= r_p,\ \hat{\boldsymbol{r}}_p)$ in the SF frame of Fig. 1.1. Expression (3.36b) means [189] that its left hand side, corresponding to the uncoupled orbital $(\boldsymbol{\ell}_p)$ and spin $(\boldsymbol{s}_p)$ angular momenta of e_p (in the presence of SOI), is expanded—with the help of the appropriate 3-j symbols—in terms of the quantities which depend upon this photoelectron's coupled angular momentum $\boldsymbol{j}_p$. Consequently, orbital (3.36a) in the j-j coupling becomes

$$\langle \boldsymbol{r}_p | \frac{1}{2}\, \nu_p\, \boldsymbol{k}_p\rangle = (-1)^{\frac{1}{2}} \sum_{\ell_{\rm p}\, m_p\, j_p\, m_{j_p}} i^{\ell_{\rm p}}\, {\rm e}^{-{\rm i}\sigma_{\ell_{\rm p}\, j_p}} (-1)^{-\ell_{\rm p} - m_{j_p}} \sqrt{2j_p+1}$$

$$\times \begin{pmatrix} \ell_{\rm p} & \frac{1}{2} & j_p \\ m_{\ell_{\rm p}} & \nu_p & -m_{j_p} \end{pmatrix} \left[Y^{m_{\ell_{\rm p}}}_{\ell_{\rm p}}(\hat{\boldsymbol{k}}_p) \right]^* |j_p\, m_{j_p}\rangle\, R^-_{j_p}(r_p; k_p), \tag{3.36c}$$

where, s_p has now been replaced by $1/2$—the spin angular momentum of e_p. In view of the right hand side of the conservation condition (3.34a), one can immediately write

$$|1^{+^*};\, \mu_p\, \hat{\boldsymbol{u}}_p\, \boldsymbol{k}_p\rangle = (-1)^{\frac{1}{2}-J_{1^{+*}}} \sum_{\substack{\ell_p\, m_{\ell_p}\, \nu_p \\ j_p\, m_{j_p}}} \sum_{J\, M_J} i^{\ell_p}\, (-1)^{\ell_p+j_p-m_{j_p}-M_J}$$
$$\times\, e^{-i\sigma_{\ell_p j_p}} \sqrt{(2j_p+1)(2J+1)} \begin{pmatrix} \ell_p & \frac{1}{2} & j_p \\ m_{\ell_p} & \nu_p & -m_{j_p} \end{pmatrix} \begin{pmatrix} J_{1^{+*}} & j_p & J \\ M_{1^{+*}} & m_{j_p} & -M_J \end{pmatrix}$$
$$\times \left[Y_{\ell_p}^{m_{\ell_p}}(\hat{\boldsymbol{k}}_p) \right]^* \left[\mathcal{D}^{\frac{1}{2}}_{\mu_p\, \nu_p}(\omega_p) \right]^* |(J_{1^{+*}}\, j_p) J\, M_J\rangle^- \tag{3.37}$$

for a state of $(\mathfrak{T}^{1+^*} + e_p)$ in j-j coupling. While writing (3.37), the quantization axis for the spin of the photoelectron e_p is changed from the polar axis of the SF in Fig. 1.1 to the direction $\hat{\boldsymbol{u}}_p\,(\vartheta_p,\, \varphi_p)$ using the Euler angles $\omega_p\,(\varphi_p,\, \vartheta_p,\, 0)$. Consequently, e_p's spin–orbital

$$\langle \boldsymbol{r}_p |\, \mu_p\, \hat{\boldsymbol{u}}_p\, \boldsymbol{k}_p\rangle = \sum_{\nu_p} \left[\mathcal{D}^{\frac{1}{2}}_{\mu_p\, \nu_p}(\omega_p) \right]^* \langle \boldsymbol{r}_p |\, \frac{1}{2}\, \nu_p\, \boldsymbol{k}_p\rangle$$
$$= (-1)^{\frac{1}{2}} \sum_{\ell_p\, m_{\ell_p}\, j_p\, m_{j_p}\, \nu_p} i^{\ell_p}\, e^{-i\sigma_{\ell_p j_p}}\, (-1)^{-\ell_p-m_{j_p}} \sqrt{2j_p+1}$$
$$\times \begin{pmatrix} \ell_p & \frac{1}{2} & j_p \\ m_{\ell_p} & \nu_p & m_{j_p} \end{pmatrix} \left[Y_{\ell_p}^{m_{\ell_p}}(\hat{\boldsymbol{k}}_p) \right]^* |j_p\, m_{j_p}\rangle\, R^-_{j_p}(r_p; k_p)$$
$$\times \left[\mathcal{D}^{\frac{1}{2}}_{\mu_p\, \nu_p}(\omega_p) \right]^*, \tag{3.38a}$$

needed in (3.37) and elsewhere in this monograph in j-j coupling, is obtained on substituting (3.36c) in (3.22b). Further in (3.37), we have defined

$$|(J_{1^{+*}}\, j_p) J\, M_J\rangle^- = \mathcal{A}\Bigg[(-1)^{-J_{1^{+*}}+j_p-M_J} \sqrt{2J+1}\, R^-_{j_p}(r_p; k_p) \sum_{M_{J_{1^{+*}}}\, m_{j_p}}$$
$$\times \begin{pmatrix} J_{1^{+*}} & j_p & J \\ M_{J_{1^{+*}}} & m_{j_p} & -M_J \end{pmatrix} |J_{1^{+*}}\, M_{J_{1^{+*}}}\rangle\, |j_p\, m_{j_p}\rangle \Bigg]. \tag{3.38b}$$

Next, a straightforward use of (3.6b), (3.35), and of (3.37) gives us the amplitude [188]

$$\mathcal{P}_{0\longrightarrow 1^{+*}} \equiv \langle 1^{+^*};\, \mu_p\, \hat{\boldsymbol{u}}_p\, \boldsymbol{k}_p | F_p\, |0;\, 1\, m_r\rangle = (-1)^{-\frac{1}{2}} \sum_{\ell_p\, m_{\ell_p}\, j_p\, m_{j_p}\, \nu_p}$$
$$\times\, (-i)^{\ell_p}\, (-1)^{\ell_p+m_{j_p}}\, e^{i\sigma_{\ell_p j_p}} \sqrt{2j_p+1} \begin{pmatrix} \ell_p & \frac{1}{2} & j_p \\ m_{\ell_p} & \nu_p & -m_{j_p} \end{pmatrix}$$
$$\times\, Y_{\ell_p}^{m_{\ell_p}}(\hat{\boldsymbol{k}}_p)\, \mathcal{D}^{\frac{1}{2}}_{\mu_p\, \nu_p}(\omega_p)\, \langle J_{1^{+*}}\, M_{J_{1^{+*}}};\, j_p\, m_{j_p} | F_p | J_0\, M_{J_0};\, 1 m_r\rangle^- \tag{3.39a}$$

for photoionization in $E1$ approximation in j-j coupling. In arriving at (3.39a), one needs to use, among other things, the definition

$$\langle J_{1^{+*}}\, M_{J_{1^{+*}}};\; j_p\, m_{j_p}|F_{\rm p}|J_0\, M_{J_0};\; 1 m_{\rm r}\rangle^- \equiv (-1)^{1-J_0-J_{1^{+*}}+j_p} \sum_{J\, M_J} (-1)^{2M_J}$$

$$\times (2J+1) \begin{pmatrix} J_0 & 1 & J \\ M_0 & m_{\rm r} & -M_J \end{pmatrix} \begin{pmatrix} J_{1^{+*}} & j_p & J \\ J_{M_{1^{+*}}} & m_{j_p} & -M_J \end{pmatrix} \langle J_{1^{+*}}\, j_p|F(J)|J_0\, 1\rangle^-. \tag{3.39b}$$

and the condition (3.34a) for the conservation of the total angular momenta in the following form

$$\begin{aligned} &\langle (J_{1^{+*}}\, j_p) J\, M_J|F_{\rm p}|(J_0\, 1) J'\, M_J'\rangle^- \\ &= \delta_{(JJ')}\, \delta_{(M_J M_J')}\, \langle J_{1^{+*}}\, j_p|F(J)|J_0\, 1\rangle^-. \end{aligned} \tag{3.39c}$$

Photoionization amplitude (3.39a) can be written in an alternative form, which is sometimes simpler to use. The angular momentum transferred from the absorbed photon γ_r to the observed e_p is readily obtained from the conservation condition (3.34a) to be [190] $\boldsymbol{j}_t = \boldsymbol{\ell}_r - \boldsymbol{j}_p = \boldsymbol{J}_{1^{+*}} - \boldsymbol{J}_0$. (Here, while the e_p is detected with its spin polarization, the excited photoion $\mathfrak{T}^{1+^*}$ remains unobserved.) In order to introduce j_t in (3.39a), expression (3.39b) should be rewritten as [190]

$$\langle J_{1^{+*}}\, M_{J_{1^{+*}}};\; j_p\, m_{j_p}|F_{\rm p}|J_0\, M_{J_0};\; 1 m_{\rm r}\rangle^- = (-1)^{-J_0-J_{1^{+*}}-M_0-m_{j_p}} \sum_{j_t\, m_t}$$

$$\times (2j_t+1) \begin{pmatrix} 1 & j_p & j_t \\ m_{\rm r} & -m_{j_p} & m_t \end{pmatrix} \begin{pmatrix} J_{1^{+*}} & J_0 & j_t \\ J_{M_{1^{+*}}} & -M_0 & m_t \end{pmatrix} \langle J_{1^{+*}}\, j_p|F(j_t)|J_0\, 1\rangle \tag{3.40a}$$

using the identity (2.19) from [187]. Here, we have defined [190]

$$\langle J_{1^{+*}}\, j_p|F(j_t)|J_0\, 1\rangle \equiv \sum_J (-1)^J\, (2J+1)$$

$$\times \begin{Bmatrix} 1 & J_0 & J \\ J_{1^{+*}} & j_p & j_t \end{Bmatrix} \langle J_{1^{+*}}\, j_p|F(J)|J_0\, 1\rangle, \tag{3.40b}$$

with the Dirac's bracket $\langle J_{1^{+*}}\, j_p|F(J)|J_0\, 1\rangle$ already given in (3.39b). On substituting (3.40a), the photoionization amplitude (3.39a) takes the following alternative form [190]

$$\mathcal{P}_{0\longrightarrow 1^{+*}} = (-1)^{-\frac{1}{2}-J_0-J_{1^{+*}}-M_{J_0}} \sum_{\substack{\ell_p\, j_p\, j_t\, \nu_p \\ m_{\ell_p}\, m_{j_p}\, m_t}} (-\mathrm{i})^{\ell_p}\,(-1)^{\ell_p}\,\mathrm{e}^{\mathrm{i}\sigma_{\ell_p j_p}}$$

$$\times (2j_t+1)\sqrt{2j_p+1}\begin{pmatrix} \ell_p & \frac{1}{2} & j_p \\ m_{\ell_p} & \nu_p & -m_{j_p}\end{pmatrix}\begin{pmatrix} 1 & j_p & j_t \\ m_r & -m_{j_p} & m_t\end{pmatrix}$$

$$\times \begin{pmatrix} J_{1^{+*}} & J_0 & j_t \\ M_{J_{1^{+*}}} & -M_0 & m_t\end{pmatrix} Y_{\ell_p}^{m_{\ell_p}}(\hat{\boldsymbol{k}}_p)\,\mathcal{D}^{\frac{1}{2}}_{\mu_p\,\nu_p}(\omega_p)\,\langle J_{1^{+*}}\, j_p|F(j_t)|J_0\, 1\rangle \quad (3.41)$$

in j-j coupling in terms of the angular momentum transfer j_t.

In analogy to (3.27) in L-S coupling, let us calculate the following quantity in the j-j coupling

$$\sum_{M_0} \mathcal{P}_{0\longrightarrow 1^{+*}}\,(\mathcal{P}_{0\longrightarrow 1'^{+*}})^{\dagger}$$

$$\equiv \sum_{M_0} \langle 1^{+^*};\ \mu_p\,\hat{\boldsymbol{u}}_p\,\boldsymbol{k}_p|F_p\,|0;\ 1\,m_r\rangle\langle 0;\ 1\,m_r|F_p^{\dagger}\,|1'^{+^*};\ \mu'_p\,\hat{\boldsymbol{u}}_p\,\boldsymbol{k}_p\rangle \qquad (3.42a)$$

$$= (-1)^{\frac{1}{2}+m_r+\mu'_p+J_0+M_{J_{1^{+*}}}}\,\frac{1}{2\sqrt{\pi}} \sum_{\substack{\ell_p\, j_p\, j_t \\ \ell'_p\, j'_p\, j'_t}}\ \sum_{\substack{L_p\, M_{L_p}\, L_r \\ S_p\, M_{S_p}\, N_{S_p}}}\ \sum_{J_p\, M_{J_p}\, J_t} (-1)^{\ell_p+j'_t}$$

$$(2L_r+1)(2S_p+1)(2J_p+1)(2J_t+1)\sqrt{2L_p+1}\begin{pmatrix} \ell_p & \ell'_p & L_p \\ 0 & 0 & 0\end{pmatrix}\begin{pmatrix} \frac{1}{2} & \frac{1}{2} & S_p \\ \mu_p & -\mu'_p & M_{S_p}\end{pmatrix}$$

$$\begin{pmatrix} 1 & 1 & L_r \\ m_r & -m_r & 0\end{pmatrix}\begin{pmatrix} L_p & S_p & J_p \\ M_{L_p} & N_{S_p} & M_{J_p}\end{pmatrix}\begin{pmatrix} J_{1^{+*}} & J_{1^{+*}} & J_t \\ M_{1^{+*}} & -M'_{1^{+*}} & M_{J_p}\end{pmatrix}\begin{pmatrix} J_p & J_t & L_r \\ M_{J_p} & -M_{J_p} & 0\end{pmatrix}$$

$$\begin{Bmatrix} J_{1^{+*}} & J_{1^{+*}} & J_t \\ j_t & j'_t & J_0\end{Bmatrix}\begin{Bmatrix} \ell_p & \ell'_p & L_p \\ \frac{1}{2} & \frac{1}{2} & S_p \\ j_p & j'_p & J_p\end{Bmatrix}\begin{Bmatrix} j_p & j'_p & J_p \\ j_t & j'_t & J_t \\ 1 & 1 & L_r\end{Bmatrix}\left[Y_{L_p}^{M_{L_p}}(\hat{\boldsymbol{k}}_p)\right]^*\left[\mathcal{D}^{S_p}_{M_{S_p}\,N_{S_p}}(\omega_p)\right]^*$$

$$\left[(-\mathrm{i})^{\ell_p}\,e^{\sigma_{\ell_p j_p}}(2j_t+1)\sqrt{(2\ell_p+1)(2j_p+1)}\langle J_{1^{+*}}\,j_p|F(j_t)|J_0\,1\rangle\right]$$

$$\left[(-\mathrm{i})^{\ell'_p}\,e^{\sigma_{\ell'_p j'_p}}(2j'_t+1)\sqrt{(2\ell'_p+1)(2j'_p+1)}\langle J_{1^{+*}}\,j'_p|F(j'_t)|J_0\,1\rangle\right]^*, \qquad (3.42b)$$

where $\langle J_{1^{+*}}\,j_p|F(j_t)|J_0\,1\rangle$ is already defined in (3.40b). Relation (3.42b) has been arrived at after substituting (3.41) and its Hermitian conjugate in (3.42a) and then simplifying the consequent expression using a procedures almost similar to that employed in deriving (3.28). At several places [e.g., in (3.42b)] in this monograph, $\begin{Bmatrix} \cdots\cdots\cdots \\ \cdots\cdots\cdots \\ \cdots\cdots\cdots \end{Bmatrix}$ is a 9-j symbol [65, 187].

Auger Decay Amplitude

The spin–orbital

$$\langle \boldsymbol{r}_a | \frac{1}{2}\, \nu_a\, \boldsymbol{k}_a \rangle = (-1)^{\frac{1}{2}} \sum_{\ell_a m_{\ell_a} j_a m_{ja}} i^{\ell_a}\, e^{-i\sigma_{\ell_a j_a}} (-1)^{-\ell_a - m_{ja}} \sqrt{2j_a+1}$$

$$\times \begin{pmatrix} \ell_a & \frac{1}{2} & j_a \\ m_{\ell_a} & \nu_a & -m_{j_a} \end{pmatrix} \left[Y_{\ell_a}^{m_{\ell_a}}(\hat{\boldsymbol{k}}_a) \right]^* | j_a\, m_{j_a} \rangle\, R^-_{j_a}(r_a; k_a) \quad (3.43a)$$

of the Auger electron e_a, with its orbital and spin angular momenta coupled according to the scheme suggested within the braces on the right-hand side of the conservation condition (3.34b), can readily be written down from that [i.e., (3.36c)] already obtained for photoelectron e_p. Analogously, following (3.37), a state of ($\mathfrak{T}^{2+}$ + e_a)-system in the j-j coupling is given by

$$|2^+;\, \mu_a\, \hat{\boldsymbol{u}}_a\, \boldsymbol{k}_a \rangle = (-1)^{\frac{1}{2} - J_{2^+}} \sum_{\substack{\ell_a m_{\ell_a} \nu_a \\ j_a m_{ja}}} \sum_{J M_J} i^{\ell_a} (-1)^{\ell_a + j_a - m_{ja} - M_J}$$

$$\times e^{-i\sigma_{\ell_a j_a}} \sqrt{(2j_a+1)(2J+1)} \begin{pmatrix} \ell_a & \frac{1}{2} & j_a \\ m_{\ell_a} & \nu_a & -m_{j_a} \end{pmatrix} \begin{pmatrix} J_{2^+} & j_a & J \\ M_{2^+} & m_{j_a} & -M_J \end{pmatrix}$$

$$\times \left[Y_{\ell_a}^{m_{\ell_a}}(\hat{\boldsymbol{k}}_a) \right]^* \left[\mathcal{D}^{\frac{1}{2}}_{\mu_a \nu_a}(\omega_a) \right]^* |(J_{2^+}\, j_a) J\, M_J\rangle^-. \quad (3.43b)$$

Quantities ν_a $(= \boldsymbol{s}_a \cdot \hat{\boldsymbol{e}}_Z)$ in (3.43a) and μ_a $(= \boldsymbol{s}_a \cdot \hat{\boldsymbol{u}}_a)$ in (3.43b) are the projections of the spin of Auger electron e_a along the polar axis of the SF and direction $\hat{\boldsymbol{u}}_a$ (ϑ_a, φ_a), respectively. Further in (3.43b),

$$|(J_{2^+}\, j_a) J\, M_J\rangle^- = \mathcal{A} \Bigg[(-1)^{-J_{2^+} + j_a - M_J} \sqrt{2J+1}\, R^-_{j_a}(r_a; k_a) \sum_{M_{J_{2^+}} m_{ja}}$$

$$\times \begin{pmatrix} J_{2^+} & j_a & J \\ M_{J_{2^+}} & m_{j_a} & -M_J \end{pmatrix} |J_{2^+} M_{J_{2^+}}\rangle\, |j_a\, m_{j_a}\rangle \Bigg], \quad (3.43c)$$

with $|j_a\, m_{j_a}\rangle$ obtained from (3.43a). Amplitude for the Auger decay in 2-DPI process (3.3) in j-j coupling is now given by [188]

$$\mathcal{A}_{0\longrightarrow 2^+} \equiv \langle 2^+;\, \mu_a\, \hat{\boldsymbol{u}}_a\, \boldsymbol{k}_a | F_{\mathrm{a}} | 1^{+^*} \rangle = (-1)^{-\frac{1}{2}+J_{2^+}+M_{1^{+*}}} \sum_{\substack{\ell_a m_{\ell_a} \nu_a \\ j_a m_{j_a}}}$$

$$\times (-\mathrm{i})^{\ell_a} (-1)^{\ell_a - j_a + m_{j_a}}\, \mathrm{e}^{\mathrm{i}\sigma_{\ell_a j_a}} \sqrt{(2j_a+1)(2J_{1^{+*}}+1)} \begin{pmatrix} \ell_a & \frac{1}{2} & j_a \\ m_{\ell_a} & \nu_a & -m_{j_a} \end{pmatrix}$$

$$\times \begin{pmatrix} J_{2^+} & j_a & J_{1^{+*}} \\ M_{2^+} & m_{j_a} & -M_{1^{+*}} \end{pmatrix} Y_{\ell_a}^{m_{\ell_a}}(\hat{\boldsymbol{k}}_a)\, \mathcal{D}^{\frac{1}{2}}_{\mu_a \nu_a}(\omega_a)\, \langle (J_{2^+}\, j_a) J_{1^{+*}} | F_{\mathrm{a}} | J_{1^{+*}} \rangle. \quad (3.44)$$

This amplitude satisfies the condition (3.34b) applicable to the Auger decay part (1.3b) of 2-DPI for the conservation of the total angular momentum in j-j coupling in the following form

$$\langle (J_{2^+}\, j_a) J\, M_J | F_{\mathrm{a}} | J_{1^{+*}}\, M_{J_{1^{+*}}} \rangle^- = \delta_{(J J_{1^{+*}})}\, \delta_{(M_J M_{J_{1^{+*}}})}\, \langle (J_{2^+}\, j_a) J_{1^{+*}} | F_a | J_{1^{+*}} \rangle^-.$$

The final calculation of the density matrix for 2-DPI (1.3) in j-j coupling is considerably simplified with a prior knowledge of the following expression

$$\sum_{M_{2^+}} \mathcal{A}_{1^{+*}\longrightarrow 2^+} (\mathcal{A}_{1'^{+*}\longrightarrow 2^+})^\dagger$$

$$\equiv \sum_{M_{2^+}} \langle 2^+;\, \mu_a\, \hat{\boldsymbol{u}}_a\, \boldsymbol{k}_a | F_{\mathrm{a}} | 1^{+^*} \rangle \langle 1'^{+^*} | F_{\mathrm{a}}^\dagger | 2^+;\, \mu_a'\, \hat{\boldsymbol{u}}_a\, \boldsymbol{k}_a \rangle \quad (3.45a)$$

$$= (-1)^{\frac{1}{2}+\mu_a'-J_{2^+}+M_{1^{+*}}} \frac{1}{2\sqrt{\pi}} (2J_{1^{+*}}+1) \sum_{\substack{\ell_a \ell_a' j_a j_a' L_a S_a J_a \\ M_{L_a} M_{S_a} N_{S_a} M_{J_a}}}$$

$$\times (-1)^{\ell_a + j_a' + J_a} (2S_a+1)(2J_a+1)\sqrt{2L_a+1} \begin{pmatrix} \ell_a & \ell_a' & L_a \\ 0 & 0 & 0 \end{pmatrix} \begin{pmatrix} \frac{1}{2} & \frac{1}{2} & S_a \\ \mu_a & -\mu_a' & M_{S_a} \end{pmatrix}$$

$$\times \begin{pmatrix} L_a & S_a & J_a \\ M_{L_a} & N_{S_a} & M_{J_a} \end{pmatrix} \begin{pmatrix} J_{1^{+*}} & J_{1^{+*}} & J_a \\ -M_{1^{+*}} & M'_{1^{+*}} & M_{J_a} \end{pmatrix} \begin{Bmatrix} J_{1^{+*}} & J_{1^{+*}} & J_a \\ j_a & j_a' & J_{2^+} \end{Bmatrix} \begin{Bmatrix} \ell_a & \ell_a' & L_a \\ \frac{1}{2} & \frac{1}{2} & S_a \\ j_a & j_a' & J_a \end{Bmatrix}$$

$$\times \left[Y_{L_a}^{M_{L_a}}(\hat{\boldsymbol{k}}_a) \right]^* \left[\mathcal{D}^{S_a}_{M_{S_a} N_{S_a}}(\omega_a) \right]^*$$

$$\times \left[(-\mathrm{i})^{\ell_a}\, \mathrm{e}^{\sigma_{\ell_a j_a}} \sqrt{(2\ell_a+1)(2j_a+1)}\, \langle (J_{2^+}\, j_a) J_{1^{+*}} | F_{\mathrm{a}} | J_{1^{+*}} \rangle \right]$$

$$\times \left[(-\mathrm{i})^{\ell_a'}\, \mathrm{e}^{\sigma_{\ell_a' j_a'}} \sqrt{(2\ell_a'+1)(2j_a'+1)}\, \langle (J_{2^+}\, j_a') J_{1^{+*}} | F_{\mathrm{a}} | J_{1^{+*}} \rangle \right]^* \quad (3.45b)$$

with the bracket $\langle (J_{2^+}\, j_a) J_{1^{+*}} | F_{\mathrm{a}} | J_{1^{+*}} \rangle$ defined in (3.44).

3.3 Amplitudes for Molecular Transitions

Molecules are known to have physical and chemical properties totally different from those of atoms. These differences arise because, unlike an atom, a molecule possesses more than one nucleus and, hence, has an extended—i.e., a linear, planar, or a three-dimensional (3-D)—structure. While electronic motion is present in both atoms and molecules; however, because of its extended structure, a molecule, unlike an atom, can vibrate and rotate as well. Presence of the extended structure plus two additional degrees of freedom in a molecule makes it necessary to have a theoretical treatment of any physical process taking place in a molecule different from that applicable to an atom. Moreover, in case of molecules with a planar or 3-D geometry, applications of the powerful group theoretical methods become necessary in order to have a proper description of molecular symmetry as well as properties (see, for example, [172, 191–211], *etc*). But, an adequate description of a linear molecule,[11] is possible even without using any group theoretical concepts. All the molecule-related discussions in this monograph will primarily be concerned, unless stated otherwise, with only the linear systems.

An energy level in a molecule is said to have positive (negative) parity[12] if its wavefunction does not (does) change its sign on inversion of its both electronic plus nuclear coordinates through the origin of the SF coordinate system $OXYZ$ shown in the Fig. 9.1 (page 207). In addition, a linear molecule [10, 68–70, 185] with a center of symmetry (e.g., H_2, N_2, CO_2, *etc*) is known to belong to the $D_{\infty h}$ point group; while the ones without such a center (e.g., CO, NO, OCS, *etc*) are classified as $C_{\infty v}$ molecules. Accordingly, an energy level in a $D_{\infty h}$ molecule possesses a gerade or ungerade character (specified, respectively, by the roman letters g or u) if it remains same or changes its sign when only its electronic coordinates are inverted through its center of symmetry.

Inclusione of nuclear vibration in a linear molecule requires merely an averaging, over the molecular vibrational wave functions, of the amplitude (calculated as a function of internuclear separation) for a physical process and, hence, does not need any extra theoretical considerations. In this monograph, therefore, vibrational degree of freedom of a molecule will not be further considered. But, nuclear rotation introduces an angular momentum $\boldsymbol{R}$ [68, 69, 185], in addition to those two which are present even in an atom—namely, electronic orbital and spin angular momenta $\boldsymbol{L}$ and $\boldsymbol{S}$, respectively. The direction of $\boldsymbol{R}$ in a linear molecule is always perpendicular to the line joining its nuclei. Further [68, 69, 185], the spin magnetic moment of an electron in a molecule can interact with the magnetic field generated, among others, by the rotation of molecular nuclei giving rise to SRI. Thus, unlike in an atom, SDIs in a linear molecule consist of not only the SOI, which is present even in an atom, but also the SRI. Hence, SDIs $\equiv$ SOI for an atom, but SDIs $\equiv$ SOI + SRI for a rotating molecule.

---

[11] A molecule with all of its nuclei in a straight line is called a linear molecule [10, 60, 68–70, 185].

[12] Parity of a molecular state is shown by a +/- superscript on the symbol representing this state.

Thus, similar to atoms, studies of the entanglement properties of the states of more than one particles generated in any of the processes (1.1) to (1.8) taking place in a rotating linear molecule (RLM) be studied both without and with SDIs. In the following two Sects. 3.3.1 and 3.3.2, respectively, we present a brief description of the calculations of photoionization and Auger decay amplitudes for a RLM in both the absence and presence of SDIs.

There [68, 69, 185] are five [called (a), (b), (c), (d), and (e)] schemes, originally suggested by Hund [212], for coupling angular momenta in a linear molecule if its rotational motion is also taken into account. For the discussions presented in this monograph, only first two [i.e., Hund's coupling schemes (a) and (b)] of the five are more relevant and, hence, sufficient. Moreover, for a proper description of the symmetry and other properties of a liner molecule, including the rotation of its nuclei, it is necessary to consider its parity adapted states [68, 69, 75].

Discussions related to a molecular target require, in general, introduction of one more coordinate system which rotates with the molecule, i.e., the molecule-fixed frame (MF) of reference $OX_mY_mZ_m$. Its polar (i.e., OZ_m)-axis is along the line[13] joining all the nuclei in a linear molecule $\mathfrak{T}$ and orientation with respect to the SF $OXYZ$ is specified by the Euler angles ω_m $(\phi_m, \theta_m, \xi_m)$. That is, a rotation [65] by ω_m brings SF in coincidence with MF.

3.3.1 *Transition Amplitudes without Spin-Dependent Interactions*

When no SDIs are taken into account, it is only the (inter-electronic, inter-nuclear, and electron-nucleus) Coulomb forces which determine properties of a molecule and those of the processes taking place in it. Unlike in an atom, electrons in a linear molecule move in an electric field which has axial symmetry. Consequently, neither of the electronic orbital and nuclear rotational angular momenta $\boldsymbol{L}$ and $\boldsymbol{R}$, respectively, is a constant of motion even in the absence of SDIs in a RLM. On the other hand, projection of the sum ($\boldsymbol{N} = \boldsymbol{L} + \boldsymbol{R}$) along the inter-nuclear axis is now a good quantum number.[14] However, electronic spin angular momentum $\boldsymbol{S}$ is still conserved. In the absence of SDIs, coupling of various angular momenta present in a rotating linear molecule is adequately described [68, 69, 185] by the Hund's scheme (b).

---

[13]This line is called also the inter-nuclear axis in a linear molecule. Its direction is given by the unit vector $\hat{\boldsymbol{e}}_{z_m}$ along the polar OZ_m-axis of the MF.

[14]The projection of $\boldsymbol{R}$ along the inter-nuclear axis is always zero. Hence, $\hat{\boldsymbol{e}}_{z_m} \cdot \boldsymbol{N} = \hat{\boldsymbol{e}}_{z_m} \cdot \boldsymbol{L} \equiv \Lambda$.

3.3.1.1 Hund's Coupling Scheme (b)

On account of the cylindrical (also called axial) symmetry of the nuclear electric field experienced by electrons in a linear molecule, the sum of the projections of all, other than spin, angular momenta along the inter-nuclear axis is conserved before and after any process taking place in a linear molecule in Hund's case (b) in the absence of SDIs. Consequently, the conservation conditions applicable to the single photoionization part of the processes (1.1), (1.3), (1.4), and (1.6)–(1.8) are

$$\begin{aligned}\lambda_r\,(=\hat{\boldsymbol{e}}_{z_m}\cdot\boldsymbol{\ell}_r)+\Lambda_0\,(=\hat{\boldsymbol{e}}_{z_m}\cdot\boldsymbol{N}_0)\\ =\Lambda_{1^{+*}}\,(=\hat{\boldsymbol{e}}_{z_m}\cdot\boldsymbol{N}_{1^{+*}})+\lambda_p\,(=\hat{\boldsymbol{e}}_{z_m}\cdot\boldsymbol{\ell}_p)\end{aligned} \tag{3.46a}$$

and

$$\boldsymbol{S}_0=\boldsymbol{S}_{1^{+*}}+\boldsymbol{s}_p\left[=\left(\frac{1}{2}\right)_p\right]; \tag{3.46b}$$

Here, angular momenta $\boldsymbol{\ell}_r$, $\boldsymbol{\ell}_p$, and $\boldsymbol{s}_p$ have meanings identical to those already introduced in (3.19) and (3.20). Further in (3.46), $\boldsymbol{N}_0=\boldsymbol{L}_0+\boldsymbol{R}_0$ with $\boldsymbol{L}_0$ and $\boldsymbol{R}_0$ being, respectively, the electronic orbital and nuclear rotational angular momenta in the pre-photoionization state $|0\rangle$ of the molecule $\mathfrak{T}$; the corresponding angular momenta for the state $|1^{+*}\rangle$ of the excited molecular photoion $\mathfrak{T}^{1+^*}$ are, respectively, $\boldsymbol{N}_{1^{+*}}=\boldsymbol{L}_{1^{+*}}+\boldsymbol{R}_{1^{+*}}$. In the conservation condition (3.46b), on the other hand, $\boldsymbol{S}_0$, $\boldsymbol{S}_{1^{+*}}$, and $\boldsymbol{s}_p$ are again the spins of $\mathfrak{T}$ in its state $|0\rangle$, of $\mathfrak{T}^{1+^*}$ in its state $|1^{+^*}\rangle$, and of the photoelectron e_p, respectively.

The conservation conditions, corresponding to (3.46) and applicable to the non-radiative Auger decay (1.3b) in a RLM are given by

$$\Lambda_{1^{+*}}\,(=\hat{\boldsymbol{e}}_{z_m}\cdot\boldsymbol{N}_{1^{+*}})=\Lambda_{2^+}\,(=\hat{\boldsymbol{e}}_{z_m}\cdot\boldsymbol{N}_{2^+})+\lambda_a\,(=\hat{\boldsymbol{e}}_{z_m}\cdot\boldsymbol{\ell}_a) \tag{3.47a}$$

and

$$\boldsymbol{S}_{1^{+*}}=\boldsymbol{S}_{2^+}+\boldsymbol{s}_a\left[=\left(\frac{1}{2}\right)_a\right] \tag{3.47b}$$

with $\boldsymbol{N}_{2^+}$ and $\boldsymbol{S}_{2^+}$ having their obvious meanings for the state $|2^+\rangle$ of the molecular dication $\mathfrak{T}^{2+}$. For the definitions of $\boldsymbol{\ell}_a$ and s_a, the reader is again referred to [(3.19), (3.20)].

A parity adapted $|0\rangle$ state of a linear molecule $\mathfrak{T}$ in Hund's coupling scheme (b) is given by [69, 75]

$$|0\rangle = \frac{1}{\sqrt{2}}\Big(|n_0\,\Lambda_0\rangle|N_0\,\Lambda_0\,M_{N_0}\rangle +$$

$$(-1)^{\mathfrak{p}_0}|n_0\,-\Lambda_0\rangle|N_0\,-\Lambda_0\,M_{N_0}\rangle\Big)|S_0\,M_{S_0}\rangle. \tag{3.48a}$$

Here, $|n_0\,\Lambda_0\rangle$ and $|S_0\,M_{S_0}\rangle$ are, respectively, the electronic spatial and spin parts of the state $|0\rangle$. The quantity n_0 collectively represents all those quantum numbers which, for brevity, have not explicitly been shown in (3.48a) as they are not relevant to the present discussion, but are needed for a complete specification of the electronic spatial part $|n_0\,\Lambda_0\rangle$ in the molecular state $|0\rangle$. $M_{S_0} = \hat{\boldsymbol{e}}_z \cdot \boldsymbol{S}_0$ is component along the polar OZ-axis of the SF in Fig. 9.1 (page 207) of electrons' spin S_0 in molecule $\mathfrak{T}$. Further in (3.48a),

$$\langle\omega_m|\,N_0\,\Lambda_0\,M_{N_0}\rangle = \sqrt{\frac{2N_0+1}{8\pi^2}}\,\mathcal{D}^{N_0}_{\Lambda_0\,M_{N_0}}(\omega_m) \tag{3.48b}$$

is a normalized function, which describes [65] rotation of nuclei in the state $|0\rangle$ of linear molecule $\mathfrak{T}$. Here and elsewhere, projection of the angular momentum $\boldsymbol{N}_0$ along the polar axis of the SF $OXYZ$ in Fig. 9.1 is $M_{N_0} = \hat{\boldsymbol{e}}_z{\cdot}\boldsymbol{N}_0$. [Euler angle ω_m, present in (3.48b), have already been defined on page 75.] Parity of the state (3.48a)[15] is [69, 75]

$$(-1)^{s+\mathfrak{p}_0+N_0} \quad \text{with} \qquad \mathfrak{p}_0 = 0 \text{ or } 1. \tag{3.48c}$$

But, for $\Lambda_0 = 0$ (i.e., for $\Sigma^{\pm}$)-states, $\mathfrak{p}_0 = 0$ [69, 75] and then the state (3.48a) reduces to

$$|0\rangle = |n_0\,0\rangle|N_0\,0\,M_{N_0}\rangle|S_0\,M_{S_0}\rangle \tag{3.48d}$$

with parity $(-1)^{s+N_0}$. In addition [69, 75], the parameter $s = 0$ always, except for Σ^--states wherein $s = 1$. Thus, when (3.48a) corresponds to a Σ^+-state, its parity is simply $(-1)^{N_0}$; whereas if (3.48a) is to represent a Σ^--state, then its parity becomes $(-1)^{1+N_0}$. In the case of non-Σ electronic states, on the other hand, the parity of $|0\rangle$ in (3.48a) is given by $(-1)^{\mathfrak{p}_0+N_0}$

Expressions similar to (3.48) in Hund's coupling scheme (b) can be written also for the state $|1^{+^*}\rangle$ of the excited molecular photoion $\mathfrak{T}^{1+^*}$, $|2^+\rangle$ of the molecular dication or, for that matter, any other states of a RLM.

[15]It, in other words, means that on the inversion through the origin "O" of the co-ordinate system in Fig. 9.1 of the position vectors of all the particles (i.e., electrons and nuclei) present in a RLM $\mathfrak{T}$, state (3.48a) does not change its sign if $s + \mathfrak{p}_0 + N_0$ is even; otherwise (i.e., for $s + \mathfrak{p}_0 + N_0$ odd), the sign of (3.48a) is reversed.

The molecular spin–orbital for the photoelectron e_p in SF in Hund's case (b) is given by [213, 214, 217]

$$|\mu_p \hat{\boldsymbol{u}}_p \boldsymbol{k}_p\rangle = \sum_{\ell_p \lambda_p m_p \nu_p} i^{\ell_p} e^{-i\sigma_{\ell_p}} \mathcal{D}^{\ell_p}_{\lambda_p m_p}(\omega_m) \times \left[\mathcal{D}^{\frac{1}{2}}_{\mu_p \nu_p}(\omega_p)\right]^* \left[Y^{m_p}_{\ell_p}(\hat{\boldsymbol{k}}_p)\right]^* |\tfrac{1}{2} \nu_p\rangle \, |F^-_{\ell_p \lambda_p}(k_p)\rangle. \quad (3.49)$$

Here, m_p $(= \hat{\boldsymbol{e}}_Z \cdot \boldsymbol{\ell}_p)$ and ν_p $(= \hat{\boldsymbol{e}}_Z \cdot \boldsymbol{s}_p)$ are, respectively, the projections of photoelectron's orbital and spin angular momenta along the OZ-axis of the SF in Fig. 9.1.[16] The spherical angles (θ_p, ϕ_p) and (ϑ_p, φ_p) specify the respective vectors $\hat{\boldsymbol{k}}_p$ and $\hat{\boldsymbol{u}}_p$ in the SF. The rest of the symbols used in (3.49) have their usual meanings explained elsewhere in the present Chap. 3, or in Chap. 1, of this monograph.

Photoionization Amplitude

If $|1, \lambda_r\rangle$ (with $\lambda_r = \hat{\boldsymbol{e}}_{Z_m} \cdot \boldsymbol{\ell}_r$) is the state of the absorbed photon γ_r in MF $OX_m X_m X_m$ in Fig. 9.1, one can then readily write [174, 213, 214]

$$|1 m_r\rangle = \sum_{\lambda_r} \mathcal{D}^1_{\lambda_r m_r}(\omega_m) \, |1 \lambda_r\rangle \quad (3.50)$$

for the state of the same photon in SF. State $|0;\ 1m_r\rangle$ of the (photon + molecule)-system in Hund's case (b), before any interaction between these two particles takes place, is then simply a product of (3.50) and (3.48a) for γ_r and $\mathfrak{T}$, respectively, and is given by

$$|0;\ 1m_r\rangle = |0\rangle \, |1m_r\rangle. \quad (3.51)$$

On the other hand, an antisymmetrized state of the $(\mathfrak{T}^{1+^*}, e_p)$-system, with the conservation conditions (3.46) taken into account, is [213, 214]

$$|1^{+^*}; e_p\rangle \equiv |N_{1^{+*}} \Lambda_{1^{+*}} M_{N_{1^{+*}}} S_{1^{+*}}; \mu_p \hat{\boldsymbol{u}}_p \boldsymbol{k}_p\rangle = \mathcal{A}\Big(|1^{+^*}\rangle \, |\mu_p \hat{\boldsymbol{u}}_p \boldsymbol{k}_p\rangle\Big). \quad (3.52a)$$

Here, $|1^{+^*}\rangle$ is the parity adapted state of the excited molecular photoion $\mathfrak{T}^{1+^*}$ in the Hund's coupling scheme (b). On substituting in (3.52a) for the state $|1^{+^*}\rangle$ of $\mathfrak{T}^{1+^*}$ and (3.49) for the photoelectron e_p, one finds [213, 214]

---

[16] In expression (3.49) and elsewhere, $\mathcal{D}^{\ell_p}_{\lambda_p m_p}(\omega_m)$ rotates photoelectron's orbital angular momentum vector $\boldsymbol{\ell}_p$ from MF to SF in Fig. 9.1; whereas $\left[\mathcal{D}^{\frac{1}{2}}_{\mu_p \nu_p}(\omega_p)\right]^*$ is used for taking e_p's spin quantization direction from OZ-axis to $\hat{\boldsymbol{u}}_p$ in SF.

$$|1^{+^*};e_p\rangle = \frac{1}{\sqrt{2}} \sum_{\substack{\ell_p\,\lambda_p\,m_p \\ \nu_p\,S\,M_S}} i^{\ell_p}\, e^{-i\sigma_{\ell_p}}\, (-1)^{\frac{1}{2}-S_{1^{+*}}-M_S} \sqrt{2S+1}$$

$$\begin{pmatrix} S_{1^{+*}} & \frac{1}{2} & S \\ M_{S_{1^{+*}}} & \nu_p & -M_S \end{pmatrix} \Big[|n_{1^{+*}}\, \Lambda_{1^{+*}};\, \ell_p\, \lambda_p\rangle |N_{1^{+*}}\, \Lambda_{1^{+*}}\, M_{N_{1^{+*}}}\rangle$$

$$+(-1)^{\mathfrak{p}_{1^{+*}}} |n_{1^{+*}}\ -\Lambda_{1^{+*}};\, \ell_p\, \lambda_p\rangle |N_{1^{+*}}\ -\Lambda_{1^{+*}}\, M_{N_{1^{+*}}}\rangle \Big] |S M_S\rangle$$

$$\mathcal{D}^{\ell_p}_{\lambda_p\, m_p}(\omega_m) \Big[\mathcal{D}^{\frac{1}{2}}_{\mu_p\,\nu_p}(\omega_p)\Big]^* \Big[Y^{m_p}_{\ell_p}(\hat{\boldsymbol{k}}_p)\Big]^* \tag{3.52b}$$

with the definitions

$$|n_{1^{+*}}\, \Lambda_{1^{+*}};\, \ell_p\, \lambda_p\rangle \ \equiv\ |n_{1^{+*}}\, \Lambda_{1^{+*}}\rangle\, |F^-_{\ell_p\,\lambda_p}(k_p)\rangle, \tag{3.52c}$$

For brevity, the antisymmetrization operator $\mathcal{A}$ has not explicitly been shown in the expression (3.52b). Expression (3.52b) conforms to the requirements of the coupling of the spin angular momenta shown on the right-hand side of the conservation condition (3.47b) applicable in the Hund's case (b).[17]

Use of the states (3.51) [along with (3.50) as well as (3.48a)] and (3.52b) gives us [215–217] following expression for the molecular photoionization amplitude in $E1$ approximation in Hund's scheme (b) in the absence of SDIs:

$$\mathcal{P}^{(m,b)}_{0\longrightarrow 1^{+*}} \equiv \langle 1^{+^*};\ \mu_p\, \hat{\boldsymbol{u}}_p\, \boldsymbol{k}_p | F_{\mathrm{p}} | 0;\ 1 m_{\mathrm{r}}\rangle \tag{3.53a}$$

$$= (-1)^{\frac{1}{2}-S_{1^{+*}}-M_{S_0}} \Big[(2S_0+1)(2N_0+1)(2N_{1^{+*}}+1)\Big]^{\frac{1}{2}} \sum_{\substack{\ell_p\, m_p\, \nu_p \\ N\, M_N}}$$

$$(2\ell_p+1)^{-\frac{1}{2}} \begin{pmatrix} S_{1^{+*}} & \frac{1}{2} & S_0 \\ M_{S_{1^{+*}}} & \nu_p & -M_{S_0} \end{pmatrix} \begin{pmatrix} N_0 & 1 & N \\ M_{N_0} & m_r & M_N \end{pmatrix} \begin{pmatrix} N_{1^{+*}} & \ell_p & N \\ M_{N_{1^{+*}}} & m_p & M_N \end{pmatrix}$$

$$\langle n_{1^{+*}}\, \Lambda_{1^{+*}}\, \mathfrak{p}_{1^{+*}};\, \big(N_{1^{+*}}\, \ell_p\big) N |\, \mathcal{F}^{(b)}_p\, | n_0\, \Lambda_0\, \mathfrak{p}_0;\, \big(N_0\, 1\big) N\rangle Y^{m_p}_{\ell_p}(\hat{\boldsymbol{k}}_p)\, \mathcal{D}^{\frac{1}{2}}_{\mu_p\,\nu_p}(\omega_p) \tag{3.53b}$$

with

---

[17] An expression for $|1^{+^*}\rangle$ in the present situation is readily obtained on the replacement $0 \longrightarrow 1^{+^*}$ everywhere in (3.48).

$$\langle n_{1^{+*}}\,\Lambda_{1^{+*}}\,\mathfrak{p}_{1^{+*}};\,\big(N_{1^{+*}}\,\ell_p\big)N|\,\mathcal{F}_p^{(b)}\,|n_0\,\Lambda_0\,\mathfrak{p}_0;\,\big(N_0\,1\big)N\rangle = \frac{1}{2}\,(-\mathrm{i})^{\ell_p}\,\mathrm{e}^{\mathrm{i}\sigma_{\ell_p}}$$
$$\times\Big[1-(-1)^{\mathfrak{p}_0+\mathfrak{p}_{1^{+*}}+\ell_p+N_0+N_{1^{+*}}}\Big](2N+1)\sqrt{2\ell_p+1}\sum_{\lambda_r\,\lambda_p\,\Lambda_N}$$
$$\times\begin{pmatrix} N_{1^{+*}} & \ell_p & N\\ \Lambda_{1^{+*}} & \lambda_p & \Lambda_N\end{pmatrix}\Bigg[\begin{pmatrix} N_0 & 1 & N\\ \Lambda_0 & \lambda_r & \Lambda_N\end{pmatrix}\langle n_{1^{+*}}\,\Lambda_{1^{+*}};\,\ell_p\,\lambda_p|\,F_p\,|n_0\,\Lambda_0;\,1\,\lambda_r\rangle$$
$$+(-1)^{\mathfrak{p}_0}\begin{pmatrix} N_0 & 1 & N\\ -\Lambda_0 & \lambda_r & \Lambda_N\end{pmatrix}\langle n_{1^{+*}}\,\Lambda_{1^{+*}};\,\ell_p\,\lambda_p|\,F_p\,|n_0-\Lambda_0;\,1\,\lambda_r\rangle\Bigg]. \quad (3.54)$$

The superscript (m, b) on $\mathcal{P}^{(m,b)}_{0\longrightarrow 1^{+*}}$ in (3.53), and hereafter, is used to indicate that it refers to photoionization in a RLM in Hund's case (b) as opposed to (3.26) meant for the same process in an atomic target.

Further in (3.54), $\langle n_{1^{+*}}\,\Lambda_{1^{+*}};\,\ell_p\,\lambda_p|\,F_p\,|n_0\,\Lambda_0;\,1\,\lambda_r\rangle$ is the matrix element of the $E1$ photoionization operator F_p [specified in (3.6)] between states (3.52c) and $|n_0\,\Lambda_0;\,1\lambda_r\rangle \equiv |n_0\,\Lambda_0\rangle\,|1\,\lambda_r\rangle$ of the (excited molecular photoion $\mathfrak{T}^{1+^*}$ + photoelectron e_p) and (molecule $\mathfrak{T}$ + ionizing photon γ_r), respectively, participating in the process (1.3a) [see, (3.48a) and (3.49)–(3.52)].

Derivation [215,217] of the photoionization amplitude (3.53) [and also of Auger amplitude (3.59)] is a lengthy and multi-step process requiring many operations in angular momentum algebra. While calculating these amplitudes, one should keep, in particular, two things in mind:
(i) Each of the matrix elements (3.53a) and (3.59a) involves integrations over the nuclear angular coordinates [i.e., Euler angles ω_m $(=\phi_m, \theta_m, \xi_m)$] which represent orientation of the inter-nuclear axis (i.e., of the MF) in SF, multiplied by integrals over spatial coordinates of all electrons present in a RLM $\mathfrak{T}$. All integrals present in (3.53a), (3.59a), and elsewhere in this monograph involving spatial coordinates of molecular electrons have to be done in MF. It becomes necessary because the wavefunction for the bound state of any molecular specie is always given in an MF. It is for this reason that the state $|1m_r\rangle$ in SF of the incident γ_r has been expanded in (3.50) in terms of the MF-state $|1\lambda_r\rangle$ of the same photon.
(ii) On account of the fact that both in (3.53a) and in (3.59a), each of the respective states $|0\rangle$, $|1^{+^*}\rangle$, and $|2^+\rangle$ of the molecular target $\mathfrak{T}$, its excited photoion $\mathfrak{T}^{1+^*}$, and the residual dication $\mathfrak{T}^{2+}$ contains sum of two terms [e.g., in (3.48a), (3.57a), *etc.*], each of the expressions (3.53a) and (3.59a) shall initially have a sum of four amplitudes. But, applying twice the well-known property [214,216]

$$\langle n_{1^{+*}}\,-\Lambda_{1^{+*}};\,\ell_p\,-\lambda_p|\,F_p\,|n_0\,-\Lambda_0;\,1\,-\lambda_r\rangle$$
$$=\langle n_{1^{+*}}\,\Lambda_{1^{+*}};\,\ell_p\,\lambda_p|\,F_p\,|n_0\,\Lambda_0;\,1\,\lambda_r\rangle, \quad (3.55)$$

in case of (3.53) [or, an equivalent relation while simplifying (3.59)], one is able to reduce the number of terms from four to those two given in the second square brackets on the right-hand side of the expressions(3.54) [or, (3.60)]. Property (3.55), or its equivalent needed for (3.59), is always applicable to photoionization, or Auger decay, of a linear molecules.

Derivation in Chap. 10 of the desired density matrix will be greatly facilitated with the help of the expression

$$\frac{1}{(2N_0+1)(S_0+1)} \sum_{\mathfrak{p}_0\, M_{N_0}\, M_{S_0}} \mathcal{P}^{(m,b)}_{0\longrightarrow 1^{+*}} \left(\mathcal{P}^{(m,b)}_{0\longrightarrow 1'^{+*}}\right)^{\dagger} \equiv \frac{1}{(2N_0+1)(S_0+1)}$$

$$\sum_{\mathfrak{p}_0\, M_{N_0}\, M_{S_0}} \langle 1^{+^*};\, \mu_p\, \hat{\boldsymbol{u}}_p\, \boldsymbol{k}_p | F_{\mathrm{p}} | 0;\, 1m_{\mathrm{r}}\rangle\, \langle 0;\, 1m_{\mathrm{r}} | F_p^{\dagger} | 1'^{+^*};\, \mu_p'\, \hat{\boldsymbol{u}}_p\, \boldsymbol{k}_p\rangle. \qquad (3.56\mathrm{a})$$

After substituting (3.53b) and its Hermitian conjugate, the resulting (3.56a) is simplified using various identities involving 3-j, 6-j, 9-j symbols and given in books like ([64, 65, 68, 187], *etc*). Expression (3.56a), in its final simplified form, can be written as [215, 217]

$$\frac{1}{(2N_0+1)(S_0+1)} \sum_{\mathfrak{p}_0\, M_{N_0}\, M_{S_0}} \mathcal{P}^{(m,b)}_{0\longrightarrow 1^{+*}} \left(\mathcal{P}^{(m,b)}_{0\longrightarrow 1'^{+*}}\right)^{\dagger} = \frac{1}{2\sqrt{\pi}}\,(2N_{1^{+*}}+1)$$

$$(-1)^{S_0+N_0+N_{1^{+*}}+M_{S_{1^{+*}}}+M'_{N_{1^{+*}}}+m_r+\mu'_p} \sum_{\substack{\ell_p\,\ell_p'\,L_p\,M_{L_p}\,L_r\,N_t\\ S_p\,M_{S_p}\,N_{S_p}\,N\,N'\,\mathfrak{p}_0}} (-1)^{N+\ell_p'}$$

$$(2L_r+1)(2N_t+1)(2S_p+1)\sqrt{2L_p+1} \begin{pmatrix} 1 & 1 & L_r\\ m_r & -m_r & 0\end{pmatrix} \begin{pmatrix} \ell_p & \ell_p' & L_p\\ 0 & 0 & 0\end{pmatrix}$$

$$\begin{pmatrix} \frac{1}{2} & \frac{1}{2} & S_p\\ \mu_p & -\mu_p' & M_{S_p}\end{pmatrix} \begin{pmatrix} L_r & L_p & N_t\\ 0 & -M_{L_p} & M_{L_p}\end{pmatrix} \begin{pmatrix} N_{1^{+*}} & N_{1^{+*}} & N_t\\ M_{N_{1^{+*}}} & -M'_{N_{1^{+*}}} & -M_{L_p}\end{pmatrix}$$

$$\begin{pmatrix} S_{1^{+*}} & S_{1^{+*}} & S_p\\ M_{S_{1^{+*}}} & -M'_{S_{1^{+*}}} & -N_{S_p}\end{pmatrix} \begin{Bmatrix} 1 & 1 & L_r\\ N & N' & N_0\end{Bmatrix} \begin{Bmatrix} \frac{1}{2} & \frac{1}{2} & S_p\\ S_{1^{+*}} & S_{1^{+*}} & S_0\end{Bmatrix} \begin{Bmatrix} \ell_p & \ell_p' & L_p\\ N & N' & L_r\\ N_{1^{+*}} & N_{1^{+*}} & N_t\end{Bmatrix}$$

$$\left[Y_{L_p}^{M_{L_p}}(\hat{\boldsymbol{k}}_p)\right]^{*} \left[\mathcal{D}^{S_p}_{M_{S_p}\,N_{S_p}}(\omega_p)\right]^{*}$$

$$\langle n_{1^{+*}}\, \Lambda_{1^{+*}}\, \mathfrak{p}_{1^{+*}};\, (N_{1^{+*}}\,\ell_p)N | \mathcal{F}_p^{(b)} | n_0\, \Lambda_0\, \mathfrak{p}_0;\, (N_0\,1)N\rangle$$

$$\langle n_{1^{+*}}\, \Lambda_{1^{+*}}\, \mathfrak{p}'_{1^{+*}};\, (N_{1^{+*}}\,\ell_p')N' | \mathcal{F}_p^{(b)} | n_0\, \Lambda_0\, \mathfrak{p}_0;\, (N_0\,1)N'\rangle^{*} \qquad (3.56\mathrm{b})$$

Auger Decay Amplitude

Calculation of the Auger amplitude for the spontaneous decay (1.3b) of the excited photoion $\mathfrak{T}^{1+^*}$ of the RLM $\mathfrak{T}$ in the absence of SDIs in the Hund's coupling scheme

(b) can be performed [214, 215, 217] along the lines used for the calculation of the photoionization amplitude (3.53). In analogy to (3.48) and (3.49), we have[18]

$$|2^+\rangle = \frac{1}{\sqrt{2}}\Big(|n_{2^+}\ \Lambda_{2^+}\rangle|N_{2^+}\ \Lambda_{2^+}\ M_{N_{2^+}}\rangle$$

$$+(-1)^{\mathfrak{p}_{2^+}}|n_{2^+}\ -\Lambda_{2^+}\rangle|N_{2^+}\ -\Lambda_{2^+}\ M_{N_{2^+}}\rangle\Big)|S_{2^+}\ M_{S_{2^+}}\rangle. \quad (3.57a)$$

with

$$\langle\omega_m|\ N_{2^+}\ \Lambda_{2^+}\ M_{N_{2^+}}\rangle = \sqrt{\frac{2N_{2^+}+1}{8\pi^2}}\ \mathcal{D}^{N_{2^+}}_{\Lambda_{2^+}\ M_{N_{2^+}}}(\omega_m) \quad (3.57b)$$

for the molecular dication $\mathfrak{T}^{2+}$ and[19]

$$|\mu_a\ \hat{\boldsymbol{u}}_a\ \boldsymbol{k}_a\rangle = \sum_{\ell_a\ \lambda_a\ m_a\ \nu_a} i^{\ell_a}\ e^{-i\sigma_{\ell_a}}\ \mathcal{D}^{\ell_a}_{\lambda_a\ m_a}(\omega_m)$$

$$\times\left[\mathcal{D}^{\frac{1}{2}}_{\mu_a\ \nu_a}(\omega_a)\right]^*\left[Y^{m_a}_{\ell_a}(\hat{\boldsymbol{k}}_a)\right]^*|\tfrac{1}{2}\ \nu_a\rangle\ |F^-_{\ell_a\ \lambda_a}(k_a)\rangle. \quad (3.58)$$

for the Auger electron e_a, respectively, in the absence of SDIs in Hund's coupling scheme (b). The various terms present in [(3.57), (3.58)] have their meanings identical to those of the corresponding terms in [(3.48), (3.49)], but now applicable to [$\mathfrak{T}^{2+}$, e_a]. The explicit forms (3.57) and (3.58) of the respective molecular state $|2^+\rangle$ and spin–orbital $|\mu_a\hat{\boldsymbol{u}}_a\boldsymbol{k}_a\rangle$ have been given for further references in this monograph.

Now, one can immediately write an expression for the state $|2^+;\ e_a\rangle \equiv |2^+;\ \mu_a\ \hat{\boldsymbol{u}}_a\ \boldsymbol{k}_a\rangle$ of the ($\mathfrak{T}^{2+} + e_a$)-system obeying the conservation conditions (3.47) in the absence of SDIs in Hund's case (b). On using properties similar to (3.55) and relevant to the present case, the Auger amplitude can be simplified to the following form [215, 217]

$$\mathcal{A}^{(m,\,b)}_{1^{+*}\longrightarrow 2^+} \equiv \langle 2^+;\ \mu_a\ \hat{\boldsymbol{u}}_a\ \boldsymbol{k}_a|F_{\rm a}|1^{+*}\rangle \quad (3.59a)$$

$$= (-1)^{\frac{1}{2}-S_{2^+}+\Lambda_{1^{+*}}-M_{N_{1^{+*}}}-M_{S_{1^{+*}}}}\sqrt{(2S_{1^{+*}}+1)(2N_{1^{+*}}+1)(2N_{2^+}+1)}$$

---

[18] In order to obtain an expression for $|2^+\rangle$ representing the dication $\mathfrak{T}^{2+}$ of the RLM $\mathfrak{T}$ in the absence of SDIs in Hund's coupling scheme (b), one merely needs to replace 0 by 2^+everywhere in (3.48).

[19] The replacement $p \longrightarrow a$ every where in (3.49) will give us a spin–orbital for the Auger electron e_a in the absence of SDIs in Hund's case (b).

$$\sum_{\ell_a\, m_a\, \nu_a} (-1)^{L_a+S_a}(2\ell_a+1)^{-\frac{1}{2}} \begin{pmatrix} N_{2^+} & \ell_a & N_{1^{+*}} \\ M_{N_{2^+}} & m_a & -M_{N_{1^{+*}}} \end{pmatrix} \begin{pmatrix} S_{2^+} & \frac{1}{2} & S_{1^{+*}} \\ M_{S_{2^+}} & \nu_a & -M_{S_{1^{+*}}} \end{pmatrix}$$

$$Y_{\ell_a}^{m_a}(\hat{\boldsymbol{k}}_a)\mathcal{D}^{\frac{1}{2}}_{\mu_a\,\nu_a}(\omega_a)\langle n_{2^+}\,\Lambda_{2^+}\,\mathfrak{p}_{2^+};\,(N_{2^+}\,\ell_a)N_{1^{+*}}|\,\mathcal{F}_a^{(b)}\,|n_{1^{+*}}\,\Lambda_{1^{+*}}\,\mathfrak{p}_{1^{+*}};\,N_{1^{+*}}\rangle, \tag{3.59b}$$

where

$$\langle n_{2^+}\,\Lambda_{2^+}\,\mathfrak{p}_{2^+};\,(N_{2^+}\,\ell_a)N_{1^{+*}}|\,\mathcal{F}_a^{(b)}\,|n_{1^{+*}}\,\Lambda_{1^{+*}}\,\mathfrak{p}_{1^{+*}};N_{1^{+*}}\rangle = \frac{1}{2}\,(-\mathrm{i})^{\ell_a}\,\mathrm{e}^{\mathrm{i}\sigma_{\ell_a}}$$
$$\times\Big[1+(-1)^{\mathfrak{p}_{1^{+*}}+\mathfrak{p}_{2^+}+\ell_a+N_{1^{+*}}+N_{2^+}}\Big]\sqrt{2\ell_a+1}$$
$$\times\sum_{\lambda_a}\Bigg[\begin{pmatrix} N_{2^+} & \ell_a & N_{1^{+*}} \\ \Lambda_{2^+} & \lambda_a & -\Lambda_{1^{+*}} \end{pmatrix}\langle n_{2^+}\,\Lambda_{2^+};\,\ell_a\,\lambda_a|\,F_a\,|n_{1^{+*}}\,\Lambda_{1^{+*}}\rangle$$
$$+(-1)^{\mathfrak{p}_{1^{+*}}}\begin{pmatrix} N_{2^+} & \ell_a & N_{1^{+*}} \\ \Lambda_{2^+} & \lambda_a & \Lambda_{1^{+*}} \end{pmatrix}\langle n_{1^{+*}}\,\Lambda_{1^{+*}};\,\ell_a\,\lambda_a|\,F_a\,|n_{1^{+*}}\,-\Lambda_{1^{+*}}\rangle\Bigg]. \tag{3.60}$$

The superscripts $(m,\,b)$ on $\mathcal{A}$ in (3.59) and elsewhere in this monograph have meanings identical to those used on $\mathcal{P}$ in the photoionization amplitude (3.53). Further in $\langle n_{2^+}\,\Lambda_{2^+};\,\ell_a\,\lambda_a|\,F_a\,|n_{1^{+*}}\,\Lambda_{1^{+*}}\rangle$ in (3.60), $|n_{2^+}\,\Lambda_{2^+};\,\ell_a\,\lambda_a\rangle \equiv |n_{2^+}\,\Lambda_{2^+}\rangle\,|F^-_{\ell_a\,\lambda_a}(k_a)\rangle$ is product of the spatial parts of the electronic wavefunction in the state $|2^+\rangle$ [(3.57)] of the molecular dication $\mathfrak{T}^{2+}$ and of the spin–orbital (3.58) of the Auger electron e_a.

For calculating the density matrix needed for studying spin-entanglement between the spins of the photoelectron e_p and Auger electron e_a, emitted sequentially from a rotating linear molecule $\mathfrak{T}$ in the process (1.3) in Hund's coupling scheme (b) with the exclusion of SDIs, one will require also the following expression [215, 217]

$$\sum_{\mathfrak{p}_{2^+}\,M_{N_{2^+}}\,M_{S_{2^+}}} \mathcal{A}^{(m,b)}_{1^{+*}\longrightarrow 2^+}\left(\mathcal{A}^{(m,b)}_{1'^{+*}\longrightarrow 2^+}\right)^{\dagger} \equiv$$

$$\sum_{\mathfrak{p}_{2^+}\,M_{N_{2^+}}\,M_{S_{2^+}}} \langle 2^+;\,\mu_a\,\hat{\boldsymbol{u}}_a\,\boldsymbol{k}_a|F_a\,|1^{+^*}\rangle\,\langle 1^{+^*}|F_a^\dagger\,|2^+;\,\mu_a'\,\hat{\boldsymbol{u}}_a\,\boldsymbol{k}_a\rangle \tag{3.61a}$$

$$=\frac{1}{2\sqrt{\pi}}(-1)^{1+\mu_a'+N_{2^+}-S_{2^+}+M_{N_{1^{+*}}}-M'_{S_{1^{+*}}}}\,(2S_{1^{+*}}+1)(2N_{1^{+*}}+1)(2N_{2^+}+1)$$

$$\sum_{\substack{\ell_a\,\ell_a'\,L_a\,M_{L_a}\\ S_a\,M_{S_a}\,N_{S_a}\,\mathrm{p}_{2+}}} (-1)^{(L_a+S_a)}(2S_a+1)\sqrt{2L_a+1}\begin{pmatrix}\ell_a & \ell_a' & L_a\\ 0 & 0 & 0\end{pmatrix}\begin{pmatrix}N_{1+*} & N_{1+*} & L_a\\ M_{N_{1+*}} & -M_{N_{1+*}}' & M_{L_a}\end{pmatrix}$$

$$\begin{pmatrix}\frac{1}{2} & \frac{1}{2} & S_a\\ \mu_a & -\mu_a' & M_{S_a}\end{pmatrix}\begin{pmatrix}S_{1+*} & S_{1+*} & S_a\\ M_{S_{1+*}} & -M_{S_{1+*}}' & N_{S_a}\end{pmatrix}\begin{Bmatrix}\ell_p & \ell_p' & L_a\\ N_{1+*} & N_{1+*} & N_{2+}\end{Bmatrix}$$

$$\begin{Bmatrix}\frac{1}{2} & \frac{1}{2} & S_a\\ S_{1+*} & S_{1+*} & S_{2+}\end{Bmatrix}\langle n_{2+}\;\Lambda_{2+}\;p_{2+};\;(N_{2+}\;\ell_a)N_{1+*}\,|\,\mathcal{F}_a^{(b)}\,|n_{1+*}\;\Lambda_{1+*}\;p_{1+*};N_{1+*}\rangle$$

$$\langle n_{2+}\;\Lambda_{2+}\;p_{2+};\;(N_{2+}\;\ell_a')N_{1+*}\,|\,\mathcal{F}_a^{(b)}\,|n_{1+*}\;\Lambda_{1+*}\;p_{1+*}';N_{1+*}\rangle^*$$

$$\left[Y_{L_p}^{M_{L_p}}(\hat{\boldsymbol{k}}_p)\right]^*\left[\mathcal{D}_{M_{S_a}\,N_{S_a}}^{S_a}(\omega_a)\right]^* \tag{3.61b}$$

3.3.2 *Transition Amplitudes with Spin-Dependent Interactions*

The two (namely, SRI and SOI) components of the SDIs in a RLM can contribute [68, 69, 185] either jointly or individually. For example, spin–orbit coupling does not exist if the molecule $\mathfrak{T}$ in its $|0\rangle$ state (3.48a) has $\Lambda_0 = 0$ and/or $S_0 = 0$ (i.e., it is in a Σ and/or singlet electronic state); moreover, even for the non-Σ and/or non-singlet states of light molecules, SOI is negligibly small.[20] In such cases, only the SRI needs to be taken into account, provided $S_0 \neq 0$ for spin-rotation interaction to be non-zero. Such kind of a RLM are properly described [68, 69, 185] by Hund's scheme (b). Those RLM for which both SRI+SOI contribute, on the other hand, are adequately represented [68, 69, 185] by Hund's case (a).

Chandra and Ghosh [213] have studied spin-entanglement of electrons, simultaneously coming out from a RLM in the 1-DPI process (1.2), taking the Coulomb + spin-rotation interactions (C+SRIs) in Hund's coupling scheme (b), as well as with C+SDIs (i.e., Coulomb + spin-rotation+ spin–orbit interactions) in Hund's case (a), in addition to that generated purely by the Coulomb forces. Reference [216] describes properties of the Coulombic entanglement between electronic qubits, emitted sequentially in the 2-DPI process (1.3) taking place in a RLM. Ghosh et al. [217] have, on the other hand, investigated the effects of SDIs on the Coulombic entanglement reported in [216]. Ghosh [215] has further studied bipartite electronic entanglement generated in (1.3) in the presence of C + SRIs in

[20]These circumstances are similar to those of atomic cases wherein spin–orbit coupling vanishes identically, or is negligibly small, whenever an atom's orbital and/or spin angular momentum is nil (i.e., atom's electronic state is S and/or has spin multiplicity one), or it has a small atomic number Z.

Hund's scheme (b). The following section describes calculations of amplitudes for photoionizing and Auger transitions in a RLM in the presence of C+SDIs in Hund's scheme (a).[21]

3.3.2.1 Hund's Coupling Scheme (a)

The dynamics of a RLM in the presence of both SRI and SOI, in addition of Coulomb forces, is adequately described in Hund's Coupling scheme (a) [68, 69, 185]. Here, one needs to define the total angular momenta [$\boldsymbol{J}_0$ (= $\boldsymbol{N}_0$ + $\boldsymbol{S}_0$), $\boldsymbol{J}_{1^{+*}}$ (= $\boldsymbol{N}_{1^{+*}}$ + $\boldsymbol{S}_{1^{+*}}$), $\boldsymbol{J}_{2^+}$ (= $\boldsymbol{N}_{2^+}$ + $\boldsymbol{S}_{2^+}$)] of molecular species [$\mathfrak{T}$, $\mathfrak{T}^{1+^*}$, $\mathfrak{T}^{2+}$] participating in the 2-step process (1.3) with their states [$|0\rangle$, $|1^{+^*}\rangle$, $|2^+\rangle$]. Let us take [68, 69, 185] [M_0 (= $\hat{\boldsymbol{e}}_z\cdot\boldsymbol{J}_0$), $M_{1^{+*}}$ (= $\hat{\boldsymbol{e}}_z\cdot\boldsymbol{J}_{1^{+*}}$), M_{2^+} (= $\hat{\boldsymbol{e}}_z\cdot\boldsymbol{J}_{2^+}$)] and [$\Omega_0$ (= $\hat{\boldsymbol{e}}_{Z_m}\cdot J_0$), $\Omega_{1^{+*}}$ (= $\hat{\boldsymbol{e}}_{Z_m}\cdot\boldsymbol{J}_{1^{+*}}$), Ω_{2^+} (= $\hat{\boldsymbol{e}}_{Z_m}\cdot J_{2^+}$)] to be the projections of [$\boldsymbol{J}_0$, $\boldsymbol{J}_{1^{+*}}$, $\boldsymbol{J}_{2^+}$] along the polar axes of the SF and MF shown in Fig. 9.1 (page 207), respectively. On defining [68, 69, 185], [Σ_0 (= $\hat{\boldsymbol{e}}_{Z_m}\cdot S_0$), $\Sigma_{1^{+*}}$ (= $\hat{\boldsymbol{e}}_{Z_m}\cdot\boldsymbol{S}_{1^{+*}}$), Σ_{2^+} (= $\hat{\boldsymbol{e}}_{Z_m}\cdot\boldsymbol{S}_{2^+}$)], one can write [$\Omega_0 = \Lambda_0 + \Sigma_0$, $\Omega_{1^{+*}} = \Lambda_{1^{+*}} + \Sigma_{1^{+*}}$, $\Omega_{2^+} = \Lambda_{2^+} + \Sigma_{2^+}$][22] for the states [$|0\rangle$, $|1^{+^*}\rangle$, $|2^+\rangle$] of [$\mathfrak{T}$, $\mathfrak{T}^{1+^*}$, $\mathfrak{T}^{2+}$].

A parity adapted state of the RLM molecule $\mathfrak{T}$ in Hund's coupling scheme (a) is given by [69, 75]

$$|0\rangle = \frac{1}{\sqrt{2}}\Big(|n_0\,\Lambda_0\rangle|S_0\,\Sigma_0\rangle|J_0\,\Omega_0\,M_0\rangle + (-1)^{p_0}|n_0\,-\Lambda_0\rangle|S_0\,-\Sigma_0\rangle|J_0\,-\Omega_0\,M_0\rangle\Big). \tag{3.62a}$$

Here, $|S_0\,\Sigma_0\rangle$ represents spin state of those molecular electrons whose spatial motion is described—as in (3.48a)—by $|n_0\,\Lambda_0\rangle$. The nuclear rotation in $\mathfrak{T}$ in Hund's case (a) is given by the normalized function [65]

$$\langle\omega_m|\,J_0\,\Omega_0\,M_0\rangle = \sqrt{\frac{2J_0+1}{8\pi^2}}\,\mathcal{D}^{J_0}_{\Omega_0\,M_0}(\omega_m). \tag{3.62b}$$

State (3.62a) has following parity [69, 75]

---

[21] Spin-entanglement generated in either of the two processes (1.2) and (1.3) taking place in a RLM with the third possibility in which only SOI is important, while SRI is negligible, has not yet probably been considered. Such situations are properly described in Hund's case (c) [68, 69, 185].

[22] For [68, 75, 185] $\Lambda_0 = 0$, spin S_0 is no longer coupled to the inter-nuclear axis. It, in other words, means, $\Sigma_0 = 0$ and, hence, $\Omega_0 = 0$. In such a situation, therefore, Hund's case (a) reduces [68, 75, 185] to its scheme (b).

$$(-1)^{s+\mathfrak{p}_0+J_0-S_0} \quad \text{with} \qquad \mathfrak{p}_0 = 0 \text{ or } 1. \tag{3.62c}$$

But, for $\Lambda_0 = 0$ (i.e., for $\Sigma^{\pm}$)-states, $\mathfrak{p}_0 = 0$ [69, 75] and the consequent state

$$|0\rangle = |n_0\,0\rangle\,|S_0\,0\rangle\,|J_0\,0\,M_0\rangle \tag{3.62d}$$

has parity $(-1)^{s+J_0-S_0}$. The parameter s in (3.62c) has meanings identical to that applicable in (3.48).

It is also the total angular momenta, in addition to the energies, which are conserved in each of the two steps of the process (1.3) in the present coupling scheme. That is

$$\boldsymbol{\ell}_r + \boldsymbol{J}_0 = \boldsymbol{J}_{1^{+*}} + \boldsymbol{j}_p \tag{3.63a}$$

and

$$\boldsymbol{J}_{1^{+*}} = \boldsymbol{J}_{2^+} + \boldsymbol{j}_a \tag{3.63b}$$

for the photoionization (1.3a) and Auger emission (1.3b), respectively. Angular momenta $\boldsymbol{j}_p$ and $\boldsymbol{j}_a$ have already been defined elsewhere in the present Chap. 3.

The spin–orbital of the photoelectron e_p in the Hund's coupling case (a) becomes [213, 215]

$$|\mu_p\,\hat{\boldsymbol{u}}_p\,\boldsymbol{k}_p\rangle = \sum_{\ell_p\,\lambda_p\,m_p\,\nu_p\,\sigma_p} i^{\ell_p}\, e^{-i\sigma_{\ell_p\sigma_p}}\, \mathcal{D}^{\ell_p}_{\lambda_p\,m_p}(\omega_m)\Big[Y^{m_p}_{\ell_p}(\hat{\boldsymbol{k}}_p)\Big]^* \times \Big[\mathcal{D}^{\frac{1}{2}}_{\mu_p\,\nu_p}(\omega_p)\Big]^* \mathcal{D}^{\frac{1}{2}}_{\sigma_p\,\nu_p}(\omega_m)|\tfrac{1}{2}\,\nu_p\rangle\,|F^-_{\ell_p\,\lambda_p\,\sigma_p}(k_p)\rangle. \tag{3.64}$$

Here, σ_p $(= \boldsymbol{s}\cdot\hat{\boldsymbol{e}}_{Z_m})$ is the projection of the spin angular momentum of e_p along the inter-nuclear axis. Thus, unlike in the spin–orbital (3.49) for Hund's case (b), both the Coulomb phase $(\sigma_{\ell_p\sigma_p})$ and the spatial part $(|F^-_{\ell_p\lambda_p\sigma_p}(k_p)\rangle)$ in the spin-orbital (3.64) depend, among other things, on σ_p as well. Further in (3.64), the harmonic $\mathcal{D}^{\frac{1}{2}}_{\sigma_p\,\nu_p}(\omega_m)$ rotates spin quantization of e_p from internuclear (i.e., OZ_m)-axis to the polar (i.e., OZ)-axis of the SF; whereas $\Big[\mathcal{D}^{\frac{1}{2}}_{\mu_p\,\nu_p}(\omega_p)\Big]^*$ brings this photoelectron's spin quantization along $\hat{\boldsymbol{u}}_p$ defined in Chap. 1.

Photoionization Amplitude

In the present Hund's coupling scheme, one now needs to use (3.50) and (3.62a) in the expression (3.51) for writing a state of the (e_p + $\mathfrak{T}$)-system. Similarly, a state of the ($\mathfrak{T}^{1+^*}$ + e_p)-system is now given by the substitution in (3.52a), of (3.64) for the photoelectron e_p and of an expression obtained by the replacement $0 \longrightarrow 1^{+^*}$ in (3.62a) for the state $|1^{+^*}\rangle$ of excited molecular photoion $\mathfrak{T}^{1^+}$ in the Hund's case (a). Calculation of the photoionization amplitude in Hund's scheme (a) is a complicated procedure already explained in brief in [215]. Using this procedure, one obtains [215, 217]

$$\begin{aligned}
\mathcal{P}^{(m,a)}_{0\longrightarrow 1^{+*}} &\equiv \langle 1^{+^*};\ \mu_p\, \hat{\boldsymbol{u}}_p\, \boldsymbol{k}_p | F_{\mathrm{p}}\, |0;\ 1m_{\mathrm{r}}\rangle \\
&= (-1)^{\Omega_0 - M_0 + J_{1^{+*}} - \Omega_{1^{+*}}} \sqrt{J_0+1} \sum_{\substack{\ell_p m_{j_p} \nu_p \\ j_p n_{j_p} j_t n_t}} (-1)^{\ell_p + j_p + j_t} \\
&\quad \times (2\ell_p+1)^{-\frac{1}{2}} \begin{pmatrix} \ell_p & \frac{1}{2} & j_p \\ -m_{j_p} & -\nu_p & n_{j_p} \end{pmatrix} \begin{pmatrix} 1 & j_p & j_t \\ m_r & -n_{j_p} & n_t \end{pmatrix} \begin{pmatrix} J_{1^{+*}} & J_0 & j_t \\ M_{1^{+*}} & -M_0 & n_t \end{pmatrix} \\
&\quad \times Y^{m_{\ell_p}}_{\ell_p}(\hat{\boldsymbol{k}}_p)\, \mathcal{D}^{\frac{1}{2}}_{\mu_p \nu_p}(\omega_p)\, \langle \mathfrak{p}_{1^{+*}} J_{1^{+*}}; \left(\ell_p \tfrac{1}{2}\right) j_p | F(j_t) | \mathfrak{p}_0 J_0;\ 1\rangle
\end{aligned} \tag{3.65a}$$

with

$$\begin{aligned}
\langle \mathfrak{p}_{1^{+*}} J_{1^{+*}}; \left(\ell_p \tfrac{1}{2}\right) j_p | F(j_t) | \mathfrak{p}_0 J_0;\ 1\rangle &\equiv \frac{1}{2} (-\mathrm{i})^{\ell_p} (2j_p+1)(2j_t+1) \\
&\quad \times \sqrt{(2J_{1^{+*}}+1)(2\ell_p+1)} \sum_{J_T} (2J_T+1) \begin{Bmatrix} j_p & 1 & j_t \\ J_0 & J_{1^{+*}} & J_T \end{Bmatrix} \\
&\quad \times \langle n_{1^{+*}} J_{1^{+*}} \Lambda_{1^{+*}} \Omega_{1^{+*}}\, p_{1^{+*}};\ \ell_p j_p |\, F_{J_T}\, | n_0 J_0 \Lambda_0 \Omega_0 p_0;\ 1\rangle
\end{aligned} \tag{3.65b}$$

and

$$\begin{aligned}
&\langle n_{1^{+*}} J_{1^{+*}} \Lambda_{1^{+*}} \Omega_{1^{+*}}\, p_{1^{+*}};\ \ell_p j_p |\, F_{J_T}\, | n_0 J_0 \Lambda_0 \Omega_0 p_0;\ 1\rangle \\
&\equiv \left[1 + (-1)^{\frac{1}{2} + p_0 + p_{1^{+*}} + \ell_p + J_0 - J_{1^{+*}}}\right] \sum_{\substack{\lambda_p \upsilon_p m_{j_p} \\ \lambda_r M_T}} (-1)^{1+\lambda_r} e^{i\sigma_{\ell_p \sigma_p}} \begin{pmatrix} \ell_p & \frac{1}{2} & j_p \\ -\lambda_p & -\sigma_p & m_{j_p} \end{pmatrix} \\
&\times \begin{pmatrix} j_p & J_{1^{+*}} & J_T \\ m_{j_p} & \Omega_{1^{+*}} & M_T \end{pmatrix} \left[\begin{pmatrix} J_0 & 1 & J_T \\ \Omega_0 & \lambda_r & M_T \end{pmatrix} \right. \\
&\times \langle n_{1^{+*}} \Lambda_{1^{+*}} \Sigma_{1^{+*}} \Omega_{1^{+*}};\ \ell_p \lambda_p \sigma_p | F_p | n_0 \Lambda_0 \Sigma_0 \Omega_0;\ 1\lambda_r\rangle \\
&+ (-1)^{p_0 + 2\Omega_0} \begin{pmatrix} J_0 & 1 & J_T \\ -\Omega_0 & \lambda_r & M_T \end{pmatrix} \\
&\times \left. \langle n_{1^{+*}} \Lambda_{1^{+*}} \Sigma_{1^{+*}} \Omega_{1^{+*}};\ \ell_p \lambda_p \sigma_p | F_p | n_0 - \Lambda_0 - \Sigma_0 - \Omega_0;\ 1\lambda_r\rangle \right].
\end{aligned} \tag{3.65c}$$

Similar to (3.53), superscripts (m, a) in $\mathcal{P}^{(m,a)}_{0\longrightarrow 1^{+*}}$ means that (3.65) is amplitude for photoionizing transition $|0\rangle \longrightarrow |1^{+*}\rangle$ in a RLM in the presence of C+SDIs in Hund's coupling scheme (a). Dirac's bracket $\langle n_{1^{+*}} \Lambda_{1^{+*}} \Sigma_{1^{+*}} \Omega_{1^{+*}};\ \ell_p \lambda_p\ \sigma_p | F_p$ $|n_0 \Lambda_0 \Sigma_0 \Omega_0;\ 1\lambda_r\rangle$ in (3.65c) stands for a matrix element of the photoionization operator (3.6a) between the states $|n_{1^{+*}} \Lambda_{1^{+*}} \Sigma_{1^{+*}} \Omega_{1^{+*}};\ \ell_p \lambda_p \sigma_p\rangle \equiv |n_{1^{+*}} \Lambda_{1^{+*}}\rangle$ $|S_{1^{+*}} \Sigma_{1^{+*}}\rangle\ |F^{-}_{\ell_p \lambda_p \sigma_p}\rangle$ of electrons in a $(\mathfrak{T}^{1^{+*}} + e_p)$-system and $|n_0 \Lambda_0 \Sigma_0 \Omega_0;\ 1\lambda_r\rangle$ $\equiv |n_0 \Lambda_0\rangle\ |S_0 \Sigma_0\rangle\ |1\lambda_r\rangle$ of absorbed photon and electrons in $\mathfrak{T}$. A reader is referred to (3.50), (3.62a), and (3.64) in this connection.

In order to obtain in Chap. 10 a density matrix for studying entanglement between the spins of photoelectron e_p and Auger electron e_a, sequentially leaving a RLM in the 2-step process (1.3) in the presence of C+SDIs in Hund's coupling (a), let us calculate [215] the following quantity using (3.65a):

$$\frac{1}{2J_0+1} \sum_{\mathfrak{p}_0\, M_0} \mathcal{P}^{(m,a)}_{0\longrightarrow 1^{+*}} \left(\mathcal{P}^{(m,a)}_{0\longrightarrow 1'^{+*}}\right)^{\dagger}$$

$$\equiv \frac{1}{2J_0+1} \sum_{\mathfrak{p}_0\, M_0} \langle 1^{+*};\ \mu_p\, \hat{\boldsymbol{u}}_p\, \boldsymbol{k}_p | F_p\, |0;\ 1m_r\rangle\ \langle 0;\ 1m_r | F_p^{\dagger}\, |1'^{+*};\ \mu'_p\, \hat{\boldsymbol{u}}_p\, \boldsymbol{k}_p\rangle \tag{3.66a}$$

$$= (-1)^{\frac{1}{2}+m_r+\mu'_p+J_0+M_{1^{+*}}} \frac{1}{2\sqrt{\pi}} \sum_{\substack{\ell_p\, j_p\, j_t\, \ell'_p\, j'_p\, j'_t\, J_p\, L_r\, K \\ \mathfrak{p}_0\, L_p\, M_{L_p}\, S_p\, M_{S_p}\, N_{S_p}\, Q}} (-1)^{\ell_p+j_p-j'_p+j'_t}$$

$$(2J_p+1)(2L_r+1)(2K+1)(2S_p+1)\sqrt{2L_p+1} \begin{pmatrix} \ell_p & \ell'_p & L_p \\ 0 & 0 & 0 \end{pmatrix} \begin{pmatrix} \frac{1}{2} & \frac{1}{2} & S_p \\ \mu_p & -\mu'_p & M_{S_p} \end{pmatrix}$$

$$\begin{pmatrix} 1 & 1 & L_r \\ m_r & -m_r & 0 \end{pmatrix} \begin{pmatrix} L_p & S_p & J_p \\ M_{L_p} & N_{S_p} & Q \end{pmatrix} \begin{pmatrix} J_{1^{+*}} & J_{1^{+*}} & K \\ M_{1^{+*}} & -M'_{1^{+*}} & Q \end{pmatrix} \begin{pmatrix} J_p & K & L_r \\ Q & -Q & 0 \end{pmatrix}$$

$$\begin{Bmatrix} J_{1^{+*}} & J_{1^{+*}} & K \\ j_t & j'_t & J_0 \end{Bmatrix} \begin{Bmatrix} \ell_p & \ell'_p & L_p \\ \frac{1}{2} & \frac{1}{2} & S_p \\ j_p & j'_p & J_p \end{Bmatrix} \begin{Bmatrix} 1 & 1 & L_r \\ j_p & j'_p & J_p \\ j_t & j'_t & K \end{Bmatrix} \left[Y^{M_{L_p}}_{L_p}(\hat{\boldsymbol{k}}_p)\right]^{*} \left[\mathcal{D}^{S_p}_{M_{S_p}\, N_{S_p}}(\omega_p)\right]^{*}$$

$$\langle \mathfrak{p}_{1^{+*}}\, J_{1^{+*}};\ (\ell_p \tfrac{1}{2}) j_p |\ F(j_t)\, |\mathfrak{p}_0 J_0;\ 1\rangle\ \langle \mathfrak{p}_{1'^{+*}}\, J_{1^{+*}};\ (\ell'_p \tfrac{1}{2}) j'_p |\ F(j_t)\, |\mathfrak{p}_0 J_0;\ 1\rangle^{*} \tag{3.66b}$$

On making $\mu_p = \mu'_p$ in (3.66), the resulting expression gives us angular distribution of spin-resolved electrons ejected in photoionization of a RLM on the inclusion of C+SDIs in Hund's case (a)—previously obtained by Chandra and Sen in their (A6) in [218].

Auger Amplitude

Lastly, let us calculate $\langle 2^+;\mu_a\,\hat{\boldsymbol{u}}_a\,\boldsymbol{k}_a|\,F_a\,|1^{+^*}\rangle$ for non-radiative, spontaneous decay of the excited photoion $\mathfrak{T}^{1+^*}$ of the RLM $\mathfrak{T}$ in the present case. This calculation requires states $|1^{+^*}\rangle$ and $|2^+\rangle$ of $\mathfrak{T}^{1+^*}$ and dication $\mathfrak{T}^{2+}$, respectively.[23] In addition, one will also require spin–orbital[24] for Auger electron e_a in Hund's scheme (a).

The required Auger amplitude for the transition (1.3b) in a RLM $\mathfrak{T}$ in Hund's case (a) can be shown to be given by [215]

$$\begin{aligned}
\mathcal{A}^{(m,a)}_{1^{+*}\longrightarrow 2^+} &\equiv \langle 2^+;\,\mu_a\,\hat{\boldsymbol{u}}_a\,\boldsymbol{k}_a|F_{\rm a}\,|1^{+^*}\rangle \;=\; (-1)^{M_{2^+}-\Omega_{2^+}}\\
&\times\sqrt{(2J_{1^{+*}}+1)(2J_{2^+}+1)}\sum_{\ell_a\,m_a\,\nu_a\,j_a\,n_{j_a}}(2\ell_a+1)^{-\frac{1}{2}}\begin{pmatrix}\ell_a & \frac{1}{2} & j_a\\ -m_a & -\nu_a & m_{j_a}\end{pmatrix}\\
&\times\begin{pmatrix}J_{2^+} & j_a & J_{1^{+*}}\\ M_{2^+} & n_{j_a} & -M_{1^{+*}}\end{pmatrix}Y_{\ell_a}^{m_a}(\hat{\boldsymbol{k}}_a)\mathcal{D}^{\frac{1}{2}}_{\mu_a\,\nu_a}(\omega_a)\\
&\times\langle \mathfrak{p}_{2^+}J_{2^+};\,(\ell_a\tfrac{1}{2})j_a|\,\mathcal{F}^{(a)}_a\,|\mathfrak{p}_{1^{+*}}J_{1^{+*}}\rangle
\end{aligned}\tag{3.67a}$$

where

$$\begin{aligned}
&\langle \mathfrak{p}_{2^+}J_{2^+};\,(\ell_a\tfrac{1}{2})j_a|\,\mathcal{F}^{(a)}_a\,|\mathfrak{p}_{1^{+*}}J_{1^{+*}}\rangle\\
&\equiv \frac{1}{2}\Big[1-(-1)^{\frac{1}{2}+\mathfrak{p}_{1^{+*}}+\mathfrak{p}_{2^+}+\ell_a+J_{1^{+*}}-J_{2^+}}\Big](-{\rm i})^{\ell_a}\\
&\times(2j_a+1)\sqrt{2\ell_a+1}\sum_{\lambda_a\,\sigma_a\,m_{j_a}}e^{i\,\sigma_{\ell_a}\,\sigma_a}\begin{pmatrix}\ell_a & \frac{1}{2} & j_a\\ -\lambda_a & -\sigma_a & m_{j_a}\end{pmatrix}\\
&\times\Bigg[\begin{pmatrix}J_{2^+} & j_a & J_{1^{+*}}\\ \Omega_{2^+} & m_{j_a} & -\Omega_{1^{+*}}\end{pmatrix}\langle n_{2^+}\Lambda_{2^+}\Sigma_{2^+}\Omega_{2^+};\ell_a\lambda_a\sigma_a|F_a|n_{1^{+*}}\Lambda_{1^{+*}}\Sigma_{1^{+*}}\Omega_{1^{+*}}\rangle\\
&+(-1)^{\mathfrak{p}_{1^{+*}}}\begin{pmatrix}J_{2^+} & j_a & J_{1^{+*}}\\ \Omega_{2^+} & m_{j_a} & \Omega_{1^{+*}}\end{pmatrix}\\
&\times\langle n_{2^+}\Lambda_{2^+}\Sigma_{2^+}\Omega_{2^+};\,\ell_a\lambda_a\sigma_a|\,F_a\,|n_{1^{+*}}-\Lambda_{1^{+*}}-\Sigma_{1^{+*}}-\Omega_{1^{+*}}\rangle\Bigg].
\end{aligned}\tag{3.67b}$$

The final expression needed for the calculation of the required density matrix in Hund's case (a) is given by [215]

---

[23] These two states are obtained on making everywhere the respective replacements $0\longrightarrow 1^{+^*}$ and $0\longrightarrow 2^+$ of the subscripts in (3.62).

[24] Replacement of the subscript p by a at every place in (3.64) will immediately provide an expression for $|\mu_a\hat{\boldsymbol{u}}_a\boldsymbol{k}_a\rangle$ in Hund's case (a).

$$\sum_{\mathfrak{p}_{2^+} M_{2^+}} \mathcal{A}^{(m,a)}_{1^{+*} \longrightarrow 2^+} \left(\mathcal{A}^{(m,a)}_{1'^{+*} \longrightarrow 2^+} \right)^{\dagger}$$

$$\equiv \sum_{\mathfrak{p}_{2^+} M_{2^+}} \langle 2^+;\ \mu_a\, \hat{\boldsymbol{u}}_a\, \boldsymbol{k}_a | F_{\mathrm{a}} | 1^{+*} \rangle \, \langle 1^{+*} | F_a^{\dagger} | 2^+;\ \mu_a'\, \hat{\boldsymbol{u}}_a\, \boldsymbol{k}_a \rangle \tag{3.68a}$$

$$= \frac{1}{2\sqrt{\pi}} (-1)^{\frac{1}{2}+\mu_a'+2J_{1^{+*}}+J_{2^+}+M_{1^{+*}}} \, (2J_{1^{+*}}+1)(2J_{2^+}+1) \sum_{\substack{\mathfrak{p}_{2^+}\, \ell_a\, j_a\, \ell_a'\, j_a'\, J_a\, M_{J_a} \\ L_a\, M_{L_a}\, S_a\, M_{S_a}\, N_{S_a}}}$$

$$\times (-1)^{\ell_a - j_a' + J_a} \, (2S_a+1)(2J_a+1)\sqrt{2L_a+1} \begin{pmatrix} \ell_a & \ell_a' & L_a \\ 0 & 0 & 0 \end{pmatrix} \begin{pmatrix} \frac{1}{2} & \frac{1}{2} & S_a \\ \mu_a & -\mu_a' & M_{S_a} \end{pmatrix}$$

$$\times \begin{pmatrix} L_a & S_a & J_a \\ M_{L_a} & N_{S_a} & M_{J_a} \end{pmatrix} \begin{pmatrix} J_{1^{+*}} & J_{1^{+*}} & J_a \\ -M_{1^{+*}} & M'_{1^{+*}} & M_{J_a} \end{pmatrix} \begin{Bmatrix} J_{1^{+*}} & J_{1^{+*}} & J_a \\ j_a & j_a' & J_{2^+} \end{Bmatrix} \begin{Bmatrix} \ell_a & \ell_a' & L_a \\ \frac{1}{2} & \frac{1}{2} & S_a \\ j_a & j_a' & J_a \end{Bmatrix}$$

$$\times \langle \mathfrak{p}_{2^+} J_{2^+};\ (\ell_a \tfrac{1}{2}) j_a | \, \mathcal{F}_a^{(a)} \, | \mathfrak{p}_{1^{+*}} J_{1^{+*}} \rangle$$

$$\times \langle \mathfrak{p}_{2^+} J_{2^+};\ (\ell_a' \tfrac{1}{2}) j_a' | \, \mathcal{F}_a^{(a)} \, | \mathfrak{p}_{1'^{+*}} J_{1^{+*}} \rangle^*$$

$$\times \left[Y_{L_a}^{M_{L_a}}(\hat{\boldsymbol{k}}_a) \right]^* \left[\mathcal{D}^{S_a}_{M_{S_a}\, N_{S_a}}(\omega_a) \right]^* . \tag{3.68b}$$

Part I
Atomic Processes

Chapter 4
Coulombic Entanglement: One-Step Single Photoionization of Atoms

Photoionization (1.1) is, probably, the simplest among the basic processes in atomic and molecular physics for generating an entangled state of two flying particles one of which is an electronic qubit. It is simplest for two reasons: First, the whole process is completed in a single step; second, it is based on the well-understood physics of photoelectric effect [1] taking place due to the absorption of a single photon γ_r in a target $\mathfrak{T}$ in the $E1$ approximation. Here, it is obvious from (1.1), the constituents of a bipartite state are the residual photoion $\mathfrak{T}^{1+}$ and the photoelectron e_p. Unlike the processes (1.2)-(1.8), the incident photon in (1.1) is absorbed by one of the electrons in the outer most shell of $\mathfrak{T}$ so that the residual photoion $\mathfrak{T}^{1+}$ is left in its ground electronic state $|1^+\rangle$, which is stable against any decay. A study of the entanglement properties of the spins of $(e_p, \mathfrak{T}^{1+})$ helps explain—in terms of very simple and familiar examples—some of the important concepts (e.g., qutrit, qudit, distillation of entanglement, bound entanglement, etc.) [43, 126, 219] developed in quantum information science during the last two decades. These examples are such that the corresponding experiments can readily be performed in a laboratory and, hence, protocols (see, for, example [43], *etc*) based on these concepts can in reality be experimentally verified.

4.1 Density Operator and Density Matrix

4.1.1 Derivation

Equation

$$\epsilon_p = h\nu_r - (E_{1^+} - E_0), \tag{4.1}$$

expressing conservation of energy in a photoionization experiment, was first written by Einstein [1] more than a century ago. Thus, in an ionization experiment, energy ϵ_p of photoelectron e_p continuously changes with every change in the frequency ν_r

N. Chandra and R. Ghosh, *Quantum Entanglement in Electron Optics*, Springer Series on Atomic, Optical, and Plasma Physics 67, DOI 10.1007/978-3-642-24070-6_4,

of the absorbed photon γ_r and/or change in energies (E_0, E_{1^+}) of either, or both, of ($\mathfrak{T}$, $\mathfrak{T}^{1+}$) participating in the 1-SPI process (1.1). In the absence of SOI, when Russell–Saunders (i.e., L-S) coupling becomes applicable [10, 60, 184, 186], both the total orbital and total spin angular momenta will also be conserved. Thus, in the present case, two more conservation conditions, in addition to (4.1), need to be used. These can directly be written from the respective (3.19a) and (3.20a) in the following form

$$\boldsymbol{\ell}_r + \boldsymbol{L}_0 = \boldsymbol{L}_{1^+} + \boldsymbol{\ell}_p \tag{4.2a}$$

and

$$\boldsymbol{S}_0 = \boldsymbol{S}_{1^+} + \boldsymbol{s}_p\Big[= \Big(\frac{1}{2}\Big)_p\Big] \tag{4.2b}$$

in slightly different notations. Here, ($\boldsymbol{L}_{1^+}$, $\boldsymbol{S}_{1^+}$) are the total (orbital, spin) angular momenta of the residual photoion $\mathfrak{T}^{1+}$ formed in the process (1.1) after the departure of e_p from the target $\mathfrak{T}$. Let ($M_{L_{1^+}} = \hat{\boldsymbol{e}}_Z \cdot \boldsymbol{L}_{1^+}$, $M_{S_{1^+}} = \hat{\boldsymbol{e}}_Z \cdot \boldsymbol{S}_{1^+}$) be the projections of ($\boldsymbol{L}_{1^+}$, $\boldsymbol{S}_{1^+}$) along the OZ-axis of the SF in Fig. 4.1. Then [10, 60, 184, 186], $|1^+\rangle \equiv |L_{1^+}\, S_{1^+}\, M_{L_{1^+}}\, M_{S_{1^+}}\rangle$ is the ground electronic state of the residual atomic photoion $\mathfrak{T}^{1+}$ possessing energy E_{1^+} in the L-S coupling.

Equation (3.5b) is the density operator for any one-step process taking place in state $|0\rangle$ of an unpolarized target $\mathfrak{T}$ due to the absorption (in $E1$ approximation) of a single photon whose state of pure polarization is represented by $|1m_r\rangle$. In L-S coupling and in the absence of any externally applied magnetic fields, all the Zeeman components (specified by M_{L_0} and M_{S_0}) of the state $|0\rangle$ are degenerate and their total number is $n_{d_0} = (2L_0{+}1)\,(2S_0{+}1)$. Consequently, the density operator (3.5b) for any photon-induced, one-step process in the L-S coupling becomes

$$\rho^{(1)} = \frac{\mathcal{K}_p}{(2L_0+1)(2S_0+1)} \sum_{M_{L_0}=-L_0}^{L_0} \sum_{M_{S_0}=-S_0}^{S_0} F_{\mathrm{p}}\Big(|0; 1m_r\rangle\langle 0; 1m_r|\Big)\, F_{\mathrm{p}}^{\dagger}. \tag{4.3}$$

This expression is completely general, independent of all dynamical models, and applicable to any process induced due to the absorption of a single photon in an unpolarized atom in L-S coupling.

In the entanglement between the spins of photoelectron e_p and of the residual photoion in the 1-SPI process (1.1), the Zeeman components ($M_{L_{1^+}} = -L_{1^+}$, $-L_{1^+}+1, \ldots, 0, \ldots, L_{1^+}-1, L_{1^+}$) of $\mathfrak{T}^{1+}$ due to the space quantization of its orbital angular momentum $\boldsymbol{L}_{1^+}$ are not observed. The required density matrix, obtained from (4.3), can, therefore, be written as

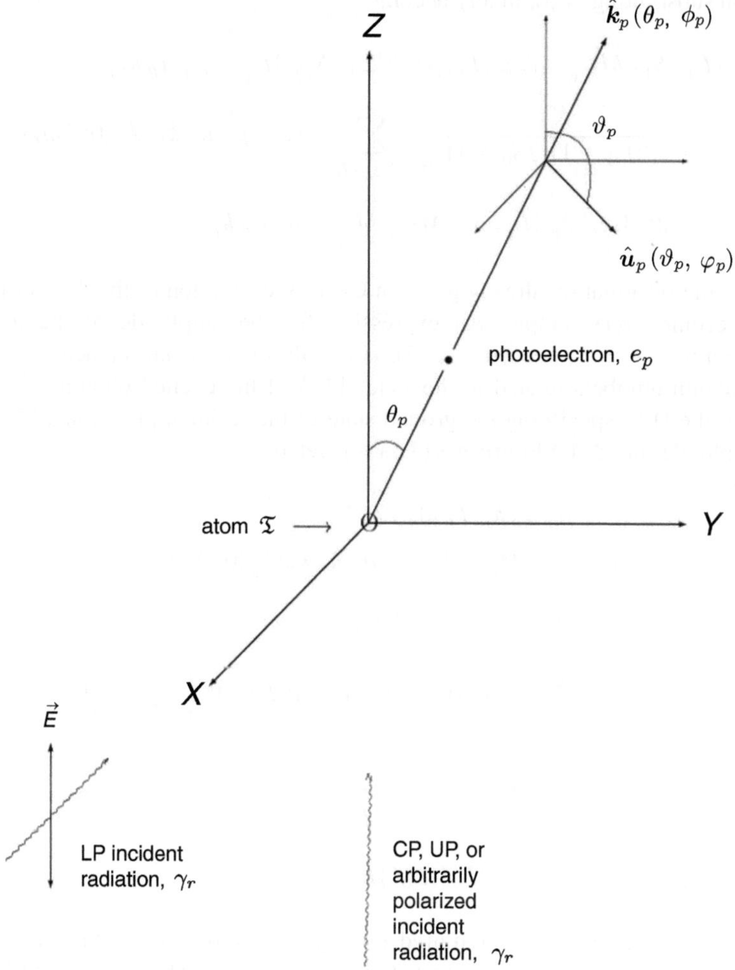

Fig. 4.1 Directions of propagation [$\hat{\boldsymbol{k}}_p$ (θ_p, ϕ_p)] and of spin quantization [$\hat{\boldsymbol{u}}_p$ (ϑ_p, φ_p)] of the photoelectron e_p emitted in the process (1.1) taking place in an atomic target $\mathfrak{T}$ in the absence of the SOI. The orientation of the SF/LF coordinate system $OXYZ$ is identical to that shown in Fig. 1.1 (page 8). (First published in [220].)

$$\langle L_{1+} S_{1+} M_{S_{1+}};\ \mu_p\, \hat{\boldsymbol{u}}_p\, \boldsymbol{k}_p |\ \rho^{(1\text{-SPI})}\ | L_{1+} S_{1+} M'_{S_{1+}};\ \mu'_p\, \hat{\boldsymbol{u}}_p\, \boldsymbol{k}_p\rangle$$

$$= \sum_{M_{L_{1+}}=-L_{1+}}^{L_{1+}} \langle L_{1+}\ S_{1+}\ M_{L_{1+}}\ M_{S_{1+}};$$

$$\mu_p\, \hat{\boldsymbol{u}}_p\, \boldsymbol{k}_p |\ \rho^{(1)}\ | L_{1+} S_{1+} M_{L_{1+}} M'_{S_{1+}};\ \mu'_p\, \hat{\boldsymbol{u}}_p\, \boldsymbol{k}_p\rangle. \qquad (4.4)$$

This, on substituting (4.3), finally becomes

$$\begin{aligned}
&\langle L_{1^+} S_{1^+} M_{S_{1^+}};\ \mu_p\,\hat{\boldsymbol{u}}_p\,\boldsymbol{k}_p|\ \rho^{(1\text{-SPI})}\ |L_{1^+} S_{1^+} M'_{S_{1^+}};\ \mu'_p\,\hat{\boldsymbol{u}}_p\,\boldsymbol{k}_p\rangle \\
&= \frac{\mathcal{K}_p}{(2L_0+1)(2S_0+1)} \sum_{M_{L_0}\,M_{S_0}\,M_{L_{1^+}}} \langle 1^+;\ \mu_p\,\hat{\boldsymbol{u}}_p\,\boldsymbol{k}_p|\ F_{\mathrm{p}}\ |0;\ 1m_r\rangle \\
&\quad \langle 0;\ 1m_r|\ F_{\mathrm{p}}^{\dagger}\ |L_{1^+} S_{1^+} M_{L_{1^+}} M'_{S_{1^+}};\ \mu'_p\,\hat{\boldsymbol{u}}_p\,\boldsymbol{k}_p\rangle. \qquad (4.5)
\end{aligned}$$

With the information already given in Chap. 3, evaluation of the density matrix (4.5) becomes very simple. An expression for the amplitude of the ionizing transition $\mathfrak{T} \longrightarrow \mathfrak{T}^{1+}$, needed in (4.5), is readily obtained on replacing in (3.26) the quantum numbers related to the state $|1^{+^*}\rangle$ of the excited photoion $\mathfrak{T}^{1+^*}$ by those of the $|1^+\rangle$ specifying the ground state of the residual photoion $\mathfrak{T}^{1+}$. Thus, the amplitude for the 1-SPI process (1.1) is given by

$$\begin{aligned}
\mathcal{P}_{0\longrightarrow 1^+} &\equiv \langle 1^+;\ \mu_p\,\hat{\boldsymbol{u}}_p\,\boldsymbol{k}_p|F_{\mathrm{p}}\ |0;\ 1\,m_r\rangle \\
&\equiv \langle L_{1^+}\ S_{1^+}\ M_{L_{1^+}}\ M_{S_{1^+}};\ \mu_p\,\hat{\boldsymbol{u}}_p\,\boldsymbol{k}_p|F_{\mathrm{p}}\ |0;\ 1\,m_r\rangle \\
&= (-1)^{\frac{1}{2}+L_0+L_{1^+}+S_{1^+}+M_{S_0}}\sqrt{2S_0+1} \\
&\quad \times \sum_{\ell_{\mathrm{p}}\,m_p\,\nu_p\,L\,M_L} (-\mathrm{i})^{\ell_{\mathrm{p}}}\,(-1)^{\ell_{\mathrm{p}}}\,\mathrm{e}^{\mathrm{i}\sigma_{\ell_{\mathrm{p}}}}\,(2L+1) \begin{pmatrix} L_{1^+} & \ell_{\mathrm{p}} & L \\ M_{L_{1^+}} & m_p & -M_L \end{pmatrix} \\
&\quad \times \begin{pmatrix} L_0 & 1 & L \\ M_{L_0} & m_r & -M_L \end{pmatrix} \begin{pmatrix} S_{1^+} & \frac{1}{2} & S_0 \\ M_{S_{1^+}} & \nu_p & -M_{S_0} \end{pmatrix} Y_{\ell_{\mathrm{p}}}^{m_p}(\hat{\boldsymbol{k}}_p)\,\mathcal{D}^{1/2}_{\mu_p\,\nu_p}(\omega_p) \\
&\quad \times \langle (L_{1^+}\ \ell_{\mathrm{p}})L|F_{\mathrm{p}}\ |(L_0 1)L\rangle. \qquad (4.6)
\end{aligned}$$

Let us now calculate the required density matrix by substituting the amplitude (4.6) and its Hermitian conjugate (with $M_{S_{1^+}}$ replaced by $M'_{S_{1^+}}$) in (4.5). The consequent expression is simplified on using, in the following order, (*a*) addition theorems for rotational and spherical harmonics given in the respective Eqs. (4.3.2) and (4.6.5) in Edmond's book [65], (*b*) identity (6.2.8) [65] for evaluating a triple sum over three 3-j symbols, (*c*) Eq. (5) from Chap. 12 in [67] for converting a single sum over two 3-j symbols into a double sum containing two 3-j and one 6-j symbols, (d) identity (6.2.8) [65] once again, and (e) the orthonormality (3.7.7) [65] of 3-j symbols after using the relation (3.7.9) for changing a phase factor into a single 3-j symbol whose one of the j's and the corresponding m are both zero. With the help of some additional minor simplifications, the density matrix (4.5) can be written as [220]

$$\langle L_{1+} S_{1+} M_{S_{1+}};\ \mu_p\, \hat{\boldsymbol{u}}_p\, \boldsymbol{k}_p |\, \rho^{(1\text{-SPI})}\, | L_{1+} S_{1+} M'_{S_{1+}};\ \mu'_p\, \hat{\boldsymbol{u}}_p\, \boldsymbol{k}_p \rangle$$
$$= \frac{\mathrm{d}^2\sigma^{(1\text{-SPI})}(m_r)}{\mathrm{d}\epsilon_p\, \mathrm{d}\hat{\boldsymbol{k}}_p}\, \sigma^{(1\text{-SPI})}(S_0;\ S_{1+};\ \hat{\boldsymbol{u}}_p)_{\mu_p, M_{S_{1+}};\ \mu'_p M'_{S_{1+}}}\,. \tag{4.7}$$

The first term on the right-hand side of (4.7) consists of [220]

$$\frac{\mathrm{d}^2\sigma^{(1\text{-SPI})}(m_r)}{\mathrm{d}\epsilon_p\, \mathrm{d}\hat{\boldsymbol{k}}_p} = (-1)^{L_0+L_{1+}+m_r}\, \frac{\mathcal{K}_p}{4\pi(2L_0+1)}$$
$$\times \sum_{\ell_p\, \ell'_p\, L_r\, L\, L'} (2L_r+1) \begin{pmatrix} \ell_p & \ell'_p & L_r \\ 0 & 0 & 0 \end{pmatrix} \begin{pmatrix} 1 & 1 & L_r \\ m_r & -m_r & 0 \end{pmatrix} \begin{Bmatrix} 1 & 1 & L_r \\ L' & L & L_0 \end{Bmatrix}$$
$$\times \begin{Bmatrix} \ell_p & \ell'_p & L_r \\ L' & L & L_{1+} \end{Bmatrix} \langle L_{1+}\ell_p|\mathcal{F}(L)|L_0 1\rangle\, \langle L_{1+}\ell'_p|\mathcal{F}(L')|L_0 1\rangle$$
$$\times\ P_{L_r}(\cos\theta_p). \tag{4.8a}$$

Here, $P_{L_r}(\cos\theta_p)$ are the Legendre polynomials [65] of order L_r and argument $\cos\theta_p$. Further, in (4.8a)

$$\langle L_{1+}\ell_p|\mathcal{F}(L)|L_0 1\rangle = (-\mathrm{i})^{\ell_\mathrm{p}}\, \mathrm{e}^{\mathrm{i}\sigma_{\ell_\mathrm{p}}}\, (-1)^{\ell_\mathrm{p}}(2L+1)$$
$$\times\ \sqrt{2\ell_\mathrm{p}+1}\langle (L_{1+}\ \ell_\mathrm{p})L|F_\mathrm{p}|(L_0\ 1)L\rangle, \tag{4.8b}$$

are the dynamical amplitudes with the bracket $\langle (L_{1+}\ \ell_\mathrm{p})L|F_\mathrm{p}|(L_0\ 1)L\rangle$ already defined in (3.25).

The second term on the right-hand side of (4.7) is, on the other hand, given by [220]

$$\sigma^{(1\text{-SPI})}(S_0;\ S_{1+};\ \hat{\boldsymbol{u}}_p)_{\mu_p, M_{S_{1+}};\ \mu'_p M'_{S_{1+}}} = (-1)^{\mu'_p+S_0+M_{S_{1+}}} \sum_{Q\, m\, n} (2Q+1)$$
$$\begin{pmatrix} \frac{1}{2} & \frac{1}{2} & Q \\ \mu_\mathrm{p} & -\mu'_\mathrm{p} & m \end{pmatrix} \begin{pmatrix} S_{1+} & S_{1+} & Q \\ M_{S_{1+}} & -M'_{S_{1+}} & -n \end{pmatrix} \begin{Bmatrix} \frac{1}{2} & \frac{1}{2} & Q \\ S_{1+} & S_{1+} & S_0 \end{Bmatrix} \left[\mathcal{D}^Q_{mn}(\omega_p)\right]^*. \tag{4.9}$$

4.1.2 Properties

Expression (4.7) is, obviously, a matrix of dimensions $[2(2S_{1+}+1) \times 2(2S_{1+}+1)]$. Here, while the photoelectron e_p is a qubit, the dimensionality of the Hilbert space of the residual photoion $\mathfrak{T}^{1+}$ in L-S coupling is equal to the spin multiplicity

$(2S_{1+} + 1)$ of its electronic state $|1^+\rangle$ formed in the 1-SPI process (1.1). Thus, the density matrix (4.7) for $S_{1+} = \frac{1}{2}, 1, \geq \frac{3}{2}$, *etc*, represents bipartite state of a (qubit, qubit)-, (qubit, qutrit)-, (qubit, qudit[1])-system, respectively. However, in order to study the spin-entanglement properties of a $(e_p, \mathfrak{T}^{1+})$ pair generated in (1.1) in L-S coupling, spin of the electronic state $|1^+\rangle$ of the residual photoion $\mathfrak{T}^{1+}$ must never be zero or, in other words,[2] $S_{1+} \geq \frac{1}{2}$.

The first term [i.e., (4.8)] in the density matrix (4.7) is simply a multiplicative factor which depends, among other things, on the orbital angular momenta of (L_0, L_{1+}) of the electronic states $(|0\rangle, |1^+\rangle)$ of $(\mathfrak{T}, \mathfrak{T}^{1+})$ participating in the 1-SPI, dynamical amplitudes (4.8b), and the Legendre polynomial specifying the direction of motion of the photoelectron e_p in the SF shown in Fig. 4.1. A familiar reader may have, indeed, noticed that (4.8) is, in fact, the well-known expression (see, for example, [221]) for angular distribution of spin-unresolved electrons in atomic photoionization in $E1$ approximation in the L-S coupling.

The angular distribution (4.8) is, obviously, independent of all the spin variables (or quantum numbers) related to the atomic target $\mathfrak{T}$, its residual photoion $\mathfrak{T}^{1+}$, or the photoelectron e_p. Nevertheless, expression (4.8) has an implicit dependence on both (S_0, S_{1+}) because, in L-S coupling, the energies (E_0, E_{1+}) of $(\mathfrak{T}, \mathfrak{T}^{1+})$ depend upon the spin multiplicities of their electronic states $(|0\rangle, |1^+\rangle)$ participating in the 1-SPI process (1.1). Consequently, according to (4.1), energy ϵ_p of the photoelectron e_p will also change with a change in the value of either of (S_0, S_{1+}), although neither of (L_0, L_{1+}) may be affected by these changes. Changes in either or both of (S_0, S_{1+}) will affect the dynamical amplitude (4.8b) as well.

The matrix character of the density matrix (4.7) is due, obviously, to the second term [i.e., (4.9)] on its right-hand side. It is independent of all those quantities which specify the photoelectron angular distribution $\mathrm{d}^2\sigma^{(1\text{-SPI})}(m_r)/\mathrm{d}\epsilon_p\, \mathrm{d}\hat{\boldsymbol{k}}_p$ in (4.8). Expression (4.9), on the other hand, is completely determined by the known values of the spin $(S_0, S_{1+}, M_{S_{1+}}, M'_{S_{1+}}, \mu_p, \mu'_p)$ and space $(\hat{\boldsymbol{u}}_p)$ variables of the particles $(\mathfrak{T}, \mathfrak{T}^{1+}, e_p)$ participating in the 1-SPI process (1.1). Expression (4.9) has, therefore, been called [220] as spin-correlation matrix. Its diagonal elements[3] [i.e., terms with $M_{S_{1+}} = M'_{S_{1+}}$ and $\mu_p = \mu'_p$ in (4.9)], multiplied by $\mathrm{d}^2\sigma^{(1\text{-SPI})}(m_r)/\mathrm{d}\epsilon_p\, \mathrm{d}\hat{\boldsymbol{k}}_p$, give us spin-resolved angular distribution of photoelectrons e_p ejected in $E1$ ionization of an atom in L-S coupling with the residual photoion $\mathfrak{T}^{1+}$ observed in its electronic state $|L_{1+}\, S_{1+}\, M_{S_{1+}}\rangle$; whereas the non-diagonal elements (i.e., those wherein $M_{S_{1+}} \neq M'_{S_{1+}}$ and/or $\mu_p \neq \mu'_p$) of the matrix (4.9) describe coherence effects.[4]

---

[1]For a definition of a qudit, see, for example, pages 22 and 26.

[2]If the spin of the residual photoion happens to be zero (i.e., $S_{1+} = 0$), the non-local correlation in a $(e_p, \mathfrak{T}^{1+})$ pair is then never possible as $\mathfrak{T}^{1+}$ can now exists only in a single spin state corresponding to $M_{S_{1+}}, M'_{S_{1+}} = 0$.

[3]See, appendix C on pages 269–271.

[4]See, appendix C on pages 269–271.

On account of the fact that the angular distribution $\mathrm{d}^2\sigma^{(1\text{-SPI})}(m_r)/\mathrm{d}\epsilon_p\, \mathrm{d}\hat{\boldsymbol{k}}_p$ in the density matrix (4.7) is simply a non-negative multiplicative factor, it is then only the spin correlation matrix $\sigma^{(1\text{-SPI})}(S_0;\ S_{1^+};\ \hat{\boldsymbol{u}}_p)$, given by (4.9), which will always determine all the entanglement properties of a spin state of the bipartite $(e_p,\ \mathfrak{T}^{1+})$-system formed in the 1-SPI process (1.1). For brevity, we, therefore, write, unless specified otherwise,

$$\langle L_{1^+} S_{1^+} M_{S_{1^+}};\ \mu_p\, \hat{\boldsymbol{u}}_p\, \boldsymbol{k}_p |\, \rho^{(1\text{-SPI})}\, | L_{1^+} S_{1^+} M'_{S_{1^+}};\ \mu'_p\, \hat{\boldsymbol{u}}_p\, \boldsymbol{k}_p\rangle$$
$$\equiv \sigma^{(1\text{-SPI})}(S_0;\ S_{1^+};\ \hat{\boldsymbol{u}}_p)_{\mu_p,\, M_{S_{1^+}};\ \mu'_p\, M'_{S_{1^+}}}. \tag{4.10}$$

One can calculate partial transpose [124] of the density matrix (4.10) with respect to either of the photoelectron e_p or photoion $\mathfrak{T}^{1+}$. A straightforward application of (2.38) immediately makes one to write

$$\sigma^{T_p}(S_0;\ S_{1^+};\ \hat{\boldsymbol{u}}_p)_{\mu_p,\, M_{S_{1^+}};\ \mu'_p\, M'_{S_{1^+}}}$$
$$= \sigma^{(1\text{-SPI})}(S_0;\ S_{1^+};\ \hat{\boldsymbol{u}}_p)_{\mu'_p,\, M_{S_{1^+}};\ \mu_p\, M'_{S_{1^+}}} \tag{4.11a}$$

and

$$\sigma^{T_{1^+}}(S_0;\ S_{1^+};\ \hat{\boldsymbol{u}}_p)_{\mu_p,\, M_{S_{1^+}};\ \mu'_p,\, M'_{S_{1^+}}}$$
$$= \sigma^{(1\text{-SPI})}(S_0;\ S_{1^+};\ \hat{\boldsymbol{u}}_p)_{\mu_p,\, M'_{S_{1^+}};\ \mu'_p,\, M_{S_{1^+}}}. \tag{4.11b}$$

Here and in the remaining part of this chapter, unless stated otherwise, superscript T_p means that (4.11a) is the partial transpose of the density matrix (4.10) with respect to the photoelectron e_p; whereas a partial transpose of (4.10) with respect to the photoion has been indicated by the superscript T_{1^+} in (4.11b).

4.2 Entanglement Between the Spins of an Electronic Qubit and an Ionic Qudit

It is obvious from the spin conservation condition (4.2b) in L-S coupling, as well as from the 6-j symbol in (4.9), that for the density matrix (4.10) not to vanish identically, the electronic states $|0\rangle$ and $|1^+\rangle$ of the atom $\mathfrak{T}$ and of its residual photoion $\mathfrak{T}^{1+}$ should, respectively, be such that the spins of these two species are related by $\left|S_0 - S_{1^+}\right| = \frac{1}{2}$. The only two possibilities are, therefore, (*i*) $S_0 = S_{1^+} - \frac{1}{2}$ and (*ii*) $S_0 = S_{1^+} + \frac{1}{2}$ with $S_{1^+} \geq \frac{1}{2}$.[5] In the following, we will see that these

[5] See, footnote (2) on page 98.

two possibilities give rise to spin states of $(e_p, \mathfrak{X}^{1+})$ which have totally different properties: Namely, a state of a $(e_p, \mathfrak{X}^{1+})$-system always has negative or positive partial transpose for all values of $S_{1+} \geq \frac{1}{2}$ depending upon whether it belongs to the case (i) or (ii), respectively.

4.2.1 *Qubit–Qudit Bipartite States with Negative Partial Transpose: Free Entanglement*

Let us specialize the density matrix (4.10) to the case (i) $S_0 = S_{1+} - \frac{1}{2}$. Expression (4.9) now simplifies to

$$\begin{aligned} &\sigma^{(\text{1-SPI})}\left(S_0 = S_{1+} - \frac{1}{2};\ S_{1+};\ \hat{\boldsymbol{u}}_p\right)_{\mu_p, M_{S_{1+}};\ \mu'_p M'_{S_{1+}}} \\ &= \frac{1}{2(2S_{1+}+1)}\,\delta_{\mu_p\mu'_p}\,\delta_{M_{S_{1+}} M'_{S_{1+}}} + (-1)^{\frac{1}{2}+\mu'_p+S_{1+}-M_{S_{1+}}}\sqrt{\frac{3(S_{1+}+1)}{2S_{1+}(2S_{1+}+1)}} \\ &\quad\times \sum_{mn} \begin{pmatrix} \frac{1}{2} & \frac{1}{2} & 1 \\ \mu_p & -\mu'_p & m \end{pmatrix} \begin{pmatrix} S_{1+} & S_{1+} & 1 \\ M_{S_{1+}} & -M'_{S_{1+}} & -n \end{pmatrix} \left[\mathcal{D}^1_{mn}(\omega_p)\right]^* \end{aligned} \tag{4.12}$$

with $\omega_p = (\varphi_p, \vartheta_p, 0)$.

Examples for Case (*i*): $S_0 = S_{1+} - \frac{1}{2}$

(*a*) $S_0 = 0$, $S_{1+} = \frac{1}{2}$

Let us begin with the simplest possible example for this case wherein photoion $\mathfrak{X}^{1+}$ too is a qubit, i.e., $S_{1+} = \frac{1}{2}$. It corresponds to taking the electronic states $(|0\rangle, |1^+\rangle)$ of $(\mathfrak{X}, \mathfrak{X}^{1+})$ to be (singlet, doublet), respectively. The density matrix (4.12) for this example becomes

$$4\,\sigma^{(\text{1-SPI})}\left(S_0 = 0;\ S_{1+} = \frac{1}{2};\ \hat{\boldsymbol{u}}_p\right)_{\mu_p, M_{S_{1+}};\ \mu'_p M'_{S_{1+}}} =$$

$$\begin{array}{lcccc} \mu'_p M'_{S_{1+}} \longrightarrow & \frac{1}{2}\ \frac{1}{2} & \frac{1}{2}\ -\frac{1}{2} & -\frac{1}{2}\ \frac{1}{2} & -\frac{1}{2}\ -\frac{1}{2} \\ \mu_p M_{S_{1+}} & & & & \\ \downarrow & & & & \\ \frac{1}{2}\ \frac{1}{2} & 1-\cos\vartheta_p & -e^{-\mathrm{i}\varphi_p}\sin\vartheta_p & \sin\vartheta_p & e^{-\mathrm{i}\varphi_p} \\ & & & & (1-\cos\vartheta_p) \end{array}$$

$$
\begin{array}{lcccc}
\frac{1}{2}\ -\frac{1}{2} & -e^{i\varphi_p}\sin\vartheta_p & 1+\cos\vartheta_p & -e^{i\varphi_p}\left(1+\cos\vartheta_p\right) & -\sin\vartheta_p \\
-\frac{1}{2}\ \frac{1}{2} & \sin\vartheta_p & -e^{-i\varphi_p}\left(1+\cos\vartheta_p\right) & 1+\cos\vartheta_p & e^{-i\varphi_p}\sin\vartheta_p \\
-\frac{1}{2}\ -\frac{1}{2} & e^{i\varphi_p}\left(1-\cos\vartheta_p\right) & -\sin\vartheta_p & e^{i\varphi_p}\sin\vartheta_p & 1-\cos\vartheta_p
\end{array}
\tag{4.13}
$$

On diagonalizing this matrix, one finds that its eigenvalues are (0, 0, 0, 1). Although each of the sixteen elements in (4.13) is a trigonometric function of one or both of the spherical angles $(\vartheta_p,\ \varphi_p)$ specifying the direction $\hat{\boldsymbol{u}}_p$ of spin quantization of photoelectron e_p in the SF in Fig. 4.1; however, none of its eigenvalues depends upon either of these angles.[6] In view of the properties of a matrix discussed on pages 253 and 254 in the Appendix A, the spin state of $(e_p,\ \mathfrak{T}^{1+})$ in the present example is pure and normalized to unit trace.

Next, we need to calculate the partial transpose of the density matrix (4.13) with respect to one of the members—say, photoelectron e_p—of this bipartite system. Using (4.11a), one readily obtains the following expression:

$$
4\,\sigma^{T_p}\left(S_0=0;\ S_{1+}=\frac{1}{2};\ \hat{\boldsymbol{u}}_p\right)_{\mu_p,\,M_{S_{1+}};\ \mu_p',\,M_{S_{1+}}'} =
$$

$$
\begin{array}{lcccc}
\mu_p' M_{S_{1+}}' \longrightarrow & \frac{1}{2}\ \frac{1}{2} & \frac{1}{2}\ -\frac{1}{2} & -\frac{1}{2}\ \frac{1}{2} & -\frac{1}{2}\ -\frac{1}{2} \\
\mu_p M_{S_{1+}} & & & & \\
\downarrow & & & & \\
\frac{1}{2}\ \frac{1}{2} & 1-\cos\vartheta_p & -e^{-i\varphi_p}\sin\vartheta_p & \sin\vartheta_p & -e^{-i\varphi_p}\left(1+\cos\vartheta_p\right) \\
\frac{1}{2}\ -\frac{1}{2} & -e^{i\varphi_p}\sin\vartheta_p & 1+\cos\vartheta_p & e^{i\varphi_p}\left(1-\cos\vartheta_p\right) & -\sin\vartheta_p \\
-\frac{1}{2}\ \frac{1}{2} & \sin\vartheta_p & e^{-i\varphi_p}\left(1-\cos\vartheta_p\right) & 1+\cos\vartheta_p & e^{-i\varphi_p}\sin\vartheta_p \\
-\frac{1}{2}\ -\frac{1}{2} & -e^{i\varphi_p}\left(1+\cos\vartheta_p\right) & -\sin\vartheta_p & e^{i\varphi_p}\sin\vartheta_p & 1-\cos\vartheta_p
\end{array}
\tag{4.14}
$$

---

[6] Spin $\boldsymbol{S}_{1+}$ of the photoion $\mathfrak{T}^{1+}$ (as well as $\boldsymbol{S}_0$ of the atom $\mathfrak{T}$) in the 1-SPI process (1.1) is quantized along the polar (i.e., OZ-) axis in Fig. 4.1. It is for this reason that no angles corresponding to the direction of quantization of the $\boldsymbol{S}_{1+}$ and/or $\boldsymbol{S}_0$ appear in the density matrix (4.7), or in the subsequent related discussions in the present Chap. 4 of this monograph.

Eigenvalues of this matrix are $(-\frac{1}{2}, \frac{1}{2}, \frac{1}{2}, \frac{1}{2})$.[7] According to the Peres [124]-Horodecki [125] condition discussed on page 40, $(e_p, \mathfrak{T}^{1+})$ are in a non-separable state. Thus, in the present example, density matrix (4.13) represents a pure, entangled, and normalized spin state of the $(e_p, \mathfrak{T}^{1+})$-system. The purity of the state (4.13) also means that its entanglement can be quantified by using any of the measures explained in the Sect. 2.4.2 on pages 41–46. Each of these measures leads to the same conclusion, i.e., state (4.13) is maximally entangled.

The best known examples of pure and maximally entangled bipartite states of qubits are the four Bell states given in (2.17) [or (A.30)]. In these states, spin of each of the qubits is quantized along the polar axis of the SF in Fig. 4.1. It is mentioned in the footnote (6) on page 101 that the spin $\boldsymbol{S}_{1+}$ of the photoion $\mathfrak{T}^{1+}$ in the density matrix (4.7) is already quantized along OZ-axis. In order to quantize also the spin of the photoelectron e_p along this axis, let us coincide the direction $\hat{\boldsymbol{u}}_p$ (ϑ_p, φ_p) with the polar axis in Fig. 4.1 by taking $\theta_p = 0$ in (4.13). This immediately reduces (up to an inconsequential phase factor of $e^{\pm\varphi_p}$) the density matrix (4.13) to the form of (A.30b), which represents a singlet state of two spin-$\frac{1}{2}$ particles.

Thus, in the present example, $(e_p, \mathfrak{T}^{1+})$ are in the pure and maximally entangled singlet spin state. It, however, is not a surprising result. The atom $\mathfrak{T}$, before its photoionization, was in a singlet electronic state; whereas the photoion $\mathfrak{T}^{1+}$, after the completion of the 1-SPI process (1.1), is left in a doublet electronic state. The conservation condition (4.2b), applicable in the presently being considered L-S coupling, demands that the total spin of $(e_p + \mathfrak{T}^{1+})$ must also be zero. Consequently, density matrix (4.13) represents a singlet state of the $(e_p, \mathfrak{T}^{1+})$-system.

There can be numerous experiments, which can realistically be performed in a laboratory based on the present example. Photoionization of any closed-shell atom (e.g., He, Be, Ne, Mg, etc.) will generate, in L-S coupling, a pure, maximally entangled, singlet spin state of the photoion (e.g., He^{1+}, Be^{1+}, Ne^{1+}, Mg^{1+}, etc.) plus photoelectron e_p. There will not be any SOI in the singlet electronic state of (e.g., He, Be, Ne, Mg, etc.). But SOI will certainly exist in the bound doublet electronic states of those photoions $\mathfrak{T}^{1+}$ (e.g., Ne^{1+}) wherein the photoelectron e_p in the 1-SPI process (1.1) is ejected from an outer most other than ns sub-shell of its parent atom $\mathfrak{T}$ (e.g., Ne); the SOI will contribute also to the continuum of e_p if it is represented by partial waves (e.g., p, d, etc.) corresponding to non-zero values of its orbital angular momentum (i.e., $\ell_p \neq 0$). However, on account of a small atomic number Z for (e.g., He, Be, Ne, Mg, etc.), the contribution of SOI will be negligibly small compared to the Coulomb interaction in these atoms.[8]

---

[7] In view of the discussion given on pages 40 and 41, the partial transpose $\sigma^{T_{1+}}(S_0 = 0; S_{1+} = \frac{1}{2}; \hat{\boldsymbol{u}}_p)$, obtained using (4.11b), of the $(e_p, \mathfrak{T}^{1+})$ state (4.13) with respect to the photoion $\mathfrak{T}^{1+}$ may, in general, be different from that given in (4.14); but its eigenvalues are necessarily identical to those of $\sigma^{T_p}(S_0 = 0; S_{1+} = \frac{1}{2}; \hat{\boldsymbol{u}}_p)$ obtained herein.

[8] See footnotes (8) and (9) on page 58, respectively.

(*b*) $S_0 = \frac{1}{2}$, $S_{1+} = 1$

In this second example for the present case (*i*) wherein $S_0 = S_{1+} - \frac{1}{2}$, the residual photoion in the 1-SPI process (1.1) has spin $S_{1+} = 1$. Hence, the multiplicities of the electronic states $(|0\rangle, |1^+\rangle)$ of $(\mathfrak{T}, \mathfrak{T}^{1+})$ in the 1-SPI process (1.1) are (2, 3). That is, photoion is now a qutrit with three simultaneously accessible spin states of this photoion being $|S_{1+}, M_{S_{1+}}\rangle = |1,1\rangle, |1,0\rangle, |1,-1\rangle$; whereas photoelectron e_p is still a qubit. Thus, one is now looking at a bipartite system whose members' Hilbert spaces are of dimensions $d_1 = 2$ (for photoelectron e_p) and $d_2 = 3$ (for photoion $\mathfrak{T}^{1+}$), respectively.

One calculates from (4.12) the following density matrix for this qubit–qutrit bipartite system with $S_0 = \frac{1}{2}$ and $S_{1+} = 1$:

$$6\sqrt{2}\,\sigma^{(1\text{-SPI})}\left(S_0 = \frac{1}{2};\ S_{1+} = 1;\ \hat{\boldsymbol{u}}_p\right)_{\mu_p, M_{S_{1+}};\ \mu_p' M_{S_{1+}}'} =$$

$$\begin{array}{l|cccccc}
\mu_p' M_{S_{1+}}' \longrightarrow & \frac{1}{2}\,1 & \frac{1}{2}\,0 & \frac{1}{2}\,-1 & -\frac{1}{2}\,1 & -\frac{1}{2}\,0 & -\frac{1}{2}\,-1 \\
\mu_p M_{S_{1+}} \downarrow & & & & & & \\
\frac{1}{2}\,1 & \sqrt{2}(1-\cos\vartheta_p) & -e^{-i\varphi_p}\sin\vartheta_p & 0 & \sqrt{2}\sin\vartheta_p & e^{-i\varphi_p}(1-\cos\vartheta_p) & 0 \\
\frac{1}{2}\,0 & -e^{i\varphi_p}\sin\vartheta_p & \sqrt{2} & -e^{-i\varphi_p}\sin\vartheta_p & -e^{i\varphi_p}(1+\cos\vartheta_p) & 0 & e^{-i\varphi_p}(1-\cos\vartheta_p) \\
\frac{1}{2}\,-1 & 0 & -e^{i\varphi_p}\sin\vartheta_p & \sqrt{2}(1+\cos\vartheta_p) & 0 & -e^{i\varphi_p}(1+\cos\vartheta_p) & -\sqrt{2}\sin\vartheta_p \\
-\frac{1}{2}\,1 & \sqrt{2}\sin\vartheta_p & -e^{-i\varphi_p}(1+\cos\vartheta_p) & 0 & \sqrt{2}(1+\cos\vartheta_p) & e^{-i\varphi_p}\sin\vartheta_p & 0 \\
-\frac{1}{2}\,0 & e^{i\varphi_p}(1-\cos\vartheta_p) & 0 & -e^{-i\varphi_p}(1+\cos\vartheta_p) & e^{i\varphi_p}\sin\vartheta_p & \sqrt{2} & e^{-i\varphi_p}\sin\vartheta_p \\
-\frac{1}{2}\,-1 & 0 & e^{i\varphi_p}(1-\cos\vartheta_p) & -\sqrt{2}\sin\vartheta_p & 0 & e^{i\varphi_p}\sin\vartheta_p & \sqrt{2}(1-\cos\vartheta_p)
\end{array} \tag{4.15}$$

Eigenvalues of this matrix are $(0, 0, 0, 0, \frac{1}{2}, \frac{1}{2})$. These eigenvalues are again independent of the angles (ϑ_p, φ_p) and further suggest that the spin state (4.15) representing the qubit–qutrit bipartite $(e_p, \mathfrak{T}^{1+})$-system is mixed, but normalized.

The following partial transpose

$$6\sqrt{2}\,\sigma^{T_p}\left(S_0 = \frac{1}{2};\ S_{1+} = 1;\ \hat{\boldsymbol{u}}_p\right)_{\mu_p, M_{S_{1+}};\ \mu_p' M_{S_{1+}}'} =$$

$$\begin{array}{c|cccccc}
\mu_p' M_{S_{1+}}' \longrightarrow & \frac{1}{2}\,1 & \frac{1}{2}\,0 & \frac{1}{2}\,-1 & -\frac{1}{2}\,1 & -\frac{1}{2}\,0 & -\frac{1}{2}\,-1 \\
\mu_p M_{S_{1+}} \downarrow & & & & & & \\
\hline
\frac{1}{2}\,1 & \sqrt{2}(1-\cos\vartheta_p) & -e^{-i\varphi_p}\sin\vartheta_p & 0 & \sqrt{2}\sin\vartheta_p & -e^{-i\varphi_p}(1+\cos\vartheta_p) & 0 \\
\frac{1}{2}\,0 & -e^{i\varphi_p}\sin\vartheta_p & \sqrt{2} & -e^{-i\varphi_p}\sin\vartheta_p & e^{i\varphi_p}(1-\cos\vartheta_p) & 0 & -e^{-i\varphi_p}(1+\cos\vartheta_p) \\
\frac{1}{2}\,-1 & 0 & -e^{i\varphi_p}\sin\vartheta_p & \sqrt{2}(1+\cos\vartheta_p) & 0 & e^{i\varphi_p}(1-\cos\vartheta_p) & -\sqrt{2}\sin\vartheta_p \\
-\frac{1}{2}\,1 & \sqrt{2}\sin\vartheta_p & e^{-i\varphi_p}(1-\cos\vartheta_p) & 0 & \sqrt{2}(1+\cos\vartheta_p) & e^{-i\varphi_p}\sin\vartheta_p & 0 \\
-\frac{1}{2}\,0 & -e^{i\varphi_p}(1+\cos\vartheta_p) & 0 & e^{-i\varphi_p}(1-\cos\vartheta_p) & e^{i\varphi_p}\sin\vartheta_p & \sqrt{2} & e^{-i\varphi_p}\sin\vartheta_p \\
-\frac{1}{2}\,-1 & 0 & -e^{i\varphi_p}(1+\cos\vartheta_p) & -\sqrt{2}\sin\vartheta_p & 0 & e^{i\varphi_p}\sin\vartheta_p & \sqrt{2}(1-\cos\vartheta_p)
\end{array} \quad (4.16)$$

with respect to the photoelectron in a $(e_p, \mathfrak{T}^{1+})$-system is obtained on applying the operation (4.11a) to the state (4.15). A diagonalization of the state (4.16) gives its eigenvalues to be $(\frac{1}{3}, \frac{1}{3}, \frac{1}{3}, \frac{1}{3}, -\frac{1}{6}, -\frac{1}{6})$. The presence of the negative eigenvalues imply [124, 125] that (4.15) is a non-separable state of a $(e_p, \mathfrak{T}^{1+})$-system wherein photoelectron is a qubit but the photoion $\mathfrak{T}^{1+}$ is a qutrit. This mixed entangled state (4.15) of a qubit–qutrit system is generated in the 1-SPI process (1.1) taking place in the $E1$ approximation in the L-S coupling in an atom $\mathfrak{T}$ in its doublet electronic state $|0\rangle$ while the photoion $\mathfrak{T}^{1+}$ is formed in a ground state $|1^+\rangle$ which is triplet.

Nitrogen atom is, probably, the lightest possible atom whose photoionization in a laboratory can generate a positive ion with spin $S_{1+} = 1$. Here, the 1—SPI process (1.1) $N(1s^2\,2s^2\,2p^3\;^2P/^2D) \longrightarrow N^{1+}(1s^2\,2s^2\,2p^2\;^3P)$ will give us a mixed entangled state of an electronic qubit and N^{1+} qutrit. Similarly, the photoionization process $P(1s^2\,2s^2\,2p^6\,3s^2\,3p^3\;^2P/^2D) \longrightarrow P^{1+}(1s^2\,2s^2\,2p^6\,3s^2 3p^2\;^3P)$ will generate a $(e_p,\, P^{1+})$-system wherein the qutrit photoion of phosphorous forms a mixed, non-separable spin state with the photoelectron. There can be several such examples for realistically generating in a laboratory mixed entangled states of an electronic qubit and an ionic qutrit in a simple process of photoionization taking place in an atom in $E1$ approximation in the L-S coupling.

(c) Thus, matrix representation (4.12) [in the $(\frac{1}{2})(S_{1+}), (\frac{1}{2})(S_{1+}-1), \cdots, (\frac{1}{2})(-S_{1+}+1), (\frac{1}{2})(-S_{1+}), (-\frac{1}{2})(S_{1+}), (-\frac{1}{2})(S_{1+}-1), \cdots, (-\frac{1}{2})(-S_{1+}+1), (-\frac{1}{2})(-S_{1+})$ basis] of the spin-state of an electronic qubit e_p and ionic qudit, generated in the $E1$ process (1.1) taking place in L-S coupling in an atom $\mathfrak{T}$ whose spin is related to that of $\mathfrak{T}^{1+}$ by $S_0 = S_{1+} - \frac{1}{2}$, is of dimensions $[2(2S_{1+}+1) \times 2(2S_{1+}+1)]$]. It can, in general, be shown [220] that the diagonalization of this matrix will give following $(4S_{1+}+2)$ eigenvaluese:

Eigenvalues of the density matrix (4.12)

$$\begin{cases} 0 & \text{there number is } 2(S_{1+}+1), \\ (2S_{1+})^{-1} & \text{there number is } 2S_{1+}. \end{cases} \tag{4.17a}$$

On the other hand, eigenvalues of the partial transpose of the density matrix (4.12) are [220]:

Eigenvalues of the partial transpose of the density matrix (4.12)

$$\begin{cases} (2S_{1+}+1)^{-1} & \text{there number is } 2(S_{1+}+1), \\ -[2S_{1+}(2S_{1+}+1)]^{-1} & \text{there number is } 2S_{1+}. \end{cases} \tag{4.17b}$$

In each of the two cases in (4.17), the total number of eigenvalues is always equal to the dimensionality $[2 \times (2S_{1+}+1)]$ of the density matrix (4.12) and there sum is equal to unity.

Thus, whatever may be the spin S_{1+} of the photoion $\mathfrak{T}^{1+}$ in the process (1.1), some of the eigenvalues of the partial transpose of the density matrix (4.12) are always negative, provided spin S_0 of the atomic target $\mathfrak{T}$ is given by $S_0 = S_{1+} - \frac{1}{2}$. This, in view of the very stringent Peres–Horodecki [124,125] condition, means that the density matrix (4.12) always represents an entangled spin-state of a (qubit, qudit) bipartite system consisting of $(e_p,\, \mathfrak{T}^{1+})$. This state is pure, if and only if, $(|0\rangle, |1^+\rangle)$ representing $(\mathfrak{T},\, \mathfrak{T}^{1+})$ in (1.1) are (singlet, doublet); otherwise, $(e_p,\, \mathfrak{T}^{1+})$ always form a mixed, but non-separable, spin state.

Let us now quantify the entanglement for these non-separable states. Out of the three entanglement measures, explained in the Sect. 2.4.2 on pages 41–46, negativity is the only measure which one can use to know the amount of entanglement present in a bipartite system composed of a qubit and a qudit.[9] According to the definition (2.45c), the negativity of each of the bipartite states (4.12) is

$$\mathcal{N}\Big[\sigma^{((1\text{-SPI}))}(S_0 = S_{1^+} - \frac{1}{2};\ S_{1^+};\ \hat{\boldsymbol{u}}_p)\Big] = \frac{2}{2S_{1^+} + 1}. \tag{4.18}$$

Thus, with increasing dimensions of the Hilbert space (i.e., value of the spin S_{1^+}) of the photoion in the 1-SPI process (1.1), the amount of entanglement in a $(e_p, \mathfrak{T}^{1+})$ pair decreases. It is maximum (i.e., equal to unity) for $S_{1^+} = \frac{1}{2}$—i.e., when the photoion is itself a qubit (e.g., in the example (*a*) discussed on pages 100–102)—and asymptotically goes to $(S_{1^+})^{-1}$ for increasing spin of the electronic state $|1^+\rangle$ of the photoion $\mathfrak{T}^{1+}$ generated in the process (1.1).

The results discussed in the present and in the following sections are completely general and valid for any dynamical models describing an atom in the L-S coupling. Their applications require a knowledge only of the spins (S_0, S_{1^+}) of the states $(|0\rangle, |1^+\rangle)$ of $(\mathfrak{T}, \mathfrak{T}^{1+})$—whatever may be their electronic configurations—participating in the 1-SPI process (1.1) in the absence of SOI. Earlier, Kim et al. [222] have considered spin entanglement of $(e_p, \mathfrak{T}^{1+})$ generated in the photoionization of only the singlet and triplet electronic states of an atom $\mathfrak{T}$ described in the independent particle model [10] by two [60] *equivalent* (i.e., ns^2) or *non-equivalent* (i.e., ns^1 $n's^1$) bound electrons in L-S coupling.

4.2.2 *Qubit–Qudit Bipartite States with Positive Partial Transpose: Bound Entanglement*

Let us now consider second of the two possibilities arising from the spin-conservation condition (4.2b) for the 1-SPI process (1.1) in the L-S coupling, i.e., the case (*ii*) mentioned on pages 99 and 100, at the beginning of the present Sect. 4.2. Specialization of the density matrix (4.10) for this case means taking $S_0 = S_{1^+} + \frac{1}{2}$ in (4.9). This gives us [220]

$$\sigma^{(1\text{-SPI})}(S_0 = S_{1^+} + \frac{1}{2};\ S_{1^+};\ \hat{\boldsymbol{u}}_p)_{\mu_p, M_{S_{1^+}};\ \mu'_p M'_{S_{1^+}}}$$

$$= \frac{1}{2(2S_{1^+} + 1)}\, \delta_{\mu_p \mu'_p}\, \delta_{M_{S_{1^+}} M'_{S_{1^+}}}$$

[9] Both [149], concurrence and entanglement of formation, defined in (2.46) and (2.47), respectively, can be calculated for states of two qubits only (see also [168–170], etc.).

$$+ (-1)^{\mu_p' - \frac{1}{2} + S_{1+} - M_{S_{1+}}} \sqrt{\frac{3S_{1+}}{2(S_{1+}+1)(2S_{1+}+1)}}$$

$$\times \sum_{mn} \begin{pmatrix} \frac{1}{2} & \frac{1}{2} & 1 \\ \mu_{\rm p} & -\mu_{\rm p}' & m \end{pmatrix} \begin{pmatrix} S_{1+} & S_{1+} & 1 \\ M_{S_{1+}} & -M_{S_{1+}}' & -n \end{pmatrix} \left[\mathcal{D}^1_{m\,n}(\omega_p)\right]^*, \tag{4.19a}$$

which can also be written as

$$\sigma^{(1\text{-SPI})}(S_0 = S_{1+} + \frac{1}{2};\ S_{1+};\ \hat{\boldsymbol{u}}_p)_{\mu_p, M_{S_{1+}};\ \mu_p' M_{S_{1+}}'}$$

$$= \frac{1}{2(S_{1+}+1)} \delta_{\mu_{\rm p}\mu_{\rm p}'} \delta_{M_{S_{1+}} M_{S_{1+}}'}$$

$$- \frac{S_{1+}}{S_{1+}+1} \sigma^{(1\text{-SPI})}(S_0 = S_{1+} - \frac{1}{2};\ S_{1+};\ \hat{\boldsymbol{u}}_p)_{\mu_p, M_{S_{1+}};\ \mu_p' M_{S_{1+}}'} \tag{4.19b}$$

in terms of the density matrix (4.12). This relationship between the density matrices for the cases (i) and (ii) will hold also for their partial transpose with respect to either of the particles in a $(e_p,\ \mathfrak{T}^{1+})$ pair.

Equation (4.19b) is applicable also to the diagonal forms of the density matrices present on both sides of this expression. Consequently, eigenvalues [220]

$$\begin{cases} \left[2(S_{1+}+1)\right]^{-1} & \text{there number is } 2(S_{1+}+1), \\ 0 & \text{there number is } 2S_{1+}. \end{cases} \tag{4.20a}$$

are readily obtained on substituting (4.17a) in (4.19b). It is immediately obvious from these that always more than one of the eigenvalues of the density matrix (4.19) is non-zero even for the smallest[10] possible (i.e., $\frac{1}{2}$) allowed value of S_{1+}. This, in other words, means that $(e_p,\ \mathfrak{T}^{1+})$, for none of the values of $S_0 = S_{1+} + \frac{1}{2}$, are in a pure spin state. Hence, in the present case, each of the possible states of a $(e_p,\ \mathfrak{T}^{1+})$-system, generated in a 1-SPI process (1.1), is always mixed. This property of the states described by the density matrix (4.19) is very different from those which belong to the case (i) discussed in the last Sect. 4.2.1 wherein at least one (with $S_0 = 0$, $S_{1+} = \frac{1}{2}$) of the states of $(e_p,\ \mathfrak{T}^{1+})$ was pure [see, example (a) on pages 100–102.]

In order to study entanglement properties of the states (4.19), one needs to diagonalize a partial transpose of this density matrix. It can be done in two ways: We first calculate partial transpose by applying either of the operations (4.11) to any of the expressions (4.19) and then diagonalize the resulting matrix. The other direct approach is to merely substitute eigenvalues (4.17b) of the partial transpose

[10]See footnote (2) on page 98.

of the density matrix (4.12) in (4.19b).[11] This second method immediately gives us following [220] eigenvalues of a partial transpose of (4.19)

$$\begin{cases} \left[2(S_{1+}+1)(2S_{1+}+1)\right]^{-1} & \text{there number is } 2(S_{1+}+1), \\ (2S_{1+}+1)^{-1} & \text{there number is } 2S_{1+}. \end{cases} \tag{4.20b}$$

The important thing to notice about these is that, unlike those given in (4.17b) wherein some of the eigenvalues were always negative, none of the eigenvalues in (4.20b) is ever less than zero. Thus, each of the bipartite states $(e_p, \mathfrak{T}^{1+})$ generated in the 1-SPI process (1.1) with $S_0 = S_{1+} + \frac{1}{2}$ is always mixed and has positive partial transpose (PPT). Some of these PPT states are separable, while others are non-separable for the following reasons.

Separable States with Positive Partial Transpose

Let us first discuss those PPT states (4.19) wherein the residual photoion $\mathfrak{T}^{1+}$ is in an electronic state $|1^+\rangle$ with its spin $S_{1+} = \frac{1}{2}$ or 1. It means, we are now considering those bipartite systems wherein $(e_p, \mathfrak{T}^{1+})$ are either (qubit, qubit) or (qubit, qutrit). According to Peres [124] and Ḥorodecki et al. [125], the existence of PPT for a state of a bipartite system is a necessary and sufficient condition for its separability if the product of the dimensions of the Hilbert spaces of its two sub-systems does not exceed six.[12] Hence, $(e_p, \mathfrak{T}^{1+})$ are always in a separable/product spin state whenever spin multiplicity of the electronic state $|1^+\rangle$ of the photoion $\mathfrak{T}^{1+}$ formed in the 1-SPI process (1.1) is either two or three (i.e., $S_{1+} = \frac{1}{2}$ or 1, respectively), provided the atom $\mathfrak{T}$ absorbs the ionizing photon γ_r in its electronic state $|0\rangle$ with spin $S_0 = 1$ or $\frac{3}{2}$, respectively. Such $(e_p, \mathfrak{T}^{1+})$ states do not have any entanglement at all.

Some of the examples of photoionizing transitions which can generate in *L*-*S* coupling in a laboratory one or the other of these two mixed PPT states without any entanglement are: $\mathrm{C}(1\mathrm{s}^2\,2\mathrm{s}^2\,2\mathrm{p}^2\,{}^3\mathrm{P}) \longrightarrow \mathrm{C}^{1+}(1\mathrm{s}^2\,2\mathrm{s}^2\,2\mathrm{p}^1\,{}^2\mathrm{P})$, $\mathrm{N}(1\mathrm{s}^2\,2\mathrm{s}^2\,2\mathrm{p}^3\,{}^4\mathrm{S}) \longrightarrow \mathrm{N}^{1+}(1\mathrm{s}^2\,2\mathrm{s}^2\,2\mathrm{p}^2\,{}^3\mathrm{P})$, *etc.* These are the respective examples of mixed, bipartite, separable PPT states of (qubit, qubit), (qubit, qutrit) systems.

Non-Separable States with Positive Partial Transpose

It has already been discussed in detail on pages 38–41 in the Sect. 2.4.1.2 that Peres [124] had shown that a separable bipartite state will necessarily have PPT with respect to either of its sub-systems (U, V) whatever may be the dimensions (d_u, d_v) of their Hilbert spaces (H_u, H_v); Subsequently, Horodecki et al. [125] proved that for any bipartite PPT state to be separable, it is sufficient that product of

[11] The partial transpose of a diagonal matrix is the matrix itself.

[12] See discussion on page 40.

the dimensions of its members' Hilbert spaces must not exceed six, i.e., $d_u \times d_v \leq 6$. But, what about those bipartite PPT states which have $d_u \times d_v > 6$? Are they, or are they not, separable? Horodecki [138] gave two explicit examples (with $d_u \times d_v = 3 \times 3$ and 2×4) showing that such a PPT state can also be non-separable and, hence, entangled. However, properties of the entanglement possessed by a PPT state with $d_u \times d_v > 6$ are very different from those present in a state with negative partial transpose (NPT). Before discussing differences between entanglements of bipartite states with NPT and PPT, let us first examine whether or not any of the PPT states (4.19) with $S_{1+} > 1$ [i.e., $d_u \times d_v = 2 \times (2S_{1+} + 1) > 6$] is non-separable?

Several workers (see, for example, [138–140, 223], *etc*) have derived criteria for checking the separability of a given bipartite state with PPT. Some of these have been reviewed in [126, 127, 132]. In order to see whether the PPT states (4.19) for $2(2S_{1+} + 1) > 6$ are separable, Parida and Chandra [220] used the criteria suggested in [138, 223]. These criteria are based upon the rank[13] and range[14] of a matrix representing a bipartite state. According to the first of these two criteria [132, 223], for the PPT state (4.19) with $S_{1+} > 1$ to be entangled, its rank must satisfy the following bound

$$\begin{aligned}&\text{Rank}\Big[\sigma^{(\text{1-SPI})}(S_0 = S_{1+} + \frac{1}{2};\ S_{1+} > 1;\ \hat{\boldsymbol{u}}_p)\Big]\\&\quad \geq \max\Big\{\text{Rank}\big[\rho^{e_p}\big],\ \text{Rank}\big[\rho^{T_{1+}}\big]\Big\},\end{aligned} \tag{4.21}$$

where ρ^{e_p} and $\rho^{T_{1+}}$ are the reduced matrices calculated from $\sigma^{(\text{1-SPI})}(S_0 = S_{1+} + \frac{1}{2};\ S_{1+} > 1;\ \hat{\boldsymbol{u}}_p)$ using the prescription (A.27). Both of these reduced matrices were found [220] to be diagonal of sizes (2×2) and $[(2S_{1+}+1)\times(2S_{1+}+1)]$, respectively. Each of the diagonal elements of ρ^{e_p} is $\frac{1}{2}$ and that of $\rho^{T_{1+}}$ is $(2S_{1+} + 1)^{-1}$, always. Hence, ranks of the two respective reduced matrices are two and $(2S_{1+} + 1)$. But, according to (4.20a), density matrix (4.19) is of rank $2(S_{1+}+1)$. Thus, in the present case, the PPT state (4.20a) always satisfies the bound (4.21) for $S_{1+} \geq \frac{3}{2}$.

In order to use the range criterion [138] for testing the non-separability, or otherwise, of the PPT state (4.19), one first needs to know its eigenvectors for each of its $2(S_{1+} + 1)$ non-zero eigenvalues with $S_{1+} > 1$. If $|\psi_i\rangle$ is one of such eigenvectors,[15] then the range of the state (4.19) is given by

---

[13]See footnote (9) on page 40.

[14]Range of a Hermitian matrix is the space spanned by its those eigenvectors which belong to the non-zero eigenvalues of the matrix ([138], footnote (10) on page 161 in [132]). The number of such eigenvectors is equal to the rank [footnote (9) on page 40] of a matrix.

[15]In the present case, each of these eigenvectors will, obviously, be a column matrix consisting of $2(2S_{1+} + 1)$ rows. Although eigenvectors belonging to non-degenerate eigenvalues of a Hermitian matrix are always orthogonal; however, it is not necessarily the case with those belonging to degenerate eigenvalues. Consequently, before using in the expression (4.22) for range,

$$\text{Range}\left[\sigma^{(1\text{-SPI})}(S_0 = S_{1+} + \frac{1}{2};\ S_{1+} > 1;\ \hat{\boldsymbol{u}}_p)\right] = \sum_{i=1}^{2(S_{1+}+1)} c_i\,|\psi_i\rangle. \tag{4.22}$$

Then according to the range criterion [138, 224], for the PPT state (4.19) to be non-separable, following $2(2S_{1+} + 1)$ coupled simultaneous equations

$$\begin{pmatrix} x_1 \\ x_2 \end{pmatrix} \otimes \begin{pmatrix} y_{S_{1+}} \\ y_{S_{1+}-1} \\ \cdots \\ \cdots \\ y_{-S_{1+}-1} \\ y_{-S_{1+}} \end{pmatrix} = \sum_{i=1}^{2(S_{1+}+1)} c_i\,|\psi_i\rangle, \tag{4.23}$$

should not have any non-trivial solutions. This, indeed, is the case [220] with the states (4.19) for $S_{1+} \geq \frac{3}{2}$.

Thus, applications of both the rank [223] and range [138] criteria, used by Parida and Chandra [220], suggest that the states (4.19) of $(e_p, \mathfrak{T}^{1+})$ generated in one of the simplest processes [i.e., 1-SPI (1.1) in $E1$ approximation] in L-S coupling in atomic and molecular physics (with spin $S_{1+} \geq \frac{3}{2}$ for the photoion $\mathfrak{T}^{1+}$ and $S_0 = S_{1+} + \frac{1}{2}$ that of the atom $\mathfrak{T}$), are entangled, though they possess positive partial transposition. There can be numerous examples of the process (1.1), which can generate non-separable PPT states of $(e_p, \mathfrak{T}^{1+})$ in $L - S$ coupling in a laboratory. Consider, for example, Cr($1s^2\,2s^2\,2p^6\,3s^2\,3p^6\,3d^5\,4s^1\ {}^5P/{}^5F/{}^5D/{}^5G$) $\longrightarrow$ Cr^{1+}($1s^2\,2s^2\,2p^6\,3s^2\,3p^6\,3d^5\ {}^4P/{}^4F/{}^4D/{}^4G$), Fe($1s^2\,2s^2\,2p^6\,3s^2\,3p^6\,3d^6\ {}^5D$) $\longrightarrow$ Fe^{1+}($1s^2\,2s^2\,2p^6\,3s^2\,3p^6\,3d^5\ {}^4P/{}^4F/{}^4D/{}^4G$), *etc.*

Free Versus Bound Entanglement

Some of the applications of entangled states in quantum information science have been discussed in Sect. 2.3 on pages 30–33. For the implementation of the corresponding protocols, it is sometimes necessary, at other times conducive, to have maximally entangled pure states. But, generation of such states is not always an easy task. Moreover, decoherence and dissipation of entanglement takes place due to noisy, ambient, external environments interacting with the system. Such decoherence and loss of entanglement converts a pure and/or maximally entangled state to a mixed state with partial entanglement. Protocols have been developed which can concentrate or purify (also called distill) entanglement. In concentration of entanglement (see, for example, [126, 132, 219, 225–230], *etc*), one generates

---

the degenerate eigenvectors need to be orthonormalized using the well-known Gram–Schmidt procedure (explained, for example, on page 62 in [110]).

a maximally entangled pure state by extracting entanglement from several pure states with partial (i.e., less than maximum possible) entanglement. Purification or distillation of entanglement ([126,132,161,219,225,227,230,231], *etc*), on the other hand, means creation of a maximally entangled pure state from several mixed states each possessing partial entanglement. Concentration or purification techniques of entanglement act on distant qubits and their implementations require only LOCC.[16] It has been suggested [161] that such ([126, 132, 219, 225–233], *etc*) distillation[17] protocols perform a task in quantum information similar to reliable transmission of data through a noisy channel in classical information.

It is obvious from the above discussion that for distillation of entanglement, one necessarily requires states which already have some amount of entanglement. A separable state contains no entanglement at all and, hence, such states are of no use as far as distillation of entanglement is concerned. As the presence of entanglement is necessary for distillation, it was, therefore, originally thought that all non-separable states must be distillable. It has already been shown ([156, 219, 225, 227, 234, 235], *etc*) that a concentration protocol will work for all pure states with entanglement. Moreover, this protocol has been shown [219, 227] to work also reversibly, i.e., one can asymptotically recover all the pure, partially entangled states used to obtain a pure state with maximum entanglement by concentration technique. Horodecki et al. [236] have shown that non-separable states of two spin-$\frac{1}{2}$ particles each possessing any amount of entanglement—however small it may be—can always be distilled to a singlet form (2.17b)/(A.30b)—a maximally entangled pure state. Subsequently, it was found [43, 126, 132, 219] that all NPT states consisting of one qubit (i.e., $d_u = 2$) and one qutrit or qudit (i.e., $d_v \geq 3$) can be distilled to a bipartite state of qubits. This, in other words, means that $(e_p, \mathfrak{T}^{1+})$ states with NPT, described on pages 100–106 in the Sect. 4.2.1, can be distilled to a maximally entangled pure state wherein both photoelectron and photoion are qubits. Entanglement which can be distilled[18] from non-separable states is known as *free* entanglement [126, 132, 219].

It was further shown by Horodecki et al. [43, 126, 132, 219] that an entangled bipartite state with PPT can never be distilled. Thus, unlike NPT states, the entanglement present in a bipartite non-separable PPT state cannot be liberated to form—by the process of distillation—a pure, maximally entangled state. Consequently, non-separability (i.e., entanglement) of a system of more than one particles associated with PPT has been called [43, 126, 132, 219] as bound entanglement and the corresponding state as a bound entangled state.

Thus, there are two kinds of entanglements [126, 132, 219]: Namely, free and bound. The former of the two is distillable and is associated with NPT of a non-separable state; whereas the latter cannot be distilled and arises from the

---

[16]For LOCC, see footnotes (12) and (13) on page 42.

[17]Henceforth, unless stated otherwise, distillation of entanglement means its either concentration or purification, as the case may be.

[18]See footnote (17).

PPT character of a non-separable state. It, however, does not mean that bound entanglement is of no use. Moreover, it is not yet conclusively known [126,132,219] whether a non-separable state with NPT can also have bound entanglement. Horodecki et al. ([126, 132, 219]) have discussed several consequences of bound entanglement. One [237] of these, for example, is the activation of bound entanglement of a non-separable PPT state in the presence of an NPT state. It is a process [219, 237] which enables one, with the support of bound entanglement, to perform quantum teleportation (discussed in the Sect. 2.3.3 on page 32) by an NPT state possessing an otherwise insufficient amount of entanglement.

These developments generated much interest in the search for entangled states with PPT (see, for example, [160,219,220,224,238–246], *etc*) and for studying their properties. However, most of these ([160, 219, 224, 238–246], *etc*) procedures are such that by using them one cannot produce PPT states with bound entanglement, probably, in a laboratory. On the other hand, in order to experimentally test the many consequences [126,132,219] of entangled PPT states in a laboratory, practical generation of such states is very much needed. The discussion presented in the current Sect. 4.2.2 shows, on the other hand, as to how a very simple and well-known procedure (1.1) of 1-SPI in L-S coupling in $E1$ approximation is capable of generating PPT states of (e_p, $\mathfrak{T}^{1+}$) with bound entanglement. These bipartite states are defined in a $2 \times (2S_{1+} + 1)$-dimensional Hilbert space wherein photoelectron e_p is a qubit and the photoion $\mathfrak{T}^{1+}$ is a qudit with $d_v = (2S_{1+} + 1)$, where $S_{1+} \geq \frac{3}{2}$.

Chapter 5
Coulombic Entanglement: One-Step Double Photoionization of Atoms

The simultaneous ejection of two electrons from a target $\mathfrak{T}$, following the absorption of a single photon γ_r in the process (1.2), has come to be known [247–251] as double photoionization, or photo-double-ionization, in atomic and molecular physics. These particles, after their emission, are herein called photoelectrons e_p and $e_{p'}$. The two electrons, before their simultaneous ejection, were part of the same bound state, say, $|0\rangle$ of an electronic configuration of the many-body system $\mathfrak{T}$. But, it is not necessary that only two *equivalent* [10, 60] electrons be ejected in (1.2). That is $(e_p, e_{p'})$, while bound in $\mathfrak{T}$, may or may not belong to the same shell and subshell (specified by the principal quantum number n and orbital angular momentum quantum number ℓ) of an atomic target. The two photoelectrons $(e_p, e_{p'})$ share between themselves energy of the absorbed γ_r in excess to the potential for double ionization of $\mathfrak{T}$. The process (1.2) has then aptly been called [252] as one-step double photoionization (1-DPI). Probably, 1-DPI is analogous to the parametric down conversion [25, 28, 29] in quantum optics wherein a single photon in a nonlinear medium decays into two new photons—called signal and idler—in a single step.

Photoionization, with or without $E1$ dipole approximation [60], is a one-body process in which absorption of a single γ_r of appropriate frequency ν_r by one of the bound electrons provides the later sufficient energy to overcome its binding to the target $\mathfrak{T}$ and become free [10, 60]. But, in 1-DPI, this one-body operator causes the simultaneous emission of two electrons $(e_p, e_{p'})$ from a single bound electronic state $|0\rangle$ of the target $\mathfrak{T}$. It can, obviously, not be possible if bound electrons in $\mathfrak{T}$ are moving independently of each other and, hence, do not have any direct correlation among themselves. This, in other words, means [247–251] that, had the independent-particle model [10, 60, 72] of an atom or a molecule been valid, 1-DPI would not have taken place. Thus, 1-DPI is, probably, one of the most direct manifestations of the existence of electron–electron correlation in any target $\mathfrak{T}$.

N. Chandra and R. Ghosh, *Quantum Entanglement in Electron Optics*, Springer Series on Atomic, Optical, and Plasma Physics 67, DOI 10.1007/978-3-642-24070-6_5,

The two mechanisms, hitherto suggested for the simultaneous emission of the second electron $e_{p'}$, in addition to the one (*i.e.*, e_p) which absorbs the incident photon, in 1-DPI in the presence of electron–electron correlation, are: "Shake-off" (see, for example, [71, 253]) and "knock-out" (also called "half-collision") [254, 255]. The shake-off [71, 253], on the one hand, of $e_{p'}$ takes place due to the sudden changes produced in the potential of $\mathfrak{T}$ following the departure of e_p. Thus, shake-off is relaxation of the suddenly changed charge distribution to an unbound state of ($\mathfrak{T}^{2+}$ + $e_{p'}$) and has been argued [256] to be a purely quantum mechanical effect. The knock-out mechanism, on the other hand, assumes [254, 255] that $e_{p'}$ is produced due to the hard binary collision suffered by one of the bound electrons in $\mathfrak{T}$ with the departing e_p. However, neither of these two mechanisms can cause ejection of $e_{p'}$ without the presence of electron–electron correlation in $\mathfrak{T}$. In addition, a recent experimental study [257] has suggested that neither shake-off nor knock-out is alone, probably, sufficient to account for 1-DPI.

Thus, (e_p, $e_{p'}$) are prepared neither independently nor in total isolation of each other. These two electrons, in addition, before becoming free, were among the correlated constituents of a bound state—in the present case $|0\rangle$—of a many-body system. According to a commonly held belief (see, for example, [17, 24, 34], *etc*), two such particles—*i.e.*, (e_p, $e_{p'}$) in the present case—must necessarily form a non-separable (*i.e.*, entangled) state. But recent studies, both for atoms [252] and for rotating linear molecules [213], have shown that it is not necessarily the case—spins of (e_p, $e_{p'}$) may or may not be entangled. Thus, electron–electron correlation, without which two particles in 1-DPI can never be simultaneously ejected, is not always sufficient to entangle spins of (e_p, $e_{p'}$). Hence, the word correlation in the context of 1-DPI has come to mean [213, 252] only those effects in a quantum system containing two or more mutually interacting particles that are not present in an independent-particle model [10, 60] of an atom or a molecule, but are essential for two electrons to come out together after the absorption of a single photon of appropriate energy. In view of this meaning of the word correlation for atoms and molecules, the two electrons in 1-DPI may or may not be entangled with respect to their spins [213, 252], but are certainly correlated (see, for example, [247, 257]) with respect to their energies, directions of motion, and polarization of their spin angular momenta [258].

This chapter presents a discussion of these features for 1-DPI in an atom in the presence of only the Coulomb forces, completely disregarding SOI among the bound electrons both in $\mathfrak{T}$ and in the residual dication $\mathfrak{T}^{2+}$, as well in the continua of the photoelectrons (e_p, $e_{p'}$). Section 9.1 on pages 202–212 in this monograph contains a similar discussion for (e_p, $e_{p'}$) produced in 1-DPI of a rotating linear molecule. Such quantum informatics properties of (e_p, $e_{p'}$) for 1-DPI in atomic and molecular targets on the inclusion of C+SDIs, on the other hand, are discussed in Sects. 7.2 (pages 160–169) and 9.2 (pages 212–217), respectively.

5.1 Density Operator and Density Matrix

5.1.1 Derivation

The conservation condition

$$h\nu_r + E_0 = E_{2+} + \epsilon_p + \epsilon_{p'} \tag{5.1a}$$

for the total energy in the 1-DPI process (1.2) suggests that energies $(\epsilon_p, \epsilon_{p'})$ of $(e_p, e_{p'})$ are simultaneously affected with a change in the frequency ν_r of absorbed photon γ_r and/or ionization potential $(= E_{2+} - E_0)$ for the simultaneous emission of two photoelectrons. The ground electronic states $(|0\rangle, |2^+\rangle)$ of the (atom $\mathfrak{T}$, residual dication $\mathfrak{T}^{2+}$) participating in the process (1.2) with energies (E_0, E_{2+}) have already been introduced in L-S coupling immediately after (3.20) on page 60. Then

$$\boldsymbol{\ell}_r + \boldsymbol{L}_0 = \boldsymbol{L}_{2+} + \boldsymbol{\ell}_{p_t} \left(= \boldsymbol{\ell}_p + \boldsymbol{\ell}_{p'}\right) \tag{5.1b}$$

and

$$\boldsymbol{S}_0 = \boldsymbol{S}_{2+} + \boldsymbol{s}_{p_t} \left\{= \boldsymbol{s}_p \left[= \left(\frac{1}{2}\right)_p\right] + \boldsymbol{s}_{p'} \left[= \left(\frac{1}{2}\right)_{p'}\right]\right\} \tag{5.1c}$$

are the required conditions for the conservation of the orbital and of the spin angular momenta, respectively, in the absence of SOI. The $(\boldsymbol{\ell}_p, \boldsymbol{\ell}_{p'})$ in (5.1b) and $(\boldsymbol{s}_p, \boldsymbol{s}_{p'})$ in (5.1c) are the orbital and spin angular momenta of electrons $(e_p, e_{p'})$ simultaneously departing from atom $\mathfrak{T}$ in the 1-DPI process (1.2).

According to the description given on page 49, (3.5b) is the density operator for any one-step process involving absorption of a single photon by an unpolarized target $\mathfrak{T}$. In the present case of 1-DPI in an atom $\mathfrak{T}$ with its initial state $|0\rangle$ in L-S coupling in the absence of SOI, the required density operator is, therefore, nothing but

$$\rho^{\text{(1-DPI)}} = \rho^{(1)} \tag{5.2}$$

with the density operator $\rho^{(1)}$ identical to that given by (4.3).

The density matrix for the one-step process (1.2) of 1-DPI, involving the transition from unpolarized state $|0\rangle$ of $\mathfrak{T}$ to state $|2^+\rangle$ of the photo-dication $\mathfrak{T}^{2+}$, is now readily obtained from (5.2) to be [252]

$$\langle L_{2^+} S_{2^+};\ \mu_p \hat{\boldsymbol{u}}_p \boldsymbol{k}_p;\ \mu_{p'} \hat{\boldsymbol{u}}_{p'} \boldsymbol{k}_{p'} |\, \rho^{(1\text{-DPI})}\, | L_{2^+} S_{2^+};\ \mu'_p \hat{\boldsymbol{u}}_p \boldsymbol{k}_p;\ \mu'_{p'} \hat{\boldsymbol{u}}_{p'} \boldsymbol{k}_{p'} \rangle$$
$$= \sum_{d_{2^+}}$$
$$\langle 2^+;\ \mu_p \hat{\boldsymbol{u}}_p \boldsymbol{k}_p;\ \mu_{p'} \hat{\boldsymbol{u}}_{p'} \boldsymbol{k}_{p'} | \rho^{(1)} | 2^+;\ \mu'_p \hat{\boldsymbol{u}}_p \boldsymbol{k}_p;\ \mu'_{p'} \hat{\boldsymbol{u}}_{p'} \boldsymbol{k}_{p'} \rangle. \tag{5.3}$$

Here, sum over d_{2^+} includes all those states of the dication $\mathfrak{T}^{2+}$, which are degenerate to $|2^+\rangle$. For brevity, the dependence on quantum numbers (L_0, S_0) of the density matrix on the left-hand side of (5.3), as well as in the following discussion of this chapter, has not explicitly been shown. Substitution of (4.3) and replacement of the sum d_{2^+} by the appropriate quantum numbers brings the density matrix (5.3) to the following form

$$\langle L_{2^+} S_{2^+};\ \mu_p \hat{\boldsymbol{u}}_p \boldsymbol{k}_p;\ \mu_{p'} \hat{\boldsymbol{u}}_{p'} \boldsymbol{k}_{p'} |\, \rho^{(1\text{-DPI})}\, | L_{2^+}, S_{2^+};\ \mu'_p \hat{\boldsymbol{u}}_p \boldsymbol{k}_p;\ \mu'_{p'} \hat{\boldsymbol{u}}_{p'} \boldsymbol{k}_{p'} \rangle$$
$$= \frac{\mathcal{K}_p}{(2L_0+1)(2S_0+1)} \sum_{M_{L_0} M_{S_0} M_{L_{2^+}} M_{S_{2^+}}} \tag{5.4}$$
$$\langle L_{2^+}\ S_{2^+}\ M_{L_{2^+}}\ M_{S_{2^+}};\ \mu_p \hat{\boldsymbol{u}}_p \boldsymbol{k}_p;\ \mu_{p'} \hat{\boldsymbol{u}}_{p'} \boldsymbol{k}_{p'} |\, F_p \,| L_0\ S_0\ M_{L_0}\ M_{S_0};\ 1 m_r \rangle$$
$$\langle L_0\ S_0\ M_{L_0}\ M_{S_0};\ 1 m_r |\, F_p^\dagger \,| L_{2^+}\ S_{2^+}\ M_{L_{2^+}}\ M_{S_{2^+}};\ \mu'_p \hat{\boldsymbol{u}}_p \boldsymbol{k}_p;\ \mu'_{p'} \hat{\boldsymbol{u}}_{p'} \boldsymbol{k}_{p'} \rangle.$$

The incoherent sums over $M_{L_{2^+}}$ and $M_{S_{2^+}}$ in (5.4) means that the dication $\mathfrak{T}^{2+}$ in our experiment remains unobserved. A diagonal term $\langle L_{2^+} S_{2^+};\mu_p \hat{\boldsymbol{u}}_p \boldsymbol{k}_p;$ $\mu_{p'} \hat{\boldsymbol{u}}_{p'} \boldsymbol{k}_{p'} | \rho^{(1\text{-DPI})} | L_{2^+} S_{2^+};\ \mu_p \hat{\boldsymbol{u}}_p \boldsymbol{k}_p;\ \mu_{p'} \hat{\boldsymbol{u}}_{p'} \boldsymbol{k}_{p'} \rangle$ in (5.4) is the probability for simultaneous ejection of the photoelectrons (e_p, $e_{p'}$) in directions ($\boldsymbol{k}_p$, $\boldsymbol{k}_{p'}$) and having their spins quantizied along ($\boldsymbol{u}_p$, $\boldsymbol{u}_{p'}$) with components (μ_p, $\mu_{p'}$). The density matrix (5.4) has the two properties discussed in (3.16) and (3.17).

Further evaluation of the density matrix (5.4) requires a knowledge of the amplitude for DPI present therein. A state of the ($\mathfrak{T}$ + γ_r)-system, obeying the left-hand sides of the angular momentum conservation conditions [(5.1b), (5.1c)] in the L-S coupling, is already given in (3.21). In order to obtain a state of the ($\mathfrak{T}^{2+}$ + e_p + $e_{p'}$)-system with the couplings of their angular momenta shown on the right-hand sides of (5.1b) and (5.1c), one first writes from (3.22) the following expression

$$\langle \boldsymbol{r}_{p_i} | \mu_{p_i} \hat{\boldsymbol{u}}_{p_i} \boldsymbol{k}_{p_i} \rangle = \sum_{\ell_{p_i} m_{p_i} \nu_{p_i}} \mathrm{i}^{\ell_{p_i}}\, \mathrm{e}^{-\mathrm{i}\sigma_{\ell_{p_i}}}\, R^-_{\ell_{p_i}}(r_{p_i}; k_{p_i}) \left[Y^{m_{p_i}}_{\ell_{p_i}}(\hat{\boldsymbol{k}}_{p_i}) \right]^*$$
$$Y^{m_{p_i}}_{\ell_{p_i}}(\hat{\boldsymbol{r}}_{p_i}) \left[\mathcal{D}^{\frac{1}{2}}_{\mu_{p_i}\, \nu_{p_i}}(\omega_{p_i}) \right]^* |\tfrac{1}{2}\, \nu_{p_i}\rangle \tag{5.5}$$

for the i ($\equiv p,\ p'$)-th photoelectron emitted in the 1-DPI process (1.2). On coupling the orbital angular momenta (L_{2^+}, ℓ_p, $\ell_{p'}$) and spin angular momenta (S_{2^+}, s_p, $s_{p'}$) according to the right-hand sides of (5.1b) and (5.1c), respectively, one obtains [252]

$$
\begin{aligned}
&|2^+;\ \mu_p \hat{\boldsymbol{u}}_p \boldsymbol{k}_p;\ \mu_{p'} \hat{\boldsymbol{u}}_{p'} \boldsymbol{k}_{p'}\rangle \equiv |L_{2^+}\ S_{2^+}\ M_{L_{2^+}}\ M_{S_{2^+}};\ \mu_p \hat{\boldsymbol{u}}_p \boldsymbol{k}_p;\ \mu_{p'} \hat{\boldsymbol{u}}_{p'} \boldsymbol{k}_{p'}\rangle \\
&= (-1)^{-L_{2^+}-S_{2^+}} \sum_{\substack{L_T\, M_{L_T} \\ S_T\, M_{S_T}}} \sum_{\substack{\nu_p\, \nu_{p'} \\ s_{p_t}\, \nu_{p_t}}} \sum_{\substack{\ell_p\, \ell_{p'}\, \ell_{p_t} \\ m_p\, m_{p'}\, m_{p_t}}} i^{(\ell_p+\ell_{p'})}\, \mathrm{e}^{-\mathrm{i}(\sigma_{\ell_p}+\sigma_{\ell_{p'}})} \\
&\times (-1)^{-\ell_p+\ell_{p'}+\ell_{p_t}-m_{p_t}+s_{p_t}-\nu_{p_t}-M_{L_T}-M_{S_T}} \sqrt{(2\ell_{p_t}+1)(2s_{p_t}+1)} \\
&\times \sqrt{(2L_T+1)(2S_T+1)} \begin{pmatrix} \ell_p & \ell_{p'} & \ell_{p_t} \\ m_p & m_{p'} & -m_{p_t} \end{pmatrix} \begin{pmatrix} \frac{1}{2} & \frac{1}{2} & s_{p_t} \\ \nu_p & \nu_{p'} & -\nu_{p_t} \end{pmatrix} \\
&\times \begin{pmatrix} L_{2^+} & \ell_{p_t} & L_T \\ M_{L_{2^+}} & m_{p_t} & -M_{L_T} \end{pmatrix} \begin{pmatrix} S_{2^+} & s_{p_t} & S_T \\ M_{S_{2^+}} & \nu_{p_t} & -M_{S_T} \end{pmatrix} \left[Y_{\ell_p}^{m_p}(\hat{\boldsymbol{k}}_p)\right]^* \left[Y_{\ell_{p'}}^{m_{p'}}(\hat{\boldsymbol{k}}_{p'})\right]^* \\
&\times \left[\mathcal{D}^{1/2}_{\mu_p \nu_p}(\omega_p)\right]^* \left[\mathcal{D}^{1/2}_{\mu_{p'} \nu_{p'}}(\omega_{p'})\right]^* |(L_{2^+}\ell_{p_t})L_T M_{L_T};\ (S_{2^+}s_{p_t})S_T M_{S_T}\rangle^-
\end{aligned}
\tag{5.6}
$$

for a state of the $(\mathfrak{T}^{2+} + e_p + e_{p'})$-system in the L-S coupling. A "–" superscript on the ket in the last line of (5.6) indicates that the continuum spin orbital (5.5) of each of the photoelectrons e_p and $e_{p'}$ individually satisfies the appropriate incoming wave boundary conditions [179].

One can now readily calculate the transition amplitude for 1-DPI present in the density matrix (5.4). On substituting the states (3.21) and (5.6), the amplitude $\langle 2^+;\ \mu_p \hat{\boldsymbol{u}}_p \boldsymbol{k}_p;\ \mu_{p'} \hat{\boldsymbol{u}}_{p'} \boldsymbol{k}_{p'}|F_p|0;\ 1m_r\rangle$ will have, in addition to other things, a Dirac bracket $\langle (L_{2^+}\ell_{p_t})L_T M_{L_T};\ (S_{2^+}s_{p_t})S_T M_{S_T}|F_p|(L_0 1)LM_L;\ S_0 M_{S_0}\rangle$. Use of the conservation conditions [(5.1b), (5.1c)], applicable in the L-S coupling, simplifies this bracket to the following form

$$
\begin{aligned}
&\langle (L_{2^+}\ell_{p_t})L_T M_{L_T};\ (S_{2^+}s_{p_t})S_T M_{S_T}|F_p|(L_0 1)LM_L;\ S_0 M_{S_0}\rangle \\
&\quad = \delta_{LL_T}\,\delta_{M_L M_{L_T}}\,\delta_{S_0 S_T}\,\delta_{M_{S_0} M_{S_T}} \langle (L_{2^+}(\ell_p \ell_{p'})\ell_{p_t})L_T|F_p|(L_0 1)L\rangle_{S_0,\, S_{2^+}}.
\end{aligned}
\tag{5.7}
$$

In L-S coupling, energy of an electronic state depends upon its spin multiplicity [10, 60]. Hence, Dirac's bracket on the right-hand side of (5.7) has explicitly been shown to have an implicit dependence on the spins $(S_0,\ S_{2^+})$ of the electronic states $(|0\rangle,\ |2^+\rangle)$ of $(\mathfrak{T},\ \mathfrak{T}^{2+})$ presently participating in the 1-DPI process (1.2). Thus, a change in either, or both, of S_0 and S_{2^+} will affect not only the value of the matrix element $\langle (L_{2^+}\ell_{p_t})L_T|F_p|(L_0 1)L\rangle$, but also the total energy $\epsilon_p + \epsilon_{p'} = h\nu_r - [E_{2^+}(L_{2^+},\ S_{2^+}) - E_0(L_0,\ S_0)]$ available to the two photoelectrons $(e_p,\ e_{p'})$ according to the conservation condition (5.1a). With some additional simplifications, the transition amplitude for 1-DPI in L-S coupling finally becomes [252]

$$
\begin{aligned}
&\langle L_{2+}\, S_{2+}\, M_{L_{2+}}\, M_{S_{2+}};\, \mu_p\, \hat{\boldsymbol{u}}_p\, \boldsymbol{k}_p;\, \mu_{p'}\, \hat{\boldsymbol{u}}_{p'}\, \boldsymbol{k}_{p'} | F_p | L_0\, S_0\, M_{L_0}\, M_{S_0};\, 1 m_r \rangle \\
&= (-1)^{1+L_0+L_{2+}+S_{2+}+M_{S_0}} \sqrt{2S_0+1} \sum_{L\, M_L} \sum_{\substack{\ell_{p_t}\, m_{p_t} \\ s_{p_t}\, \nu_{p_t}}} \sum_{\substack{\ell_p\, m_p\, \nu_p \\ \ell_{p'}\, m_{p'}\, \nu_{p'}}} (-\mathrm{i})^{\ell_p+\ell_{p'}} \\
&\times \mathrm{e}^{\mathrm{i}(\sigma_{\ell_p}+\sigma_{\ell_{p'}})} (-1)^{\ell_p+\ell_{p'}+\ell_{p_t}+s_{p_t}+m_{p_t}+\nu_{p_t}} (2L+1) \sqrt{(2\ell_{p_t}+1)(2s_{p_t}+1)} \\
&\times \begin{pmatrix} \ell_p & \ell_{p'} & \ell_{p_t} \\ m_p & m_{p'} & -m_{p_t} \end{pmatrix} \begin{pmatrix} \frac{1}{2} & \frac{1}{2} & s_{p_t} \\ \nu_p & \nu_{p'} & -\nu_{p_t} \end{pmatrix} \begin{pmatrix} L_{2+} & \ell_{p_t} & L \\ M_{L_{2+}} & m_{p_t} & -M_L \end{pmatrix} \begin{pmatrix} S_{2+} & s_{p_t} & S_0 \\ M_{S_{2+}} & \nu_{p_t} & -M_{S_0} \end{pmatrix} \\
&\times \begin{pmatrix} L_0 & 1 & L \\ M_{L_0} & m_r & -M_L \end{pmatrix} Y_{\ell_p}^{m_p}(\hat{\boldsymbol{k}}_p)\, Y_{\ell_{p'}}^{m_{p'}}(\hat{\boldsymbol{k}}_{p'})\, \mathcal{D}^{1/2}_{\mu_p \nu_p}(\omega_p)\, \mathcal{D}^{1/2}_{\mu_{p'} \nu_{p'}}(\omega_{p'}) \\
&\times \langle (L_{2+}(\ell_p \ell_{p'})\ell_{p_t})L | F_p | (L_0 1)L \rangle_{S_0, S_{2+}}.
\end{aligned}
\tag{5.8}
$$

For the calculation of the required density matrix, one needs to substitute (5.8) and its Hermitian conjugate on the right-hand side of (5.4). The consequent expression is considerably simplified by analytically performing as many sums over the magnetic quantum numbers (e.g., M_{L_0}, M_{S_0}, $M_{L_{2+}}$, $M_{S_{2+}}$, M_L, m_p, $m_{p'}$, m_{p_t}, ν_p, $\nu_{p'}$, ν_{p_t}, *etc*) as possible.

These simplifications require a heavy use of Racah algebra by applying, in various orders, several identities [*e.g.*, (3.7.8), (3.7.9), (4.3.2), (4.6.5), (6.2.8), *etc.*] given in [65]. Some of these identities have to be used more than once. In addition, (14.42) given in [64] needs to be used twice. A single application of (14.42) [64] reduces quadruple sums over the product of four 3-j symbols into a double sum over the product of two 3-j and one 9-j symbols [64].

With the help of these and some other simplifications, one can finally reduce the desired density matrix to a product of two independent terms of the following form [252]

$$
\begin{aligned}
&\langle L_{2+} S_{2+};\, \mu_p\, \hat{\boldsymbol{u}}_p\, \boldsymbol{k}_p;\, \mu_{p'}\, \hat{\boldsymbol{u}}_{p'}\, \boldsymbol{k}_{p'} |\, \rho^{\text{(1-DPI)}}\, | L_{2+} S_{2+};\, \mu'_p\, \hat{\boldsymbol{u}}_p\, \boldsymbol{k}_p;\, \mu'_{p'}\, \hat{\boldsymbol{u}}_{p'}\, \boldsymbol{k}_{p'} \rangle \\
&= \frac{\mathrm{d}^3\sigma^{\text{(1-DPI)}}(m_r)}{\mathrm{d}\epsilon_1\, \mathrm{d}\hat{\boldsymbol{k}}_p\, \mathrm{d}\hat{\boldsymbol{k}}_{p'}}\, \sigma^{\text{(1-DPI)}}(S_0;\, S_{2+};\, \hat{\boldsymbol{u}}_p;\, \hat{\boldsymbol{u}}_{p'})_{\mu_p, \mu_{p'}; \mu'_p, \mu'_{p'}}.
\end{aligned}
\tag{5.9}
$$

Here, we have defined [252]

$$
\begin{aligned}
\frac{\mathrm{d}^3\sigma^{\text{(1-DPI)}}(m_r)}{\mathrm{d}\epsilon_1\, \mathrm{d}\hat{\boldsymbol{k}}_p\, \mathrm{d}\hat{\boldsymbol{k}}_{p'}} &= (-1)^{m_r+L_0+L_{2+}} \frac{\mathcal{K}_p}{4\pi(2L_0+1)} \sum_{\substack{L_p\, L_{p'}\, M \\ L\, L'\, L_r}} \sum_{\substack{\ell_p\, \ell_{p'}\, \ell_{p_t} \\ \ell'_p\, \ell'_{p'}\, \ell'_{p_t}}} \\
&\times (-1)^{\ell'_p+\ell'_{p'}+\ell'_{p_t}} (2L_r+1) \sqrt{(2L_p+1)(2L_{p'}+1)}
\end{aligned}
$$

$$\times \begin{pmatrix} \ell_p & \ell'_p & L_p \\ 0 & 0 & 0 \end{pmatrix} \begin{pmatrix} \ell_{p'} & \ell'_{p'} & L_{p'} \\ 0 & 0 & 0 \end{pmatrix}$$

$$\times \begin{pmatrix} 1 & 1 & L_r \\ m_r & -m_r & 0 \end{pmatrix} \begin{pmatrix} L_p & L_{p'} & L_r \\ M & -M & 0 \end{pmatrix} \begin{Bmatrix} 1 & 1 & L_r \\ L' & L & L_0 \end{Bmatrix} \begin{Bmatrix} \ell_{p_t} & \ell'_{p_t} & L_r \\ L' & L & L_{2+} \end{Bmatrix} \begin{Bmatrix} \ell_p & \ell_{p'} & \ell_{p_t} \\ \ell'_p & \ell'_{p'} & \ell'_{p_t} \\ L_p & L_{p'} & L_r \end{Bmatrix}$$

$$\times Y^M_{L_p}(\hat{\boldsymbol{k}}_p)\, Y^{-M}_{L_{p'}}(\hat{\boldsymbol{k}}_{p'})\, \langle L_{2+}(\ell_p \ell_{p'})\ell_{p_t}|F(L)|L_0 1\rangle \, \langle L_{2+}(\ell'_p \ell'_{p'})\ell'_{p_t}|F(L')|L_0 1\rangle^* \tag{5.10a}$$

with

$$\langle L_{2+}(\ell_p \ell_{p'})\ell_{p_t}|F(L)|L_0 1\rangle = (-\mathrm{i})^{\ell_p + \ell_{p'}}\, \mathrm{e}^{\mathrm{i}(\sigma_{\ell_p} + \sigma_{\ell_{p'}})}\,(2L+1)$$
$$\times \sqrt{(2\ell_p+1)(2\ell_{p'}+1)(2\ell_{p_t}+1)}\langle (L_{2+}(\ell_p \ell_{p'})\ell_{p_t})L|F_p|(L_0 1)L\rangle_{S_0,\,S_{2+}}, \tag{5.10b}$$

and [252]

$$\sigma^{(\text{1-DPI})}(S_0;\, S_{2+};\, \hat{\boldsymbol{u}}_p;\, \hat{\boldsymbol{u}}_{p'})_{\mu_p,\mu_{p'};\mu'_p,\mu'_{p'}} = (-1)^{\mu'_p + \mu'_{p'}} \sum_{\substack{s\, s_{p_t} \\ \mu\, \mu_1\, \mu_2}} (-1)^{s_{p_t} - \mu}(2s+1)$$

$$\times \begin{pmatrix} 1/2 & 1/2 & s \\ \mu_p & -\mu'_p & \mu_1 \end{pmatrix} \begin{pmatrix} 1/2 & 1/2 & s \\ \mu_{p'} & -\mu'_{p'} & \mu_2 \end{pmatrix} \begin{Bmatrix} 1/2 & 1/2 & s \\ 1/2 & 1/2 & s_{p_t} \end{Bmatrix} \left[\mathcal{D}^s_{\mu_1\,\mu}(\omega_p)\right]^* \left[\mathcal{D}^s_{\mu_2,-\mu}(\omega_{p'})\right]^* \tag{5.11}$$

5.1.2 Properties

The density matrix in (5.9) has been arrived at by taking into account all those effects (including electron–electron correlation), which arise due only to the Coulomb forces in an interacting, many-body quantum system. SOI neither in the bound states ($|0\rangle, |2^+\rangle$) of ($\mathfrak{T}, \mathfrak{T}^{2+}$), nor in the continua of any of the two photoelectrons (e_p, $e_{p'}$), has ever been considered in the derivation of (5.9). Thus, (5.9) can appropriately be called a Coulombic density matrix for the 1-DPI process (1.2) in an atom.

It is obvious from (5.10) that the first term [i.e., $\mathrm{d}^3\sigma^{(\text{1-DPI})}(m_r)/\mathrm{d}\epsilon_1\, \mathrm{d}\hat{\boldsymbol{k}}_p\, \mathrm{d}\hat{\boldsymbol{k}}_{p'}$] on the right-hand side of the Coulombic density matrix (5.9) depends both on the kinematics and on the dynamics of 1-DPI. $\mathrm{d}^3\sigma^{(\text{1-DPI})}(m_r)/\mathrm{d}\epsilon_1\, \mathrm{d}\hat{\boldsymbol{k}}_p\, \mathrm{d}\hat{\boldsymbol{k}}_{p'}$ involves, in addition to the energy and polarization of the absorbed photon γ_r, momentum vectors (i.e., $\boldsymbol{k}_p$, $\boldsymbol{k}_{p'}$) of the escaped electrons (e_p, $e_{p'}$) as well as the transition amplitudes (5.10b) for 1-DPI, which contain, among other things, total orbital angular momenta (L_0, L_{2+}) of the electronic states ($|0\rangle$, $|2^+\rangle$) of ($\mathfrak{T}$, $\mathfrak{T}^{2+}$) participating in the 1-DPI process (1.2). Further, $\mathrm{d}^3\sigma^{\text{(1-DPI)}}(m_r)/\mathrm{d}\epsilon_1\, \mathrm{d}\hat{\boldsymbol{k}}_p\, \mathrm{d}\hat{\boldsymbol{k}}_{p'}$ in (5.10) does not explicitly include spins of either of the two photoelectrons, the target atom, or of its residual photo-dication $\mathfrak{T}^{2+}$. Hence, in L-S coupling, $\mathrm{d}^3\sigma^{\text{(1-DPI)}}(m_r)/\mathrm{d}\epsilon_1\, \mathrm{d}\hat{\boldsymbol{k}}_p\, \mathrm{d}\hat{\boldsymbol{k}}_{p'}$ represents purely angular correlation between the photoelectrons (e_p, $e_{p'}$). It gives us angular distribution of two simultaneously ejected electrons following the absorption of a single photon by an atom in L-S coupling. Its value is always *positive* and acts merely as a multiplicative factor in the density matrix (5.9) for a given process of 1-DPI.

The second term [*i.e.*, $\sigma^{\text{(1-DPI)}}(S_0;\, S_{2+};\, \hat{\boldsymbol{u}}_p;\, \hat{\boldsymbol{u}}_{p'})_{\mu_p,\mu_{p'};\mu'_p,\mu'_{p'}}$], on the right-hand side of (5.9)—specified in (5.11)—is a (4 × 4) matrix. Unlike the angular correlation (5.10)—also occurring in (5.9)—expression (5.11) is completely independent both of the kinematics and dynamics of the 1-DPI process (1.2). Moreover, (5.11) contains none of those physical variables (*e.g.*, m_r, L_0, L_{2+}, $\boldsymbol{k}_p$, $\boldsymbol{k}_{p'}$, *etc.*) which are present in the angular correlation (5.10). On the other hand, matrix $\sigma^{\text{(1-DPI)}}(S_0;\, S_{2+};\, \hat{\boldsymbol{u}}_p;\, \hat{\boldsymbol{u}}_{p'})_{\mu_p,\mu_{p'};\mu'_p,\mu'_{p'}}$ is readily determined analytically from (5.11) from a knowledge of the directions ($\hat{\boldsymbol{u}}_p$, $\hat{\boldsymbol{u}}_{p'}$) of spin quantization of (e_p, $e_{p'}$) and of the components (μ_p, $\mu_{p'}$) of the two photoelectrons' spins along these unit vectors. Hence, expression $\sigma^{\text{(1-DPI)}}(S_0;\, S_{2+};\, \hat{\boldsymbol{u}}_p;\, \hat{\boldsymbol{u}}_{p'})_{\mu_p,\mu_{p'};\mu'_p,\mu'_{p'}}$ on the right-hand side of the Coulombic density matrix (5.9) and determined from (5.11), represents purely spin correlation between the photoelectrons (e_p, $e_{p'}$).

Neither of the respective expressions (5.10) and (5.11) for the angular and spin correlations explicitly contains either or both of the total spins (S_0, S_{2+}) of the electronic states ($|0\rangle$, $|2^+\rangle$) of ($\mathfrak{T}$, $\mathfrak{T}^{2+}$) participating in the 1-DPI process (1.2). It will, however, be wrong to conclude from this explicit absence of (S_0, S_{2+}) in (5.10) and (5.11) that the respective correlations represented by them are also independent of the spins of the atomic target and/or of its residual photodication. Both angular and spin correlations, on the other hand, implicitly depend in important ways on (S_0, S_{2+}).

In L-S coupling [10, 60], electronic state [$|0\rangle$, $|2^+\rangle$] of [$\mathfrak{T}$, $\mathfrak{T}^{2+}$] are called *term*, which are written as $\left[{}^{2S_0+1}L_0,\ {}^{2S_{2+}+1}L_{2+}\right]$. Each term has a specific energy [E_0, E_{2+}] for definite values of [(L_0, S_0), (L_{2+}, S_{2+})]. Consequently, in view of the conservation conditions (5.1), the linear momenta ($\boldsymbol{k}_p$, $\boldsymbol{k}_{p'}$) of (e_p, $e_{p'}$) and the transition amplitudes (5.10b) depend on the spins (S_0, S_{2+}) of ($\mathfrak{T}$, $\mathfrak{T}^{2+}$). This, in turn, means that the angular correlation $\mathrm{d}^3\sigma^{\text{(1-DPI)}}(m_r)/\mathrm{d}\epsilon_1\, \mathrm{d}\hat{\boldsymbol{k}}_p\, \mathrm{d}\hat{\boldsymbol{k}}_{p'}$ between (e_p, $e_{p'}$) in the Coulombic density matrix (5.9) has an implicit dependence on the spins (S_0, S_{2+}), although neither of these two quantum numbers is present in its expressions (5.10). However, this implicit dependence of the angular correlation on S_0 and S_{2+} will have no effect, whatsoever, on the spin correlation (5.11) between (e_p, $e_{p'}$).

According to the conservation condition (5.1c), applicable in L-S coupling in the present case, spin angular momenta present therein must satisfy the triangular condition [65]

$$\Delta(S_0, S_{2+}, s_{p_t}) \equiv |S_0 - S_{2+}| \leq s_{p_t} \leq S_0 + S_{2+}. \tag{5.12}$$

But, the total spin s_{p_t} of two photoelectrons $(e_p, e_{p'})$ emitted in the 1-DPI (1.2) can be 0 and/or 1. Thus, only those states $(|0\rangle, |2^+\rangle)$ of $(\mathfrak{T}, \mathfrak{T}^{2+})$ will participate in the present 1-DPI process whose total spin angular momenta (S_0, S_{2+}) are either equal or differ by unity. For example, one needs to perform incoherent sum over s_{p_t} in (5.11) over both of its allowed values (of 0 and 1) if the spin multiplicities of the electronic states of $(\mathfrak{T}, \mathfrak{T}^{2+})$ are same and more than one, *ie*, $S_0 = S_{2+} \geq \frac{1}{2}$. Whenever, each of the atom $\mathfrak{T}$ and its residual photodication $\mathfrak{T}^{2+}$ are in their singlet electronic states (i.e., S_0, $S_{2+} = 0$), (5.11) is to be summed over only for $s_{p_t} = 0$. If $|S_0 - S_{2+}| = 1$, on the other hand, then $s_{p_t} = 1$ will only contribute to (5.11). One, therefore, concludes that, although, neither S_0 of the target atom $\mathfrak{T}$ nor S_{2+} of the residual photodication $\mathfrak{T}^{2+}$ explicitly appear in the spin correlation matrix (5.11), the latter's value is, nevertheless, very much governed by these two quantum numbers on account of the spin conservation condition (5.1c). The value of s_{p_t} to be used in (5.11) is determined by the triangular condition (5.12). It is for this reason that the spin correlation matrix $\sigma^{(1\text{-DPI})}(S_0;\, S_{2+};\, \hat{\boldsymbol{u}}_p;\, \hat{\boldsymbol{u}}_{p'})_{\mu_p, \mu_{p'}; \mu'_p, \mu'_{p'}}$ has been shown to have both S_0 and S_{2+} among its arguments.

Thus, the density matrix needed for studying entanglement between the spins of two photoelectrons produced in 1-DPI (1.2) in an atom $\mathfrak{T}$ in the absence of SOI is a product of an angular correlation term, which is always positive and a (4×4) matrix representing the spin correlation between the two ejected electrons. The determination of these two correlations requires entirely different kinds of information. The calculation of the angular correlation (5.10) needs a complete knowledge of the kinematics as well as of the dynamics of 1-DPI—in addition to the energy and polarization of the absorbed photon. The latter, i.e., (5.11), on the other hand, does not require any of this information and is completely determined analytically by a knowledge of the spins of the four particles (i.e., $\mathfrak{T}$, $\mathfrak{T}^{2+}$, e_p, $e_{p'}$) participating in the 1-DPI process (1.2), in addition to the quantization directions $(\hat{\boldsymbol{u}}_p, \hat{\boldsymbol{u}}_{p'})$. As a consequence, the spin correlation matrix remains unaffected whatever may be the velocities of two photoelectrons as well as energy and polarization of the ionizing radiation γ_r in (1.2). However, both angular correlation (5.10) and spin correlation matrix (5.11) have implicit dependence, albeit of different kinds, on the spin quantum numbers S_0 and S_{2+}.

5.2 Spin-Entanglement Properties of Two Photoelectrons

In view of the investigations reported, for example, in [128–130] showing that there may be mixed non-separable states of two qubits which do not violate the hidden-variable inequalities [12, 90, 92, 98], it becomes necessary that one should study the spin-entanglement properties of photoelectrons $(e_p, e_{p'})$, emitted simultaneously in the 1-DPI process (1.2) in an atom in the absence of SOI, using the conditions, which are equally applicable to both pure and mixed bipartite states of qubits. Chandra and Chakraborty [252] used the condition whose satisfaction is both necessary [124] and sufficient [125] for a (2×2) or (2×3) state—pure or mixed—to be separable. Herein, therefore, the presence or absence of spin-entanglement in a pair of photoelectrons $(e_p,\ e_{p'})$, whose state is represented by the Coulombic density matrix (5.9), is investigated using the stringent Peres–Horodecki condition [124, 125].

It has already been discussed elsewhere in this chapter that $d^3\sigma^{\text{(1-DPI)}}(m_r)/ d\epsilon_1\, d\hat{\boldsymbol{k}}_p\, d\hat{\boldsymbol{k}}_{p'}$ in (5.9) describes purely angular distribution in L-S coupling of simultaneously observed photoelectrons $(e_p,\ e_{p'})$. Its value is, therefore, always positive. Moreover, it does not depend on any of the properties associated with the spins of the two observed photoelectrons. In the L-S coupling, all the spin-entanglement properties of a pair of $(e_p,\ e_{p'})$ will always be determined by the spin correlation matrix $\sigma^{\text{(1-DPI)}}(S_0;\ S_{2^+};\ \hat{\boldsymbol{u}}_p;\ \hat{\boldsymbol{u}}_{p'})_{\mu_p,\mu_{p'};\mu'_p,\mu'_{p'}}$ given in (5.11). The angular correlation term [given by (5.10)] in (5.9) simply multiplies this matrix by a positive factor. In the remaining part of our discussion of the spin-entanglement properties of photoelectrons $(e_p,\ e_{p'})$, we, therefore, do not write the angular correlation part (5.10) and represent, for brevity, the Coulombic density matrix (5.9) simply by its spin correlation part. That is, unless specified otherwise, we write

$$\begin{aligned} &\langle L_{2^+} S_{2^+};\ \mu_p\, \hat{\boldsymbol{u}}_p\, \boldsymbol{k}_p;\ \mu_{p'}\, \hat{\boldsymbol{u}}_{p'}\, \boldsymbol{k}_{p'} |\, \rho^{\text{(1-DPI)}}\, | L_{2^+} S_{2^+};\ \mu'_p\, \hat{\boldsymbol{u}}_p\, \boldsymbol{k}_p;\ \mu'_{p'}\, \hat{\boldsymbol{u}}_{p'}\, \boldsymbol{k}_{p'} \rangle \\ &\qquad \equiv \sigma^{\text{(1-DPI)}}(S_0;\ S_{2^+};\ \hat{\boldsymbol{u}}_p;\ \hat{\boldsymbol{u}}_{p'})_{\mu_p,\mu_{p'};\mu'_p,\mu'_{p'}}. \end{aligned} \tag{5.13}$$

In view of the spin conservation condition (5.1c) [or, the requirement (5.12)], applicable in the L-S coupling, there are only five possibilities [252] wherein the Coulombic density matrix (5.13) is different from being a null matrix [259]. In the following, we investigate the spin-entanglement properties of $(e_p,\ e_{p'})$ in all of these possible five cases.

5.2.1 $S_0 = S_{2^+} = 0$

This corresponds to the situation when each of the electronic states $|0\rangle$ and $|2^+\rangle$ of the target atom $\mathfrak{T}$ before its 1-DPI and that of the residual photodication $\mathfrak{T}^{2+}$ are singlet, respectively. The conservation condition (5.1c) [or, the requirement (5.12)] now suggests that $s_{p_t} = 0$ is the only allowed possibility. On substituting $S_0,\ S_{2^+},\ s_{p_t} = 0$ in (5.11), the Coulombic density matrix (5.13) becomes

$$4\,\sigma^{\text{(1-DPI)}}(S_0 = 0;\ S_{2^+} = 0;\ \hat{\boldsymbol{u}}_p;\ \hat{\boldsymbol{u}}_{p'})_{\mu_p,\mu_{p'};\mu'_p,\mu'_{p'}} =$$

$$\mu_p, \mu_{p'} / \mu'_p, \mu'_{p'} \Longrightarrow \quad \tfrac{1}{2}, \tfrac{1}{2} \quad \tfrac{1}{2}, -\tfrac{1}{2} \quad -\tfrac{1}{2}, \tfrac{1}{2} \quad -\tfrac{1}{2}, -\tfrac{1}{2}$$

$$
\begin{array}{lcccc}
\Downarrow & & & & \\
\frac{1}{2},\frac{1}{2} \longrightarrow & 1-\hat{\boldsymbol{u}}_p\cdot\hat{\boldsymbol{u}}_{p'} & \begin{array}{c} c_p s_{p'} \\ -s_p c_{p'} c \\ -\mathrm{i} s_p s \end{array} & \begin{array}{c} s_p c_{p'} \\ -c_p s_{p'} c \\ +\mathrm{i} s_{p'} s \end{array} & \begin{array}{c} -s_p s_{p'} \\ +(1-c_p c_{p'})c \\ -\mathrm{i}(c_p - c_{p'})s \end{array} \\
\frac{1}{2},-\frac{1}{2} \longrightarrow & \begin{array}{c} c_p s_{p'} \\ -s_p c_{p'} c \\ +\mathrm{i} s_p s \end{array} & 1+\hat{\boldsymbol{u}}_p\cdot\hat{\boldsymbol{u}}_{p'} & \begin{array}{c} -s_p s_{p'} \\ -(1+c_p c_{p'})c \\ +\mathrm{i}(c_p + c_{p'})s \end{array} & \begin{array}{c} -s_p c_{p'} \\ +c_p s_{p'} c \\ -\mathrm{i} s_{p'} s \end{array} \\
-\frac{1}{2},\frac{1}{2} \longrightarrow & \begin{array}{c} s_p c_{p'} \\ -c_p s_{p'} c \\ -\mathrm{i} s_{p'} s \end{array} & \begin{array}{c} -s_p s_{p'} \\ -(1+c_p c_{p'})c \\ -\mathrm{i}(c_p + c_{p'})s \end{array} & 1+\hat{\boldsymbol{u}}_p\cdot\hat{\boldsymbol{u}}_{p'} & \begin{array}{c} -c_p s_{p'} \\ +s_p c_{p'} c \\ +\mathrm{i} s_p s \end{array} \\
-\frac{1}{2},-\frac{1}{2} \longrightarrow & \begin{array}{c} -s_p s_{p'} \\ +(1-c_p c_{p'})c \\ +\mathrm{i}(c_p - c_{p'})s \end{array} & \begin{array}{c} -s_p c_{p'} \\ +c_p s_{p'} c \\ +\mathrm{i} s_{p'} s \end{array} & \begin{array}{c} -c_p s_{p'} \\ +s_p c_{p'} c \\ -\mathrm{i} s_p s \end{array} & 1-\hat{\boldsymbol{u}}_p\cdot\hat{\boldsymbol{u}}_{p'}
\end{array}
$$

$$\equiv\; 4\,\sigma_1(\hat{\boldsymbol{u}}_p;\,\hat{\boldsymbol{u}}_{p'})_{\mu_p,\mu_{p'};\mu'_p,\mu'_{p'}} \tag{5.14}$$

Here and elsewhere, unless specified otherwise, the definitions used are

$$\left.\begin{array}{ll} s \equiv \sin(\varphi_{p'} - \varphi_p), & c \equiv \cos(\varphi_{p'} - \varphi_p) \\ i \equiv \sqrt{(-1)}, & \hat{\boldsymbol{u}}_p\cdot\hat{\boldsymbol{u}}_{p'} = c_p c_{p'} + s_p s_{p'}\, c, \\ s_p \equiv \sin\vartheta_p, \quad s_{p'} \equiv \sin\vartheta_{p'}, & c_p \equiv \cos\vartheta_p, \quad c_{p'} \equiv \cos\vartheta_{p'}. \end{array}\right\} \tag{5.15}$$

The density matrix (5.14) represents the required spin state [in the basis $(|\frac{1}{2}\rangle\, |\frac{1}{2}\rangle)$, $(|\frac{1}{2}\rangle\, \text{-}|\frac{1}{2}\rangle)$, $(\text{-}|\frac{1}{2}\rangle\, |\frac{1}{2}\rangle)$, $(\text{-}|\frac{1}{2}\rangle\, \text{-}|\frac{1}{2}\rangle)$] of photoelectrons $(e_p,\, e_{p'})$ produced in the 1-DPI process (1.2) in L-S coupling with both the target atom $\mathfrak{T}$ and its residual photodication in their singlet electronic state. On confining [with $\hat{\boldsymbol{u}}_p$ $(\vartheta_p = \frac{\pi}{2},\, \varphi_p)$, $\hat{\boldsymbol{u}}_{p'}$ $(\vartheta_{p'} = \frac{\pi}{2},\, \varphi_{p'})$] the spin quantization directions of the photoelectrons $[e_p,\, e_{p'}]$ to the $X - Y$ plane in Fig. 5.1, the density matrix (5.14) is readily shown to reduce to that originally derived by Chandra and Chakraborty [252]. Matrix (5.14) is obviously Hermitian. The interesting thing about it is that its eigenvalues (EVs)

$$\mathrm{EVs}\left[\sigma^{(\text{1-DPI})}(S_0 = 0;\, S_{2+} = 0;\, \hat{\boldsymbol{u}}_p;\, \hat{\boldsymbol{u}}_{p'})\right] \;=\; (0,0,0,1) \tag{5.16}$$

are totally independent of the angles present in (5.14), and, hence, of the directions $(\hat{\boldsymbol{u}}_p,\, \hat{\boldsymbol{u}}_{p'})$ of quantization of the spins of the photoelectrons $(e_p,\, e_{p'})$ shown in the Fig. 5.1. According to the properties of a density matrix, discussed in the Appendix A to this monograph, (5.14) represents a normalized state of photoelectrons $(e_p,\, e_{p'})$, which is both pure and positive. The pure and normalized nature of the state (5.14) is verified from the condition (A.17a) as well.

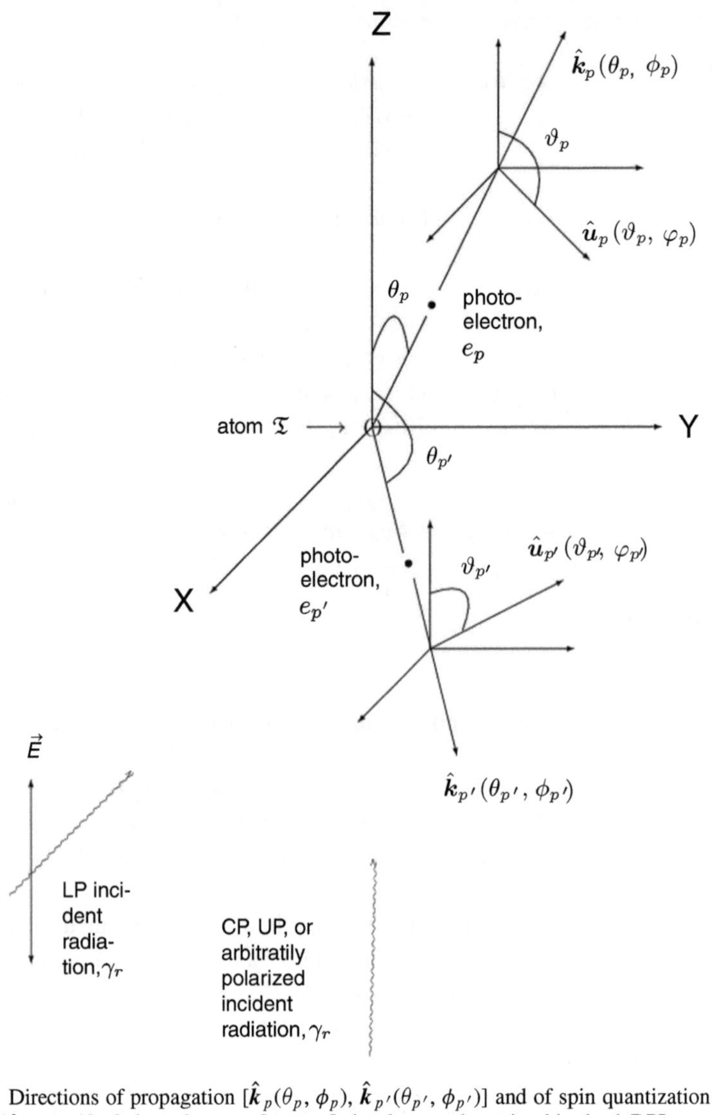

Fig. 5.1 Directions of propagation $[\hat{\boldsymbol{k}}_p(\theta_p, \phi_p), \hat{\boldsymbol{k}}_{p'}(\theta_{p'}, \phi_{p'})]$ and of spin quantization $[\hat{\boldsymbol{u}}_p(\vartheta_p, \varphi_p), \hat{\boldsymbol{u}}_{p'}(\vartheta_{p'}, \varphi_{p'})]$ of photoelectrons $[e_p, e_{p'}]$ simultaneously emitted in the 1-DPI process (1.2) taking place in an atomic target $\mathfrak{T}$ following the absorption of a single photon γ_r in the absence of the SOI. The orientation of the SF/LF coordinate system $OXYZ$ is identical to that shown in Fig. 1.1

In order to test the non-separability of the state (5.14) using the stringent Peres–Horodecki [124,125] condition, one needs to calculate its partial transpose (PT). The density matrix (5.13) can be transposed with respect to either of the photoelectrons e_p and $e_{p'}$ obtaining

$$\sigma^{T_p}(\hat{\boldsymbol{u}}_p;\,\hat{\boldsymbol{u}}_{p'})_{\mu_p,\,\mu_{p'};\,\mu'_p,\,\mu'_{p'}} \;=\; \sigma^{(1\text{-DPI})}(\hat{\boldsymbol{u}}_p;\hat{\boldsymbol{u}}_{p'})_{\mu'_p,\mu_{p'};\,\mu_p,\mu'_{p'}} \tag{5.17}$$

and

$$\sigma^{T_{p'}}(\hat{\boldsymbol{u}}_p;\,\hat{\boldsymbol{u}}_{p'})_{\mu_p,\,\mu_{p'};\,\mu'_p,\,\mu'_{p'}} \;=\; \sigma^{(1\text{-DPI})}(\hat{\boldsymbol{u}}_p;\hat{\boldsymbol{u}}_{p'})_{\mu_p,\mu'_{p'};\,\mu'_p,\mu_{p'}} \tag{5.18}$$

Here, (5.17) and (5.18) are obtained by transposing the density matrix (5.13) with respect to the indices $(\mu_p,\ \mu'_p)$ of e_p and $(\mu_{p'},\ \mu'_{p'})$ of $e_{p'}$, respectively. Using this procedure, one readily finds the following partial transpose of the state (5.14).

$$4\,\sigma^{T_{p'}}(S_0=0;\ S_{2+}=0;\ \hat{\boldsymbol{u}}_p;\,\hat{\boldsymbol{u}}_{p'})_{\mu_p,\,\mu_{p'};\,\mu'_p,\,\mu'_{p'}} =$$

$$\begin{array}{lcccc}
\mu_p\mu_{p'}/\mu'_p\mu'_{p'} \Longrightarrow & \frac{1}{2},\frac{1}{2} & \frac{1}{2},-\frac{1}{2} & -\frac{1}{2},\frac{1}{2} & -\frac{1}{2},-\frac{1}{2} \\
\Downarrow & & & & \\
\frac{1}{2},\frac{1}{2} \longrightarrow & 1-\hat{\boldsymbol{u}}_p\cdot\hat{\boldsymbol{u}}_{p'} & \begin{array}{c} c_p s_{p'} \\ -s_p c_{p'} c \\ +\mathrm{i} s_p s \end{array} & \begin{array}{c} s_p c_{p'} \\ -c_p s_{p'} c \\ +\mathrm{i} s_{p'} s \end{array} & \begin{array}{c} -s_p s_{p'} \\ -(1+c_p c_{p'})c \\ +\mathrm{i}(c_p+c_{p'})s \end{array} \\
\frac{1}{2},-\frac{1}{2} \longrightarrow & \begin{array}{c} c_p s_{p'} \\ -s_p c_{p'} c \\ -\mathrm{i} s_p s \end{array} & 1+\hat{\boldsymbol{u}}_p\cdot\hat{\boldsymbol{u}}_{p'} & \begin{array}{c} -s_p s_{p'} \\ +(1-c_p c_{p'})c \\ -\mathrm{i}(c_p-c_{p'})s \end{array} & \begin{array}{c} -s_p c_{p'} \\ +c_p s_{p'} c \\ -\mathrm{i} s_{p'} s \end{array} \\
-\frac{1}{2},\frac{1}{2} \longrightarrow & \begin{array}{c} s_p c_{p'} \\ -c_p s_{p'} c \\ -\mathrm{i} s_{p'} s \end{array} & \begin{array}{c} -s_p s_{p'} \\ +(1-c_p c_{p'})c \\ +\mathrm{i}(c_p-c_{p'})s \end{array} & 1+\hat{\boldsymbol{u}}_p\cdot\hat{\boldsymbol{u}}_{p'} & \begin{array}{c} -c_p s_{p'} \\ +s_p c_{p'} c \\ -\mathrm{i} s_p s \end{array} \\
-\frac{1}{2},-\frac{1}{2} \longrightarrow & \begin{array}{c} -s_p s_{p'} \\ -(1+c_p c_{p'})c \\ -\mathrm{i}(c_p+c_{p'})s \end{array} & \begin{array}{c} -s_p c_{p'} \\ +c_p s_{p'} c \\ +\mathrm{i} s_{p'} s \end{array} & \begin{array}{c} -c_p s_{p'} \\ +s_p c_{p'} c \\ +\mathrm{i} s_p s \end{array} & 1-\hat{\boldsymbol{u}}_p\cdot\hat{\boldsymbol{u}}_{p'}
\end{array}$$

$$\equiv 4\,\sigma_1^{T_{p'}}(\hat{\boldsymbol{u}}_p;\,\hat{\boldsymbol{u}}_{p'})_{\mu_p,\,\mu_{p'};\,\mu'_p,\,\mu'_{p'}} \tag{5.19}$$

The EVs of (5.19) are

$$\mathrm{EVs}\left[\sigma_1^{T_{p'}}(\hat{\boldsymbol{u}}_p;\,\hat{\boldsymbol{u}}_{p'})_{\mu_p,\,\mu_{p'};\,\mu'_p,\,\mu'_{p'}}\right] = \left(\frac{1}{2},\,\frac{1}{2},\,\frac{1}{2},\,-\frac{1}{2}\right). \tag{5.20}$$

Similar to those given in (5.16) for the density matrix (5.14), the EVs of its PT (5.19) too are totally independent of all the angles specifying directions $(\hat{\boldsymbol{u}}_p,\ \hat{\boldsymbol{u}}_{p'})$ of spin quantization of photoelectrons $(e_p,\, e_{p'})$.

Had one taken partial transpose of the density matrix (5.14) with respect to e_p in place of $e_{p'}$, following the prescription specified in (5.17), though the state $\sigma_1^{T_p}$ would have probably been different from that given in (5.19), but the eigenvalues would have remained unchanged, i.e., identical to (5.20).

According to the Peres–Horodecki condition [124, 125], (5.14) represents a non-separable (i.e., entangled) state of $(e_p, e_{p'})$. Furthermore, each of the three measures of entanglement, namely negativity $\mathcal{N}\big(\sigma_1(\hat{\boldsymbol{u}}_p;\, \hat{\boldsymbol{u}}_{p'})\big)$, concurrence $\mathcal{C}\,\big(\sigma_1(\hat{\boldsymbol{u}}_p;\, \hat{\boldsymbol{u}}_{p'})\big)$, and entanglement of formation $\mathcal{E}_F\big(\sigma_1(\hat{\boldsymbol{u}}_p;\, \hat{\boldsymbol{u}}_{p'})\big)$, defined by the respective (2.45c), (2.46), and (2.47), have unit value for the state (5.14). Taking into account also the EVs (5.16) and the condition (A.16b), $(e_p, e_{p'})$ in the present case are in a pure and maximally entangled state.

Thus, two photoelectrons simultaneously ejected in 1-DPI process (1.2) in the L-S coupling are always in a pure state, which possesses maximum entanglement if each of the target atom $\mathfrak{T}$ and its residual photodication $\mathfrak{T}^{2+}$ is in its singlet electronic state. This entanglement is generated purely by the Coulomb forces experienced inside $\mathfrak{T}$ by the two photoelectrons, before their emission. It can, therefore, be appropriately called as "Coulombic entanglement." Moreover, these properties of the spin state (5.14) are totally independent of the characteristics (e.g., energy, polarization) of the ionizing radiation γ_r as well as of the kinematics and dynamics of any 1-DPI process (1.2). In addition, the Coulombic entanglement generated in the present case is totally uncoupled to the environment outside its parent atom $\mathfrak{T}$. It is, therefore, immune to the harmful effects of decoherence and dissipation as the photoelectrons $(e_p, e_{p'})$ recede in directions $(\hat{\boldsymbol{k}}_p, \hat{\boldsymbol{k}}_{p'})$ from each other and from the photo-dication residue $\mathfrak{T}^{2+}$ with speeds allowed by the energy conservation condition (5.1a)

Furthermore, the reduced density matrix, defined in (A.24), is readily calculated for the state (5.14). One finds

$$\mathrm{Tr}_p\left[\sigma_1(\hat{\boldsymbol{u}}_p;\, \hat{\boldsymbol{u}}_{p'})\right] = \mathrm{Tr}_{p'}\left[\sigma_1(\hat{\boldsymbol{u}}_p;\, \hat{\boldsymbol{u}}_{p'})\right] = \begin{pmatrix} \frac{1}{2} & 0 \\ 0 & \frac{1}{2} \end{pmatrix} \tag{5.21}$$

with respect to either of the photoelectrons e_p and $e_{p'}$. This gives a degree of non-separability (participation ratio or Schmidt number) [260], defined only for a pure state, $K = [(1/2)^2 + (1/2)^2]^{-1} = 2$. This means [260], pure state represented by the density matrix (5.14) is a Bell state. In order to find out as to which of the four Bell states (2.17) the two photoelectrons $(e_p, e_{p'})$ in (5.14) belong to, we specialize the directions $(\boldsymbol{u}_p, \boldsymbol{u}_{p'})$ of their spin quantization to those of the Bell states, i.e., to the OZ-axis of the space frame in Fig. 5.1 by taking $\vartheta_p = \vartheta_{p'} = 0$. This immediately reduces the density matrix (5.14) to (A.30b), which corresponds to the singlet state (2.17b) of two spin-$\frac{1}{2}$ particles. Moreover, on coinciding $\hat{\boldsymbol{u}}_{p'}$ with OZ-axis in Fig. 5.1, the DMs (5.14) and (5.19) reduce to those given in (4.13) and (4.14), respectively, for obvious reasons.

For (5.14) to represent a pure, maximally entangled, normalized, singlet spin state of $(e_p, e_{p'})$ is, probably, an obvious result. In the present case, each of $\mathfrak{T}$ and

$\mathfrak{T}^{2+}$ has $S_0 = S_{2+} = 0$. Then according to the conservation condition (5.1c), the total spin of two photoelectrons $(e_p, e_{p'})$—in the absence of any spin-dependent interactions—$s_{p_t} = 0$ as well. Consequently, two such electrons are always in their singlet spin state, i.e., the Bell state (2.17b), for all values of the total orbital angular momenta (L_0, L_{2+}) of the electronic states $(|0\rangle, |2^+\rangle)$ of $(\mathfrak{T}, \mathfrak{T}^{2+})$ in L-S coupling, any frequency (ν_r) and polarization (m_r) of the ionizing radiation γ_r, as well as for the linear momenta $(\boldsymbol{k}_p, \boldsymbol{k}_{p'})$ and directions $(\hat{\boldsymbol{u}}_p, \hat{\boldsymbol{u}}_{p'})$ of the spin quantization of the photoelectrons $(e_p, e_{p'})$.

The best known and, probably, the simplest and most widely—both theoretically as well as experimentally—studied example of this result is 1-DPI of atomic He in its ground electronic state ^{1}S. The two photoelectrons $(e_p, e_{p'})$ which together leave the atom in this process are both correlated and entangled in the Bell state (2.17b). Some of the other simple examples of the results discussed in the present section are for $[\mathfrak{T}, \mathfrak{T}^{2+}]$ being [Be($1s^2 2s^2\ {}^1S$), Be^{2+} ($1s^2 2s^0\ {}^1S$)], [C($1s^2 2s^2 2p^2\ {}^1S/{}^1D$), C^{2+}($1s^2 2s^2 2p^0\ {}^1S$)], *etc.*

Thus, $\sigma_1(\hat{\boldsymbol{u}}_p;\, \hat{\boldsymbol{u}}_{p'})$ in (5.14) represents a maximally entangled, pure state of two qubits. For two spin-$\frac{1}{2}$ particles, it is their singlet state with spins quantized in the directions $\hat{\boldsymbol{u}}_p$ $(\vartheta_p,\, \varphi_p)$ and $\hat{\boldsymbol{u}}_{p'}$ $(\vartheta_{p'},\, \varphi_{p'})$. Hence, $\sigma_1(\hat{\boldsymbol{u}}_p;\, \hat{\boldsymbol{u}}_{p'})$ in (5.14) is the generalized form of the Bell state $|\psi^-\rangle$ given by (2.17b), or (A.30b). The examples discussed herein show as to how such states can readily be generated in the 1-DPI (1.2) of an atomic target $\mathfrak{T}$ in a laboratory. Thus, $\sigma_1(\hat{\boldsymbol{u}}_p;\, \hat{\boldsymbol{u}}_{p'})$ is one (i.e., pure and maximally entangled σ_1) of those two (i.e., σ_1, σ_0) states which form a Werner [121] state ρ_{W} in (2.28). The respective (5.16) and (5.20) are the EVs of $\sigma_1(\hat{\boldsymbol{u}}_p;\, \hat{\boldsymbol{u}}_{p'})$ and of its partial transpose (5.19).

The other part of ρ_{W} in (2.28) is the maximally chaotic [22] (also called white noise [122]) state σ_0. It is represented by the constant matrix (2.28b) with each of its diagonal elements equal to 1/4 and non-diagonal elements zero. This, in other words, means

$$\sigma_0 = \sigma_0^{T_p} = \sigma_0^{T_{p'}} = \frac{1}{4}\mathbb{I} \tag{5.22a}$$

and

$$\mathrm{EVs}[\sigma_0] = \mathrm{EVs}[\sigma_0^{T_p}] = \mathrm{EVs}[\sigma_0^{T_{p'}}] = \left(\frac{1}{4}, \frac{1}{4}, \frac{1}{4}, \frac{1}{4}\right). \tag{5.22b}$$

That is, the partial transpose of σ_0 is this state itself and each of the four eigenvalues of σ_0, or of its partial transpose, is equal to 1/4.

For further use in this monograph, we define, following (2.28), a Werner [121] state in terms of the states (5.14) and (5.22a) as follows

$$\sigma_{\mathrm{w}}(\hat{\boldsymbol{u}}_p;\, \hat{\boldsymbol{u}}_{p'}) = p\,\sigma_1(\hat{\boldsymbol{u}}_p;\, \hat{\boldsymbol{u}}_{p'}) + (1-p)\,\sigma_0; \tag{5.23}$$

whereas

$$\sigma_{\mathrm{W}}^{T_p}(\hat{\boldsymbol{u}}_p; \hat{\boldsymbol{u}}_{p'}) = p\,\sigma_1^{T_p}(\hat{\boldsymbol{u}}_p; \hat{\boldsymbol{u}}_{p'}) + (1-p)\,\sigma_0 \tag{5.24a}$$

is its partial transpose with respect to the first particle (i.e., photoelectron e_p) and

$$\sigma_{\mathrm{W}}^{T_{p'}}(\hat{\boldsymbol{u}}_p; \hat{\boldsymbol{u}}_{p'}) = p\,\sigma_1^{T_{p'}}(\hat{\boldsymbol{u}}_p; \hat{\boldsymbol{u}}_{p'}) + (1-p)\,\sigma_0 \tag{5.24b}$$

with respect to the second particle (i.e., photoelectron $e_{p'}$). Here, while state $\sigma_1^{T_{p'}}(\hat{\boldsymbol{u}}_p; \hat{\boldsymbol{u}}_{p'})$ in (5.24b) is nothing but (5.19); $\sigma_1^{T_p}(\hat{\boldsymbol{u}}_p; \hat{\boldsymbol{u}}_{p'})$ for (5.24a) can, on the other hand, be gotten by applying the procedure (5.17) to (5.14) [or, equivalently, by taking the full transpose of (5.19)]. The EVs [188]

$$\frac{1}{4}(1-p),\ \frac{1}{4}(1-p),\ \frac{1}{4}(1-p),\ \frac{1}{4}(1+3p) \tag{5.25}$$

of the Werner state (5.23) as well as

$$\frac{1}{4}(1+p),\ \frac{1}{4}(1+p),\ \frac{1}{4}(1+p),\ \frac{1}{4}(1-3p) \tag{5.26}$$

of its partial transpose (5.24) are, as usual, independent of all the angles [$\hat{\boldsymbol{u}}_p$ (ϑ_p, φ_p), $\hat{\boldsymbol{u}}_{p'}$ ($\vartheta_{p'}$, $\varphi_{p'}$)] but they do, of course, depend upon the mixing parameter p. The requirement that none of the eigenvalues (5.25) of a physically acceptable state can be negative suggests that

$$-\frac{1}{3} \le p \le 1 \tag{5.27a}$$

for (5.23). A straightforward application of the Pares–Horodecki [124, 125] condition to the eigenvalues (5.26) of the partial transpose (5.24) for the non-separability of the state (5.23), on the other hand, requires that

$$\frac{1}{3} < p \le 1. \tag{5.27b}$$

For a state of two qubits to be qualified to be a Werner [121] state, the mixing parameter in (5.23) must never be negative, i.e., $0 < p < 1$. Thus, whenever, $p < 0$, (5.23) ceases to represent a Werner state. We will see in this monograph that it indeed is possible in a real experiment in a laboratory to generate bipartite states (5.23) of electronic qubits with different values of the mixing parameter, including $p < 0$, $p = 0$, 1, *etc.*

The amount of entanglement present in the state (5.23) is quantified by calculating any one or more of the three measures given on pages 44–46. In order to calculate the negativity [142, 159, 160] of this bipartite state, one needs to substitute in (2.45c) only those eigenvalues of its partial transpose (5.24a) [or, (5.24b)], which are negative. Out of the four eigenvalues (5.26) of a partial transpose of (5.23), only the last one can be less than zero (for $p > 1/3$), while the remaining three can never become negative for the whole range ($-1/3 \leq p \leq 1$) of the allowed values of the mixing parameter. Substitution of the eigenvalue $(1 - 3p)/4$ in (2.45c), therefore, immediately determines negativity of the state (5.23).

Concurrence [149, 163] is another measure of entanglement of a bipartite state of qubits. For an arbitrary (i.e., mixed/pure) state of two qubits, concurrence is obtained from (2.46) derived by Wootters [149, 163]. Although, in general, negativity and concurrence may differ for a mixed state of two qubits, these two measures of entanglement, nevertheless, have same values if the state of two qubits happens to be pure. But, for the bipartite state (5.23), both negativity and concurrence were found to have identical values given by [188]

$$\mathcal{N}\Big(\sigma_w(\hat{\boldsymbol{u}}_p;\, \hat{\boldsymbol{u}}_{p'})\Big) = \mathcal{C}\Big(\sigma_w(\hat{\boldsymbol{u}}_p;\, \hat{\boldsymbol{u}}_{p'})\Big) = \max\left(0,\ \frac{3p-1}{2}\right). \tag{5.28}$$

The third measure of non-local correlation in a state of two qubits is the entanglement of formation $\mathcal{E}_F$ [161–163]. For an arbitrary state of two qubits, $\mathcal{E}_F$ is given by (2.47). Unlike the other two measures (i.e., negativity and concurrence), the $\mathcal{E}_F$ in (2.47) cannot be expressed in an analytical form even for the state (5.23) of two qubits.

5.2.2 $|S_0 - S_{2+}| = 1$

It includes three of the remaining four of the total five possibilities in which (5.13) [i.e., (5.11)] does not become a null matrix. These are: ($S_0 = 0, S_{2+} = 1$); or ($S_0 = 1, S_{2+} = 0$); or $S_{2+} = \left|S_0 \pm 1\right|$ with $S_0 \geq 1/2$, excluding both S_0 and S_{2+} to be simultaneously 1/2. These, in other words, mean that the electronic states ($|0\rangle$, $|2^+\rangle$) of ($\mathfrak{T}$, $\mathfrak{T}^{2+}$) in the 1-DPI process (1.2) are (singlet, triplet), or (triplet, singlet), or their spin multiplicities differ by two while neither of them has spin zero. Equation (5.1c) now imposes the condition that the total spin angular momentum s_{p_t} of (e_p, $e_{p'}$) can only be unity. On taking the triad (S_0, S_{2+}, s_{p_t}) to be (0, 1, 1), or (1, 0, 1), or (1/2, 3/2, 1), or ($S_0 \geq 1/2$, $S_{2+} = \left|S_0 \pm 1\right|$, 1) with $S_0 \neq S_{2+}$ in (5.11), the density matrix (5.13) was found [252] to be identical in each of these cases. It can beautifully be written in terms of the state (5.23) in the following form [252]

$$\sigma^{\text{(1-DPI)}}(S_0;\, S_{2+};\, \hat{\boldsymbol{u}}_p;\, \hat{\boldsymbol{u}}_{p'})\Big|_{\left|S_0 - S_{2+}\right|=1} = -\frac{1}{3}\,\sigma_1(\hat{\boldsymbol{u}}_p,\, \hat{\boldsymbol{u}}_{p'}) + \frac{4}{3}\,\sigma_0. \tag{5.29}$$

Similarly, its partial transpose with respect to the photoelectron e_p and $e_{p'}$ are given by

$$\sigma^{(1\text{-DPI})}(S_0;\ S_{2+};\ \hat{\boldsymbol{u}}_p;\ \hat{\boldsymbol{u}}_{p'})\Bigg|^{(\mathrm{PT})_{e_p}}_{|S_0-S_{2+}|=1} = -\frac{1}{3}\,\sigma_1^{T_p}(\hat{\boldsymbol{u}}_p;\ \hat{\boldsymbol{u}}_{p'}) + \frac{4}{3}\,\sigma_0 \qquad (5.30a)$$

and

$$\sigma^{(1\text{-DPI})}(S_0;\ S_{2+};\ \hat{\boldsymbol{u}}_p;\ \hat{\boldsymbol{u}}_{p'})\Bigg|^{(\mathrm{PT})_{e_{p'}}}_{|S_0-S_{2+}|=1} = -\frac{1}{3}\,\sigma_1^{T_{p'}}(\hat{\boldsymbol{u}}_p;\ \hat{\boldsymbol{u}}_{p'}) + \frac{4}{3}\,\sigma_0, \qquad (5.30b)$$

respectively.

With these simple relationships, one needs diagonalize none of (5.29) and (5.30) for calculating their eigenvalues. These are readily obtained either by using those given in [(5.16), (5.22b)] and [(5.20), (5.22b)] in the respective relationships (5.29) and in either of (5.30); or, more simply, by substituting the mixing parameter $p = -1/3$ in the eigenvalues (5.25) and (5.26). One, subsequently, finds

$$\left(\frac{1}{3},\ \frac{1}{3},\ \frac{1}{3},\ 0\right) \qquad (5.31a)$$

and

$$\left(\frac{1}{2},\ \frac{1}{6},\ \frac{1}{6},\ \frac{1}{6}\right) \qquad (5.31b)$$

for the eigenvalues of the state (5.29) and of its partial transpose (5.30), respectively.

Eigenvalues (5.31) immediately suggest that, unlike in the previous Sect. 5.2.1, photoelectrons $(e_p,\ e_{p'})$ in the present case are in a mixed spin state. Moreover, according to the Peres [124]–Horodecki [125] condition, these two electrons are presently not entangled as well. Thus, the photoelectrons produced in the 1-DPI process (1.2) with $|S_0 - S_{2+}| = 1$ in the L-S coupling and in the absence of SOI, are always in a mixed separable state. The twin electrons $(e_p,\ e_{p'})$ were integral parts of the same atom and were emitted together—in a process carried out by a one-body operator—because of their correlated motions inside $\mathfrak{T}$. Moreover, they share between themselves energy of the absorbed photon γ_r in excess to the potential for double ionization of $\mathfrak{T}$ and continue to interact with each other and with the residual photo-dication $\mathfrak{T}^{2+}$ even after leaving their common parent. However, $(e_p,\ e_{p'})$ in the present case are neither in a pure nor in an entangled spin-state. Thus, the electron–electron correlation, without which $(e_p,\ e_{p'})$ could not have simultaneously left the atomic target $\mathfrak{T}$ following the absorption of a single photon, is not capable of entangling these two electrons in the present case.

It has elsewhere been pointed out in this section that total spin $s_{p_t} = 1$ of $(e_p, e_{p'})$ in the present case. Thus, the state (5.29) of the two freely moving photoelectrons is a mixed, separable superposition of the three spin states given in (2.17a) and (2.18). Furthermore, the mixing parameter p in (5.29) in the present case is negative. Hence, the density matrix (5.29), describing the spin-correlation between $(e_p, e_{p'})$ in the present case, does not qualify to represent a Werner [121] state either.

The results presented herein are completely general, independent of any dynamical models one may like to use in an experimental or theoretical study of the 1-DPI process (1.2) in L-S coupling [10, 60] in the absence of SOI in the bound electronic states $(|0\rangle, |2^+\rangle)$ of $(\mathfrak{T}, \mathfrak{T}^{2+})$ as well as in the continua of $(e_p, e_{p'})$. These two photoelectrons are always correlated, but never entangled with respect to their spin angular momenta and, hence, do not form an EPR pair of electronic qubits.

The two lightest possible atoms in which 1-DPI can take place according to the present requirements are $C(1s^2 2s^2 2p^2\,{}^3P)$ and $N(1s^2 2s^2 2p^3\,{}^4S)$. After the simultaneous emission in 1-DPI of two electrons from, say, the outermost sub-shells of each of these two atoms, the photo-dications formed are $C^{2+}(1s^2 2s^2 2p^0\,{}^1S)$ and $N^{2+}(1s^2 2s^2 2p^1\,{}^2P)$, respectively. The two ejected p-electrons were certainly correlated within their parent atom; otherwise, they could have left together neither a C nor an N atom following the absorption of a single photon. But their spin states for each of these two atoms is mixed and separable. $C(1s^2 2s^2 2p^2\,{}^3P)$ and $N(1s^2 2s^2 2p^3\,{}^4S)$ atoms are two of the many possible examples wherein pairs of correlated electrons are produced in a separable spin state (5.29), provided 1-DPI takes place within the conditions discussed in this section.

5.2.3 $S_0 = S_{2^+} \geq \frac{1}{2}$

When spins of each of the electronic states $(|0\rangle, |2^+\rangle)$ of $(\mathfrak{T}, \mathfrak{T}^{2+})$ are equal and neither of them zero, then, according to the conservation condition (5.1c), the total spin of photoelectrons $(e_p, e_{p'})$ can simultaneously be $s_{p_t} = 0$ and 1. In the calculation of the density matrix (5.13), one needs to sum over both of these allowed values of s_{p_t} in the expression (5.11). On taking $S_0 = S_{2^+}$ in (5.11), the density matrix for the spin-state of the photoelectrons $(e_p, e_{p'})$, simultaneously emitted in the 1-DPI (1.2), is now given by [252]

$$\sigma^{\text{(1-DPI)}}(S_0;\, S_{2^+};\, \hat{\boldsymbol{u}}_p;\, \hat{\boldsymbol{u}}_{p'})\Big|_{S_0 \,=\, S_{2^+} \,\geq\, \frac{1}{2}} = \frac{1}{3}\,\sigma_1(\hat{\boldsymbol{u}}_p;\, \hat{\boldsymbol{u}}_{p'}) + \frac{2}{3}\,\sigma_0; \tag{5.32a}$$

whereas its partial transpose with respect to the photoelectron, say, $e_{p'}$ is

$$\sigma^{(1\text{-DPI})}(S_0;\, S_{2+};\, \hat{\boldsymbol{u}}_p;\, \hat{\boldsymbol{u}}_{p'})\Big|^{(\mathrm{PT})_{e_{p'}}}_{S_0 = S_{2+} \geq \frac{1}{2}} = \frac{1}{3}\sigma_1^{T_{p'}}(\hat{\boldsymbol{u}}_p;\, \hat{\boldsymbol{u}}_{p'}) + \frac{2}{3}\sigma_0. \tag{5.32b}$$

The eigenvalues

$$\left(\frac{1}{6},\, \frac{1}{6},\, \frac{1}{6},\, \frac{1}{2}\right) \tag{5.33a}$$

and

$$\left(0,\, \frac{1}{3},\, \frac{1}{3},\, \frac{1}{3}\right). \tag{5.33b}$$

of (5.32) are readily obtained using the procedure already described herein. Thus, none of the expressions (5.32) and (5.33) depends upon the spins $(S_0,\ S_{2+})$ of $(\mathfrak{T},\ \mathfrak{T}^{2+})$, participating in a 1-DPI processes (1.2), provided electronic states $(|0\rangle,\ |2^+\rangle)$ of both of these species have spin multiplicities equal and more than one.

From the eigenvalues (5.33), one readily concludes that the bipartite state (5.32a) of $(e_p,\ e_{p'})$ is not only mixed, but also separable. The correlation among the constituents of $\mathfrak{T}$, responsible for the simultaneous emission of $(e_p,\ e_{p'})$, is not sufficient for entangling these two electrons. In the present case, two simultaneously ejected electrons never form an EPR pair of particles. However, (5.32a) qualifies to be a Werner state. In the generation of the state (5.32a), both of the allowed values (i.e., $s_{p_t} = 0$ and 1) of the total spins of two photoelectrons $(e_p,\ e_{p'})$ contribute to (5.11). Hence, (5.32a) is a separable linear superposition of one singlet and three triplet spin states of $(e_p,\ e_{p'})$.

The two simplest possible experiments, meeting the present requirements, are 1-DPI of B $(1s^2 2s^2 2p^1\ {}^2P)$ and of N $(1s^2 2s^2 2p^3\ {}^2P/{}^2D)$. After the simultaneous emission of a pair of photoelectrons from each of these atoms, their photo-dications are left in the states B^{2+} $(1s^2 2s^1 2p^0\ {}^2S)$ and N^{2+} $(1s^2 2s^2 2p^1\ {}^2P)$, respectively. The twin photoelectrons from each atom is in the Werner state (5.32a) which, with a mixing parameter $p = 1/3$, is mixed and separable.

In the absence of SOI, bound electrons in an atom are correlated with each other due only to the inter-electronic Coulomb interaction. This electron–electron correlation causes simultaneous emission of two of the constituents from a single bound electronic state of their parent atom in 1-DPI. The analysis presented in this chapter shows that this Coulombic, inter-electronic correlation is capable of entangling the two emitted electrons when both the parent atom and its residual photo-dication are in their singlet electronic states. For other possible electronic states of these two participating species, the twin electrons—although correlated—are never entangled. These results are totally independent of the kinematics and dynamics of the “Shake-off” [71, 253] and “knock-out” [254, 255] mechanisms hitherto proposed for explaining this process.

Thus, the electron–electron correlation, in the context of quantum information, means only those effects in a many-body interacting system which are not present in an independent-particle model of an atom but are necessary for the simultaneous emission of two electrons in a process performed by a one-body operator which, in the present case, is the absorption of a single photon. Moreover, as the properties of the spin-state—entangled or otherwise—of the twin photoelectrons are due completely to the Coulombic interactions experienced by them inside an atom, they are in no way coupled to the environment outside their common parent. These properties are, therefore, least prone to the harmful effects of dissipation and decoherence when the two emitted particles recede from the residual photo-dication.

Chapter 6
Coulombic Entanglement: Two-Step Double Photoionization of Atoms

Unlike in the previous Chap. 5, an atom or a molecule may lose two of its electrons in two consecutive steps as well (e.g., Fig. 3.1). While its primary ionization (i.e., loss of first electron) is caused from the energy supplied by some external source, the secondary ionization (i.e., emission of second electron) takes place in the rearrangement of electrons resulting from the primary ionization. Such a rearrangement of electrons becomes necessary whenever an inner-shell electron leaves the target in latter's primary ionization. However, it is equally possible that such a rearrangement may lead to the emission of a photon in place of an electron (e.g., Fig. 3.2). This loss of energy in the form of a photon by, or departure of an electron from, an atom or a molecule due to rearrangement of its constituents is known as spontaneous radiative or non-radiative decay, respectively. The non-radiative decay was first observed by Auger [73, 74, 175, 176, 261]. The secondary ionization is, therefore, also known as Auger decay and the emitted particle called an Auger electron, say, e_a. In this monograph, we will consider the primary ionization due only to the absorption of a single photon and, hence, the emitted particle will be known as a photoelectron e_p (say).

There are many important physical differences between 1-DPI (1.2) and the present two-step process (1.3), although both are initiated due to the absorption of a single photon by the target $\mathfrak{T}$ and end up with the formation of the same residue, i.e., dication $\mathfrak{T}^{2+}$. [However, it is not necessary that the electronic states $|2^+\rangle$ of $\mathfrak{T}^{2+}$ generated in (1.2) and (1.3) are identical.] Unlike in 1-DPI, electrons (e_p, e_a) in 2-DPI are emitted from two different electronic states $(|0\rangle, |1^{+^*}\rangle)$ belonging to $(\mathfrak{T}, \mathfrak{T}^{1+*})$. Here, $\mathfrak{T}^{1+^*}$ is the excited photoion formed after primary ionization of $\mathfrak{T}$, i.e., after the departure of e_p from an inner-shell of $\mathfrak{T}$ in first [i.e., (1.3a)] of the two steps of the process (1.3). Consequently, (e_p, e_a) belonged to the electronic configurations of their parent species $(\mathfrak{T}, \mathfrak{T}^{1+^*})$ and, hence, can never be *equivalent* electrons [10, 60]. The process (1.3) of sequential emission of two electrons has, therefore, been rightly called [188] as two-step double photoionization (2-DPI).

The simultaneous emission of $(e_p, e_{p'})$ in 1-DPI is caused by a single one-body interaction (3.6) assisted [257], of course, by the mechanisms of shake-off [71, 253]

N. Chandra and R. Ghosh, *Quantum Entanglement in Electron Optics*, Springer Series on Atomic, Optical, and Plasma Physics 67, DOI 10.1007/978-3-642-24070-6_6,

and/or knock-out [254, 255]. In 2-DPI, on the other hand, emission of the second (i.e., Auger) electron takes place due to the two-body, inter-electronic Coulomb interaction (3.8). In addition, while a single condition (5.1a) determines energy of both of the electrons $(e_p, e_{p'})$ in 1-DPI, two different conservation rules are applicable in 2-DPI. Namely, energy of the photoelectron e_p is given by

$$\epsilon_{\mathrm{p}} = h\nu_{\mathrm{r}} - (E_{1^{+*}} - E_0); \tag{6.1a}$$

whereas

$$\epsilon_{\mathrm{a}} = E_{1^{+*}} - E_{2^{+}} \tag{6.1b}$$

determines that of the Auger electron e_a. Here, E_{2^+} is the energy of the electronic state $|2^+\rangle$ of the dication $\mathfrak{T}$ formed after the departure of e_a from $\mathfrak{T}^{1+^*}$. Thus, in the present case, it is only the energy of e_p which varies with that of the ionizing radiation in (1.3a); but, ϵ_{a} is completely independent of ν_{r} and remains same until energy of at least one of the states $(|1^{+^*}\rangle, |2^+\rangle)$ of $(\mathfrak{T}^{1+^*}, \mathfrak{T}^{2+})$, participating in the non-radiative decay (1.3), changes.

We will see in this chapter that these and some other differences between the processes (1.2) and (1.3) of 1-DPI and 2-DPI, respectively, caused due only to the participation of the intermediate state $|1^{+^*}\rangle$ of $\mathfrak{T}^{1+^*}$ in the later, significantly change the entanglement properties of the spin state of (e_p, e_a) compared to those of the state of $(e_p, e_{p'})$ studied in the previous Chap. 5. Here, with suitably chosen electronic states $(|0\rangle, |1^{+^*}\rangle, |2^+\rangle)$ of $(\mathfrak{T}, \mathfrak{T}^{1+^*}, \mathfrak{T}^{2+})$, participating in a 2-DPI process (1.3), one can generate almost any kind of bipartite state of electronic qubits. In the present analysis, no SDIs are taken into account. The study of the entanglement properties of a spin state of (e_p, e_a) on the inclusion of SOI are differed to Sect. 7.3 on pages 169–178.

The space-fixed coordinate system $OXYZ$ in Fig. 6.1 shows, among other things, propagation directions $(\hat{\boldsymbol{k}}_p, \hat{\boldsymbol{k}}_a)$ and spin quantization directions $(\hat{\boldsymbol{u}}_p, \hat{\boldsymbol{u}}_a)$ for electrons (e_p, e_a).

6.1 Density Matrix

6.1.1 Derivation

Equations [(3.19a), (3.20a)] and [(3.19b), (3.20b)] are the conditions for the conservation of the total [orbital, spin] angular momenta applicable to each of the [photoionization, Auger decay] parts [(1.3a), (1.3b)] of the 2-DPI process (1.3) taking place without SOI in the L-S coupling. Equations (3.12) and (3.15) are the desired expressions for the respective density operator and density matrix for both steps of (1.3). Explicitly specifying all the sums present in (3.15), the desired expression for the density matrix becomes

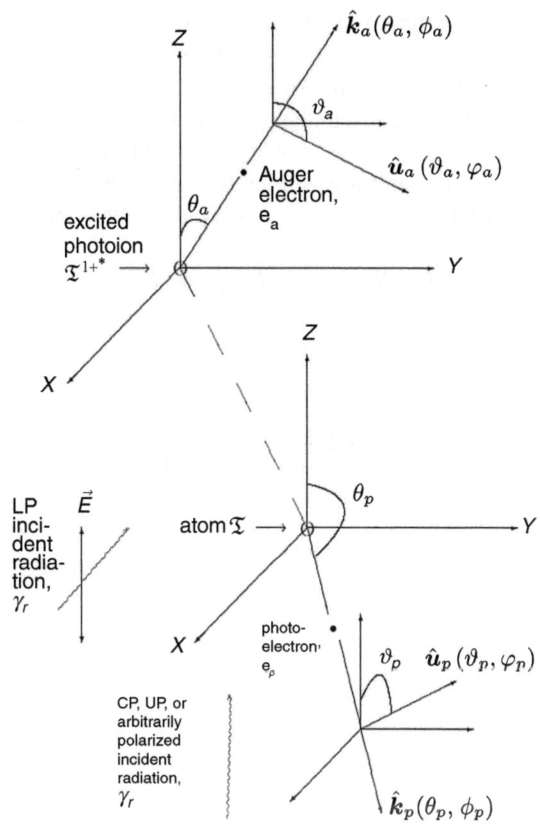

Fig. 6.1 Directions of propagation [$\hat{\boldsymbol{k}}_p(\theta_p, \phi_p)$, $\hat{\boldsymbol{k}}_a(\theta_a, \phi_a)$] and of spin quantization [$\hat{\boldsymbol{u}}_p\,(\vartheta_p, \varphi_p)$, $\hat{\boldsymbol{u}}_a\,(\vartheta_a, \varphi_a)$] of [photoelectron e_p, Auger electron e_a] sequentially emitted from the atom $\mathfrak{T}$ and from its excited photoion $\mathfrak{T}^{1+^*}$—placed at the origin of our co-ordinate system $OXYZ$—in the 2-DPI process (1.3) following the absorption of a single photon γ_r by the target (i.e., $\mathfrak{T}$) in the absence of the SOI. The orientation of the SF/LF coordinate system $OXYZ$ is identical to that shown in Fig. 1.1

$$
\begin{aligned}
\langle \mu_p\, \hat{\boldsymbol{u}}_p\, \boldsymbol{k}_p;\ \mu_a\, \hat{\boldsymbol{u}}_a\, \boldsymbol{k}_a |\, \rho^{(\text{2-DPI})}\, | \mu_p'\, \hat{\boldsymbol{u}}_p\, \boldsymbol{k}_p;\ \mu_a'\, \hat{\boldsymbol{u}}_a\, \boldsymbol{k}_a \rangle &= \frac{\mathcal{K}_{ps}}{(2L_0+1)(2S_0+1)} \\
\times \sum_{\substack{M_{L_0} M_{L_{1^{+*}}} M_{L_{1'^{+*}}} M_{L_{2^+}} \\ M_{S_0} M_{S_{1^{+*}}} M_{S_{1'^{+*}}} M_{S_{2^+}}}} \langle 2^+;\ \mu_a\, \hat{\boldsymbol{u}}_a\, \boldsymbol{k}_a |\, F_{\text{s}}\, | 1^{+^*} \rangle \langle 1^{+^*};\ \mu_p\, \hat{\boldsymbol{u}}_p\, \boldsymbol{k}_p |\, F_{\text{p}}\, | 0;\ 1m_r \rangle \\
\times \langle 0;\ 1m_r |\, F_{\text{p}}^\dagger\, | 1'^{+^*};\ \mu_p'\, \hat{\boldsymbol{u}}_p\, \boldsymbol{k}_p \rangle \langle 1'^{+^*} |\, F_{\text{s}}^\dagger\, | 2^+;\ \mu_a'\, \hat{\boldsymbol{u}}_a\, \boldsymbol{k}_a \rangle. \qquad (6.2)
\end{aligned}
$$

Here, we have defined, for brevity, $\mathcal{K}_{ps} \equiv \mathcal{K}_p \mathcal{K}_{\text{s}}$ and averaged over all the Zeeman components (in L-S coupling) [10, 60] of $|0\rangle$ assuming that the atom $\mathfrak{T}$ is unpolarized before absorbing a photon γ_r in this electronic state.

For evaluating (6.2), we first substitute in it relations (3.28) and (3.33). The subsequent expression is simplified in such a way that as many sums over the projections of the angular momenta are analytically evaluated as possible. The important steps in this simplification procedure are use of the identity (14.42) [64] for converting a quadruple sum of product of four 3-j symbols to a double sum of the product of two 3-j and one 9-j symbols, orthogonality (3.7.8) [65] of 3-j symbols, and identity (6.2.8) [65] for reducing a triple sum consisting of product of three 3-j symbols to the product of one 3-j and one 6-j symbols. After some more simplifications, one is able to write [188] the resulting expression for the density matrix (6.2) in forms similar to that obtained earlier in (4.7) for 1-SPI or in (5.9) for 1-DPI, i.e., as a product of two independent terms. Namely [188],

$$\begin{aligned}
&\langle L_{2^+} S_{2^+};\ \mu_p\, \hat{\boldsymbol{u}}_p\, \boldsymbol{k}_p;\ \mu_a\, \hat{\boldsymbol{u}}_a\, \boldsymbol{k}_a |\, \rho^{\text{(2-DPI)}}\, | L_{2^+} S_{2^+};\ \mu'_p\, \hat{\boldsymbol{u}}_p\, \boldsymbol{k}_p;\ \mu'_a\, \hat{\boldsymbol{u}}_a\, \boldsymbol{k}_a \rangle \\
&\quad = \frac{\mathrm{d}^3\sigma^{\text{(2-DPI)}}(m_r)}{\mathrm{d}\epsilon_1\, \mathrm{d}\hat{\boldsymbol{k}}_p\, \mathrm{d}\hat{\boldsymbol{k}}_a}\, \sigma^{\text{(2-DPI)}}(S_0;\ S_{1^{+*}};\ S_{2^+};\ \hat{\boldsymbol{u}}_p;\ \hat{\boldsymbol{u}}_a)_{\mu_p,\mu_a;\,\mu'_p,\mu'_a}.
\end{aligned} \tag{6.3}$$

with, of course, expressions for each term totally different from those to be used in (4.7) or (5.9). One now, instead, has [188]

$$\begin{aligned}
\frac{\mathrm{d}^3\sigma^{\text{(2-DPI)}}(m_r)}{\mathrm{d}\epsilon_1\, \mathrm{d}\hat{\boldsymbol{k}}_p\, \mathrm{d}\hat{\boldsymbol{k}}_a} &= (-1)^{m_r+L_0+L_{1^{+*}}+L_{2^+}}\, \frac{(2L_{1^{+*}}+1)\,\mathcal{K}_{ps}}{4\pi(2L_0+1)} \sum_{\substack{\ell_p\, \ell'_p\, L_p\, L\, L' \\ \ell_a\, \ell'_a\, L_a\, L_r\, M}} \\
&\times (-1)^{\ell'_a+L+L_a}(2L_r+1)\sqrt{(2L_p+1)(2L_a+1)} \begin{pmatrix} \ell_p & \ell'_p & L_p \\ 0 & 0 & 0 \end{pmatrix} \begin{pmatrix} \ell_a & \ell'_a & L_a \\ 0 & 0 & 0 \end{pmatrix} \\
&\times \begin{pmatrix} 1 & 1 & L_r \\ m_r & -m_r & 0 \end{pmatrix} \begin{pmatrix} L_a & L_p & L_r \\ M & -M & 0 \end{pmatrix} \begin{Bmatrix} 1 & 1 & L_r \\ L' & L & L_0 \end{Bmatrix} \begin{Bmatrix} \ell_a & \ell'_a & L_a \\ L_{1^{+*}} & L_{1^{+*}} & L_{2^+} \end{Bmatrix} \\
&\times \begin{Bmatrix} \ell_p & \ell'_p & L_p \\ L & L' & L_r \\ L_{1^{+*}} & L_{1^{+*}} & L_a \end{Bmatrix} P_e(L_{1^{+*}}\, \ell_p;\ L_0\, 1;\ L) \left[P_e(L_{1^{+*}}\, \ell'_p;\ L_0\, 1;\ L') \right]^* \\
&\times A_e(L_{2^+}\, \ell_a;\ L_{1^{+*}};\ L_{1^{+*}}) \left[A_e(L_{2^+}\, \ell'_a;\ L_{1^{+*}};\ L_{1^{+*}}) \right]^* \\
&\times \left[Y_{L_p}^{-M}(\hat{\boldsymbol{k}}_p) \right]^* \left[Y_{L_a}^{M}(\hat{\boldsymbol{k}}_a) \right]^*,
\end{aligned} \tag{6.4a}$$

where

$$\begin{aligned}
P_e(L_{1^{+*}}\, \ell_p;\ L_0\, 1;\ L) \equiv\ & (-\mathrm{i})^{\ell_p}\, \mathrm{e}^{\mathrm{i}\sigma_{\ell_p}}\, (-1)^{\ell_p} \\
& \times (2L+1)\sqrt{2\ell_p+1}\, \langle (L_{1^{+*}}\, \ell_p)L | F_p | (L_0\, 1)L \rangle_{S_0,\, S_{1^{+*}}}
\end{aligned} \tag{6.4b}$$

and

$$A_e(L_{2^+}\,\ell_a;\;L_{1^{+*}};\;L_{1^{+*}}) \;\equiv\; (-\mathrm{i})^{\ell_a}\,\mathrm{e}^{\mathrm{i}\sigma_{\ell_a}}\,(-1)^{\ell_a}$$
$$\times\sqrt{2\ell_a+1}\,\langle (L_{2^+}\,\ell_a)L_{1^{+*}}|F_a|L_{1^{+*}}\rangle_{S_{1^{+*}},\,S_{2^+}} \qquad (6.4c)$$

have already been introduced in (3.28) and (3.33), respectively. The second term—present on the right hand side of (6.3)—is , on the other hand, given by [188, 262]

$$\sigma^{(2\text{-DPI})}(S_0;\;S_{1^{+*}};\;S_{2^+};\;\hat{\boldsymbol{u}}_p;\;\hat{\boldsymbol{u}}_a)_{\mu_p,\mu_a;\mu_p',\mu_a'} = (-1)^{S_0+2S_{1^{+*}}+S_{2^+}+\mu_p'+\mu_a'}$$
$$\times(2S_{1^{+*}}+1)\sum_{s\,\mu\,\mu_1\,\mu_2}(-1)^{s+\mu}\,(2s+1)\begin{pmatrix}1/2 & 1/2 & s\\ \mu_p & -\mu_p' & \mu_1\end{pmatrix}\begin{pmatrix}1/2 & 1/2 & s\\ \mu_a & -\mu_a' & \mu_2\end{pmatrix}$$
$$\times\begin{Bmatrix}1/2 & 1/2 & s\\ S_{1^{+*}} & S_{1^{+*}} & S_0\end{Bmatrix}\begin{Bmatrix}1/2 & 1/2 & s\\ S_{1^{+*}} & S_{1^{+*}} & S_{2^+}\end{Bmatrix}\left[\mathcal{D}^s_{\mu_1\,\mu}(\omega_p)\right]^*\left[\mathcal{D}^s_{\mu_2,-\mu}(\omega_a)\right]^* \qquad (6.5)$$

6.1.2 Properties

Derivation of the density matrix (6.3) is independent of all dynamical models which one may like to use in a theoretical and/or experimental study of the 2-DPI process (1.3). Russell–Saunders scheme for the coupling of the angular momenta of the participating particles has been used as SDIs neither in the bound states ($|0\rangle$, $|1^{+*}\rangle$, $|2^+\rangle$) of ($\mathfrak{T}$, $\mathfrak{T}^{1+*}$, $\mathfrak{T}^{2+}$) nor in the continua of (e_p, e_a) are taken into account. Hence, similar to (4.7) or (5.9), expression (6.3) too is a Coulombic density matrix, which splits into the angular- and spin-correlation parts (6.4) and (6.5), respectively. The term (6.4), which represents in (6.3) angular-correlation between (e_p, e_a), contains all the kinematical plus dynamical information about both the photoionization as well as Auger decay steps (1.3a) and (1.3b), respectively, of the 2-DPI process. It implicitly depends upon the spins [shown in (6.4b) and (6.4c)] only to the extent that the energies [E_0, $E_{1^{+*}}$, E_{2^+}] of [$\mathfrak{T}$, $\mathfrak{T}^{1+*}$, $\mathfrak{T}^{2+}$] in L-S coupling have different values for different spin multiplicities [$(2S_0+1)$, $(2S_{1^{+*}}+1)$, $(2S_{2^+}+1)$] even for the same values of the orbital angular momenta [L_0, $L_{1^{+*}}$, L_{2^+}] arising from electronic configuration of the states [$|0\rangle$, $|1^{+*}\rangle$, $|2^+\rangle$]. Although, $d^3\sigma^{(2\text{-DPI})}(m_r)/d\epsilon_1\,d\hat{\boldsymbol{k}}_p\,d\hat{\boldsymbol{k}}_a$ in (6.3) too contains, among other things, information about the directions of propagation ($\hat{\boldsymbol{k}}_p$, $\hat{\boldsymbol{k}}_a$) of electrons (e_p, e_a) and the ionizing radiation γ_r in (1.3a), its expression in (6.4) is, of course, very different from (4.8) or (5.10) describing angular correlation between (e_p, $\mathfrak{T}^{1+}$) in a 1-SPI process (1.1), or between (e_p, $e_{p'}$) in a 1-DPI precess (1.2), respectively.

Unlike (5.11) applicable to 1-DPI, the spin-correlation part (6.5) of the density matrix (6.3) explicitly contains spins (S_0, $S_{1^{+*}}$, S_{2^+}) of the electronic states ($|0\rangle$, $|1^{+*}\rangle$, $|2^+\rangle$) of the atomic species ($\mathfrak{T}$, $\mathfrak{T}^{1+*}$, $\mathfrak{T}^{2+}$) participating in the 2-DPI process (1.3). It, of course, involves also the spins of $(e_p,\ e_a)$ as well as their quantization directions $(\omega_p,\ \omega_a)$, and does not contain any of those quantities, which determine the kinematical and/or dynamical part of (1.3) and, hence, are present in the angular correlation (6.4).

Thus, in each of the 1-SPI, 1-DPI, and 2-DPI processes (1.1)-(1.3), respectively, the corresponding density matrix (4.7), (5.9), and (6.3) is a product of two independent terms one of which represents the angular correlation while the other term describes correlation between the spins of the concerned particles [i.e., $(e_p,\ \mathfrak{T}^{1+})$, $(e_p,\ e_{p'})$, $(e_p,\ e_a)$]. Indeed, such separation of correlation between the angular and spin parts can be shown to be present in the density matrices also of some of the other processes mentioned on pages 5 and 6 in Chap. 1 if their studies too are performed in the absence of SDIs—no matter, whether the target which absorbs photon γ_r in $E1$ approximation is an atom (e.g., [56, 57, 263]) or a rotating linear molecule (e.g., [213, 216, 217], *etc*). It, in other words, means that the properties of the entanglement between those observables which have discrete values (e.g., spins of an electron and an ion, spin of an electron and polarization of a photon, spins of two or more electrons) for particles generated, for example, in (1.1)–(1.8) are totally independent of the kinematics and dynamics of these processes if only the Coulombic interactions experienced by the particles inside an atom/molecule are taken into account.

6.2 Spin-Entanglement Properties of a Photo-Auger Electron Pair

In view of the fact that the angular correlation $\mathrm{d}^3\sigma^{\text{(2-DPI)}}(m_r)/\mathrm{d}\epsilon_1\,\mathrm{d}\hat{\boldsymbol{k}}_p\,\mathrm{d}\hat{\boldsymbol{k}}_a$ in (6.3) is merely a multiplicative constant whose value is always positive, the spin-entanglement properties of a $(e_p,\ e_a)$ pair produced in the 2-DPI process (1.3) will always be determined in the absence of SOI by the (4×4) spin-correlation part (6.5) of the present density matrix [i.e., (6.3)]. Unless stated otherwise, in the remaining part of the present Chap. 6 we, therefore, write

$$\langle L_{2^+} S_{2^+};\ \mu_p\,\hat{\boldsymbol{u}}_p\,\boldsymbol{k}_p;\ \mu_a\,\hat{\boldsymbol{u}}_a\,\boldsymbol{k}_a|\,\rho^{\text{(2-DPI)}}\,|L_{2^+} S_{2^+};\ \mu_p'\,\hat{\boldsymbol{u}}_p\,\boldsymbol{k}_p;\ \mu_a'\,\hat{\boldsymbol{u}}_a\,\boldsymbol{k}_a\rangle$$
$$\longrightarrow \sigma^{\text{(2-DPI)}}(S_0;\ S_{1^{+*}};\ S_{2^+};\ \hat{\boldsymbol{u}}_p;\ \hat{\boldsymbol{u}}_a)_{\mu_p,\mu_a;\,\mu_p',\mu_a'}. \tag{6.6}$$

It is obvious from the spin conservation conditions (3.20a) and (3.20b), applicable to the respective photoionization (1.3a) and Auger decay (1.3b) steps of a 2-DPI process, that $S_0 = \left|S_{1^{+*}} \pm \frac{1}{2}\right|$ and $S_{2^+} = \left|S_{1^{+*}} \pm \frac{1}{2}\right|$. Exactly, the same restrictions are imposed on these spin angular momenta by the triangular conditions $\Delta(S_0,\ S_{1^{+*}},\ \frac{1}{2})$ and $\Delta(S_{1^{+*}},\ S_{2^+},\ \frac{1}{2})$ needed to be satisfied [65] by the angular momenta present in the two 6-j symbols in (6.5) for neither of them to vanish identically. These requirements, in other words, mean that the respective spins $(S_0,\ S_{1^{+*}},\ S_{2^+})$ of the electronic states $(|0\rangle,\ |1^{+^*}\rangle,\ |2^+\rangle)$ of the species $(\mathfrak{T},\ \mathfrak{T}^{1+^*},\ \mathfrak{T}^{2+})$ participating in the 2-DPI are not independent at all; they, instead, are related by the conditions $S_0 = S_{2+} = \left|S_{1+^*} \pm \frac{1}{2}\right|$, or $\left|S_0 - S_{2+}\right| = 1$ with $S_0 = \left|S_{1+^*} \pm \frac{1}{2}\right|$ and $S_{2+} = |S_{1+^*} \mp \frac{1}{2}|$. The remaining part of the present chapter contains a discussion of the spin-entanglement properties of a (e_p, e_a) pair for each of these possibilities.

6.2.1 $S_0 = S_{2+} = \left|S_{1+^*} \pm \frac{1}{2}\right|$

On substituting these values of the spins in (6.5), it was found that the density matrix (6.6) for each of the allowed values of S_{1+^*} can be written in the form of the Werner state (5.23) as follows [188, 262]

$$\begin{aligned}\sigma^{(2\text{-DPI})}&\left(S_0 = S_{2+} = \left|S_{1+^*} \pm \frac{1}{2}\right|;\ S_{1+^*};\ \hat{\boldsymbol{u}}_p;\ \hat{\boldsymbol{u}}_a\right)\\ &= p\left(S_0 = S_{2+} = \left|S_{1+^*} \pm \frac{1}{2}\right|; S_{1+^*}\right)\sigma_1(\hat{\boldsymbol{u}}_p;\ \hat{\boldsymbol{u}}_a) + (1-p)\sigma_0, \quad (6.7)\end{aligned}$$

where matrices $\sigma_1(\hat{\boldsymbol{u}}_p;\ \hat{\boldsymbol{u}}_a)$ and σ_0 have been already introduced in (5.14) and (5.22a), respectively; the mixing parameter, on the other hand, is given by [188]

$$p\left(S_0 = S_{2+} = \left|S_{1+^*} \pm \frac{1}{2}\right|; S_{1+^*}\right) = 0 \quad \text{for} \quad S_{1+^*} = 0, \quad (6.8a)$$

$$\begin{aligned}p&\left(S_0 = S_{2+} = \left|S_{1+^*} \pm \frac{1}{2}\right|; S_{1+^*}\right)\\ &= \frac{\left[\frac{3}{4} + S_{1+^*}(S_{1+^*}+1) - S_0(S_0+1)\right]^2}{3S_{1+^*}(S_{1+^*}+1)} \quad \text{for} \quad S_{1+^*} > 0; \quad (6.8b)\end{aligned}$$

or, equivalently [188],

$$p\left(S_{1+^*}\right) = \frac{S_{1+^*}}{3(S_{1+^*}+1)} \quad \text{for} \quad S_0 = S_{2+} = S_{1+^*} + \frac{1}{2} \quad \text{and} \quad S_{1+^*} \geq 0, \quad (6.9a)$$

$$= \frac{S_{1+^*}+1}{3S_{1+^*}} \quad \text{for} \quad S_0 = S_{2+} = S_{1+^*} - \frac{1}{2} \quad \text{and} \quad S_{1+^*} \geq \frac{1}{2}. \quad (6.9b)$$

Thus, the whole dependence on the spins of the atomic species participating in the 2-DPI process (1.3) is now contained in the mixing parameter.

It is obvious from both of the (6.8) and (6.9) that $0 \leq p \leq 1$ always for all those values of the spins (S_0, S_{1+^*}, S_{2+}) of the electronic states of $(\mathfrak{T}, \mathfrak{T}^{1+^*}, \mathfrak{T}^{2+})$, which participate in the 2-DPI process (1.3) and satisfy the conditions mentioned

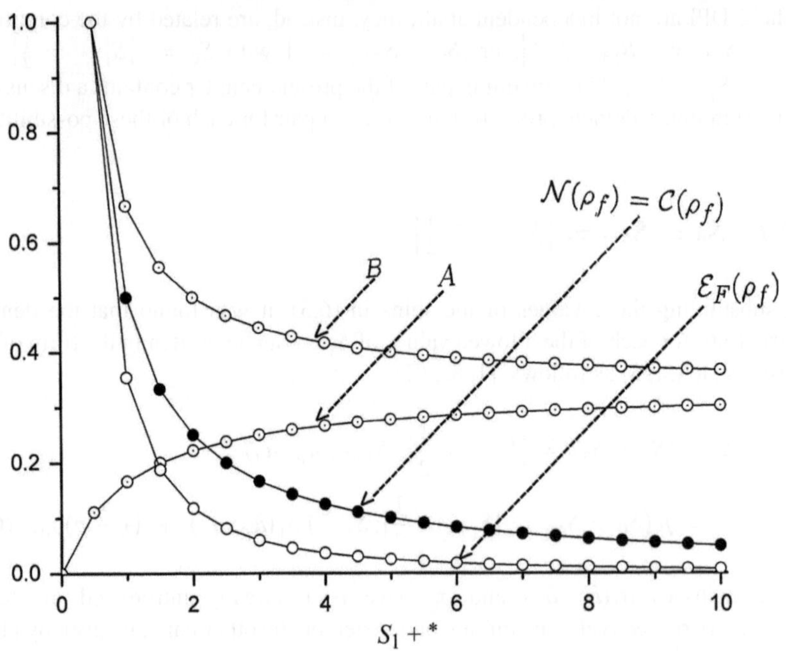

Fig. 6.2 Variations of the mixing parameter p [(6.8) or (6.9)], negativity $\mathcal{N}(\sigma^{(2\text{-DPI})}(\hat{\boldsymbol{u}}_p;\, \hat{\boldsymbol{u}}_a))$ (6.14), concurrence $\mathcal{C}(\sigma^{(2\text{-DPI})}(\hat{\boldsymbol{u}}_p;\, \hat{\boldsymbol{u}}_a))$ (6.14), and of the entanglement of formation $\mathcal{E}(\sigma^{(2\text{-DPI})}(\hat{\boldsymbol{u}}_p;\, \hat{\boldsymbol{u}}_a))$ (2.47) with the spin $S_{1^{+*}}$ of the excited photoion $\mathfrak{T}^{1+^*}$. As explained in (6.14), here we have $\mathcal{N}(\sigma^{(2\text{-DPI})}(\hat{\boldsymbol{u}}_p;\, \hat{\boldsymbol{u}}_a)) = \mathcal{C}(\sigma^{(2\text{-DPI})}(\hat{\boldsymbol{u}}_p;\, \hat{\boldsymbol{u}}_a))$ for all those values of $S_{1^{+*}}$, which are consistent with the conservation conditions (3.20). Curve A: p [from (6.9a)] $< \frac{1}{3}$, always; curve B: p [from (6.9b)] $> \frac{1}{3}$, always. The remaining two curves in this figure show negativity as well as concurrence calculated from (6.14), and entanglement of formation from [(2.47), (6.7)]. (First published in [188].)

in either of these two sets of equations. It means, in view of the discussions given on pages 35 and 128, the spin-state (6.7) of a photo-Auger electron pair $(e_p,\, e_a)$ is always of the form of a Werner [121] state (5.23).

Curves A and B in Fig. 6.2 show the variations of p with the spin $S_{1^{+*}}$ of the excited photoion $\mathfrak{T}^{1+^*}$ in the state $|1^{+^*}\rangle$ for requirements (6.9a) and (6.9b), respectively.[1] But, $p < 1/3$ always for values of $S_{1^{+*}}$ meeting the requirement (6.9a); whereas (6.9b) suggests that $p > 1/3$. However, for increasing $S_{1^{+*}}$, p in both of these cases asymptotically approaches to 1/3. Hence, depending upon the spins $(S_0,\, S_{1^{+*}},\, S_{2^+})$ of $(\mathfrak{T},\, \mathfrak{T}^{1+^*},\, \mathfrak{T}^{2+})$ participating in (1.3), density matrix (6.7) can

---

[1] According to (5.26), while three of the four eigenvalues of the partial transpose of the Werner state (6.7) are always more than zero for the allowed values $(-1/3 \le p \le 1)$ of the mixing parameter, the remaining fourth exactly vanishes for $p = 1/3$. This later one is the same eigenvalue which becomes negative (positive) for p greater (smaller) than 1/3, i.e., decides the entangled/separable character of Werner state.

represent either of a separable [for $0 \leq p < \frac{1}{3}$ from (6.9a)] or an entangled [for $\frac{1}{3} < p \leq 1$ from (6.9b)] state of (e_p, e_a).

Thus, a separable spin-state of (e_p, e_a) is

$$\sigma^{(2\text{-DPI})}(\hat{\boldsymbol{u}}_p; \hat{\boldsymbol{u}}_a)\Big|_{S_0=S_{2+}=S_{1+^*}+\frac{1}{2}} = \frac{S_{1+^*}}{3(S_{1+^*}+1)}\,\sigma_1(\hat{\boldsymbol{u}}_p; \hat{\boldsymbol{u}}_a) + \frac{2S_{1+^*}+3}{3(S_{1+^*}+1)}\,\sigma_0 \quad \text{for } S_{1+^*} \geq 0; \tag{6.10a}$$

whereas, an entangled state formed by these two electronic qubits is, on the other hand, given by

$$\sigma^{(2\text{-DPI})}(\hat{\boldsymbol{u}}_p; \hat{\boldsymbol{u}}_a)\Big|_{S_0=S_{2+}=S_{1+^*}-\frac{1}{2}} = \frac{S_{1+^*}+1}{3S_{1+^*}}\,\sigma_1(\hat{\boldsymbol{u}}_p; \hat{\boldsymbol{u}}_a) + \frac{2S_{1+^*}-1}{3S_{1+^*}}\,\sigma_0 \quad \text{for } S_{1+^*} \geq \frac{1}{2}. \tag{6.10b}$$

The $(4 \otimes 4)$ matrices $\sigma_1(\hat{\boldsymbol{u}}_p; \hat{\boldsymbol{u}}_a)$ and σ_0, present in both of (6.10), are already defined in the respective (5.14) and (5.22a).

Further,

$$\left[\frac{2S_{1+^*}+3}{12(S_{1+^*}+1)},\ \frac{2S_{1+^*}+3}{12(S_{1+^*}+1)},\ \frac{2S_{1+^*}+3}{12(S_{1+^*}+1)},\ \frac{2S_{1+^*}+1}{4(S_{1+^*}+1)}\right] \tag{6.11a}$$

are the eigenvalues of the state (6.10a) for $S_0 = S_{2+} = S_{1+^*} + \frac{1}{2}$ with $S_{1+^*} \geq 0$. While those of the state (6.10b), applicable whenever $S_0 = S_{2+} = S_{1+^*} - \frac{1}{2}$ with $S_{1+^*} \geq \frac{1}{2}$, are given by

$$\left[\frac{2S_{1+^*}-1}{12S_{1+^*}},\ \frac{2S_{1+^*}-1}{12S_{1+^*}},\ \frac{2S_{1+^*}-1}{12S_{1+^*}},\ \frac{2S_{1+^*}+1}{4S_{1+^*}}\right]. \tag{6.11b}$$

Eigenvalues [(6.11a), (6.11b)] of the bipartite states [(6.10a), (6.10b)] are obtained by substituting p from [(6.9a), (6.9b)] in (5.25). From these, one concludes in a straightforward manner, that the separable spin-state (6.10a) of (e_p, e_a) is always mixed as none of its four eigenvalues (6.11a) is ever zero; (e_p, e_a), on the other hand, are in a pure entangled state (6.10b) as the eigenvalues (6.11b) become (0, 0, 0, 1) for the spin-triad ($S_0 = 0$, $S_{1+^*} = 1/2$, $S_{2+} = 0$). In all other cases, (6.10b) is a mixed entangled state.

On combining the mixing parameter p in (6.9a) with (5.24b), one obtains the partial transpose

$$\sigma^{(2\text{-DPI})}(\hat{\boldsymbol{u}}_p;\,\hat{\boldsymbol{u}}_a)\Big|^{T_a}_{S_0=S_{2^+}=S_{1^{+*}}+\frac{1}{2}}$$
$$=\frac{S_{1^{+*}}}{3(S_{1^{+*}}+1)}\,\sigma_1^{T_a}(\hat{\boldsymbol{u}}_p;\,\hat{\boldsymbol{u}}_a)+\frac{2S_{1^{+*}}+3}{3(S_{1^{+*}}+1)}\,\sigma_0\quad\text{for } S_{1^{+*}}\geq 0;\tag{6.12a}$$

of the mixed separable state (6.10a). Use of (6.9b) in (5.24b), on the other hand, gives the following partial transpose

$$\sigma^{(2\text{-DPI})}(\hat{\boldsymbol{u}}_p;\,\hat{\boldsymbol{u}}_a)\Big|^{T_a}_{S_0=S_{2^+}=S_{1^{+*}}-\frac{1}{2}}$$
$$=\frac{S_{1^{+*}}+1}{3S_{1^{+*}}}\,\sigma_1^{T_a}(\hat{\boldsymbol{u}}_p;\,\hat{\boldsymbol{u}}_a)+\frac{2S_{1^{+*}}-1}{3S_{1^{+*}}}\,\sigma_0\quad\text{for } S_{1^{+*}}\geq \frac{1}{2}.\tag{6.12b}$$

of the entangled state (6.10b) of $(e_p,\, e_a)$. The superscript T_a in each of (6.12) means that these are partial transpose of density matrices (6.10) with respect to the Auger electron e_a. Here, $\sigma_1^{T_a}(\hat{\boldsymbol{u}}_p;\,\hat{\boldsymbol{u}}_a)$ is the partial transpose of $\sigma_1(\hat{\boldsymbol{u}}_p;\,\hat{\boldsymbol{u}}_a)$ (present in the states (6.10)) and is obtained from (5.19). Further, substitution in (5.26) of the values of p from (6.9a) gives the following eigenvalues

$$\left[\frac{4S_{1^{+*}}+3}{12(S_{1^{+*}}+1)},\,\frac{4S_{1^{+*}}+3}{12(S_{1^{+*}}+1)},\,\frac{4S_{1^{+*}}+3}{12(S_{1^{+*}}+1)},\,\frac{1}{4(S_{1^{+*}}+1)}\right]\tag{6.13a}$$

of the partial transpose (6.12a) for $S_0 = S_{2^+} = S_{1^{+*}} + \frac{1}{2}$ and $S_{1^{+*}} \geq 0$; whereas

$$\left[\frac{4S_{1^{+*}}+1}{12S_{1^{+*}}},\,\frac{4S_{1^{+*}}+1}{12S_{1^{+*}}},\,\frac{4S_{1^{+*}}+1}{12S_{1^{+*}}},\,-\frac{1}{4S_{1^{+*}}}\right],\tag{6.13b}$$

for $S_0 = S_{2^+} = S_{1^{+*}} - \frac{1}{2}$ and $S_{1^{+*}} \geq \frac{1}{2}$, are the eigenvalues of (6.12b) obtained on combining p from (6.9b) with (5.26).

In (6.13a), none of the eigenvalues is ever less than zero; but in (6.13b), one eigenvalue is always negative. In view of the necessary (Peres [125]) and sufficient (Horodecki [124]) condition for the partial transpose of a separable bipartite (mixed/pure) state of two qubits to be positive, the eigenvalues given in (6.13a) and (6.13b) reasserts that (6.10a) and (6.10b) are indeed the separable and entangled parts of the spin-state (6.7) of a pair of photoelectron e_p and Auger electron e_a, generated sequentially in the 2-DPI process (1.3) in the absence of SOI in L-S coupling.

The amount of entanglement present in (6.7) is quantified by calculating any of the three measures discussed on pages 44–46. For the bipartite state (6.7), two [i.e., negativity (2.45c) and concurrence (2.46)] of these measures were found [188] to have identical values given by the expression (5.28). Negativity for the states (6.10) is readily calculated by substituting either the mixing parameter (6.9) in (5.28) or, equivalently, eigenvalues (6.13) of the partial transpose (6.12) in (2.45c). One then finds [188]

$$\mathcal{N}\left(\sigma^{(2\text{-DPI})}(\hat{\boldsymbol{u}}_p;\,\hat{\boldsymbol{u}}_a)\right) = \mathcal{C}\left(\sigma^{(2\text{-DPI})}(\hat{\boldsymbol{u}}_p;\,\hat{\boldsymbol{u}}_a)\right)^2$$

$$= \begin{cases} 0 & \text{for } S_0 = S_{2+} = S_{1+^*} + \frac{1}{2} \text{ and } S_{1+^*} \geq 0, \\ \left(2S_{1+^*}\right)^{-1} & \text{for } S_0 = S_{2+} = S_{1+^*} - \frac{1}{2} \text{ and } S_{1+^*} \geq \frac{1}{2}. \end{cases} \qquad (6.14)$$

These entanglement measures further show that while the spin-state (6.10a) of $(e_p,\, e_a)$ is always separable, the amount of entanglement in the non-separable state (6.10b) continuously decreases with increasing values of the spin S_{1+^*} of the electronic state $|1^{+^*}\rangle$ of the excited photoion $\mathfrak{T}^{1+^*}$ participating in the 2-DPI process (1.3) in L-S coupling in the absence of SOI. This variation of entanglement in the state (6.10b) with the values of the spin S_{1+^*} is shown in Fig. 6.2. This figure also contains entanglement of formation $\mathcal{E}_F$ (2.47) calculated [188] for the spin-state (6.7) of $(e_p,\, e_a)$ for different values of the mixing parameter given by (6.8) [or, (6.9)].[3] Here again, one finds that while $\mathcal{E}_F\left(\sigma^{(2\text{-DPI})}(\hat{\boldsymbol{u}}_p;\,\hat{\boldsymbol{u}}_a)\right)$ is always zero for the state (6.10a), its value for the state (6.10b) continuously decreases with increasing S_{1+^*}. Another thing one observes in Fig. 6.2 is that, while all the three measures of entanglement varies between (0, 1)—zero being for a separable state and unity for a maximally entangled pure state—the values of $\mathcal{E}_F$ are, otherwise, always less than those of $\mathcal{N}$ and $\mathcal{C}$ for mixed, entangled states of two qubits.

Thus, (6.7)–(6.14) determine almost every property pertaining to quantum information of a spin state of $(e_p,\, e_a)$ whenever multiplicities of the electronic states $(|0\rangle,\, |2^+\rangle)$ of $(\mathfrak{T},\, \mathfrak{T}^{2+})$, participating in the 2-DPI (1.3), are equal and differ from that of the excited photoion $\mathfrak{T}^{1+^*}$ in its electronic state $|1^{+^*}\rangle$ by one. For spin angular momenta of $(e_p,\, e_a)$ to be entangled in this case, it is necessary that the spins $(S_0,\, S_{2+})$ of $(\mathfrak{T},\, \mathfrak{T}^{2+})$ are not only equal but also less than S_{1+^*} of $\mathfrak{T}^{1+^*}$ by one-half; otherwise, $(e_p,\, e_a)$ will always be in a separable spin state. Moreover, higher the value of S_{1+^*} in a non-separable state of $(e_p,\, e_a)$, lower will be their degree of entanglement with its maximum value of unity being when each of the species $(\mathfrak{T},\, \mathfrak{T}^{2+})$ is in a singlet and $\mathfrak{T}^{1+^*}$ in a doublet electronic state.

Thus, purely Coulombic interaction experienced by electrons inside an atom is capable of generating in the 2-DPI process (1.3) in L-S coupling, states of electronic qubits $(e_p,\, e_a)$ starting from maximally chaotic state σ_0 (for $S_0 = S_{1+} = \frac{1}{2}$, $S_{1+^*} = 0$) to maximally entangled pure state $\sigma_1(\hat{\boldsymbol{u}}_p;\hat{\boldsymbol{u}}_a)$ (for $S_0 = S_{1+} = 0$, $S_{1+^*} = \frac{1}{2}$). It all depends upon the spins $(S_0,\, S_{1+^*},\, S_{2+})$ of the electronic

[2]Concurrence can be calculated using the density matrices (6.10) in (2.46).

[3]Calculation of $\mathcal{E}_F$ in the present case requires [188] merely substitution of the concurrence (6.14) in (2.47).

states $(|0\rangle, |1^{+^*}\rangle, |2^{+}\rangle)$ of the species $(\mathfrak{T}, \mathfrak{T}^{1+^*}, \mathfrak{T}^{2+})$ participating in the 2-DPI process (1.3). Spins $(S_0, S_{1^{+*}}, S_{2^+})$, in addition, completely characterize this Coulombic entanglement.

Examples

One can think of numerous examples for performing real experiments on 2-DPI in a laboratory with an available Auger spectrometer. With the analysis given herein, it is possible to generate states of electronic qubits (e_p, e_a) with a degree of entanglement shown in Fig. 6.2. In order to illustrate the above discussion, let us consider following examples.

$S_0 = S_{2^+} = \frac{1}{2}$, $S_{1^{+*}} = 0$.

The lightest atom in which a 2-DPI process (1.3) with negligible SOI is possible is boron involving its daughter species and their electronic states as $[\mathrm{B}(1s^2 2s^2 2p^1\, {}^2P)$, $\mathrm{B}^{1+^*}(1s^1 2s^2 2p^1\, {}^1P)$, $\mathrm{B}^{2+}(1s^2 2s^1 2p^0\, {}^2S)]$. Thus, the spins $(S_0, S_{1^{+*}}, S_{2^+})$ of the participating electronic states of $(\mathrm{B}, \mathrm{B}^{1+^*}, \mathrm{B}^{2+})$ are (1/2, 0, 1/2). This example belongs to (6.9a) giving mixing parameter $p = 0$ and reducing the corresponding separable state (6.10a) to σ_0. Hence, the photoelectron e_p and Auger electron e_a, emitted from B in two sequential steps of the 2-DPI process (1.3a) in the absence of SOI, are not only never entangled, but are also in the maximally chaotic state [22, 122] σ_0.

Let us compare it with an experiment wherein the same doubly ionized state of $\mathrm{B}^{2+}(1s^2 2s^1 2p^0\, {}^2S)$ is formed in the 1-DPI process (1.2). According to the discussion given in the Sect. 5.2.3 on pages 131 & 132, the two emitted electrons are now in the Werner state (5.32a), which is separable as well mixed with $p = 1/3$. Thus, a change in the mechanisms of emission of two electrons has changed the nature of their spin-state from being a mixed, separable, Werner in 1-DPI to maximally chaotic in 2-DPI, although both of the respective processes (1.2) and (1.3) are caused by the absorption of a single photon γ_r in $E1$ approximation and the two emitted electrons experience exactly the same Coulomb interactions inside a boron atom.

$S_0 = S_{2^+} = 0$, $S_{1^{+*}} = \frac{1}{2}$.

Among the lightest possible atoms, suitable for such a 2-DPI process, are C, O, Ne, *etc*. One, therefore, has $[\mathrm{C}(1s^2 2s^2 2p^2\, {}^1S/{}^1D)$, $\mathrm{C}^{1+^*}(1s^1 2s^2 2p^2\, {}^2S/{}^2D)$, $\mathrm{C}^{2+}(1s^2 2s^2 2p^0\, {}^1S)]$, $[\mathrm{O}(1s^2 2s^2 2p^4\, {}^1S/{}^1D)$, $\mathrm{O}^{1+^*}(1s^1 2s^2 2p^4\, {}^2S/{}^2D)$, $\mathrm{O}^{2+}(1s^2 2s^2 2p^2\, {}^1S/{}^1D)]$, and $[\mathrm{Ne}(1s^2 2s^2 2p^6\, {}^1S)$, $\mathrm{Ne}^{1+^*}(1s^1 2s^2 2p^6\, {}^2S)$, $\mathrm{Ne}^{2+}(1s^2 2s^2 2p^4\, {}^1S/{}^1D)]$ for the C, O, and Ne atoms, respectively. Although SOI in these atoms is not as negligible as in B, nevertheless, the spin-state of a photo-Auger electron pair (e_p, e_a) formed in 2-DPI in each of these atoms is (6.10b) with the mixing parameter $p = 1$ obtained from (6.9b). Thus, (e_p, e_a) in each of these cases are in the state $\sigma_1(\hat{\boldsymbol{u}}_p; \hat{\boldsymbol{u}}_a)$

defined in (5.14). It is a pure state for which each of the three measures (i.e., negativity, concurrence, entanglement of formation) have their maximum possible value of one [see, for example (6.14) and Fig. 6.2]. This is a natural consequence of spin conservation as none of the atoms [C, O, Ne] suffer any loss in their total spin angular momenta in reducing to their dications [C^{2+}, O^{2+}, Ne^{2+}] in a 2-DPI process (1.3) in L-S coupling in the absence of SOI. It then means that the total spin of a pair of (e_p, e_a) in each of these cases must also be zero, i.e., the two emitted electrons be in their singlet spin state $\sigma_1(\hat{\boldsymbol{u}}_p; \hat{\boldsymbol{u}}_a)$.

Suppose two electrons depart from each of the atoms [C(^{1}S/^{1}D), O(^{1}S/^{1}D), Ne(^{1}S)] in a 1-DPI, rather than in a 2-DPI, process leaving behind the dication residues [C^{2+}(^{1}S), O^{2+}(^{1}S/^{1}D), O^{2+}(^{1}S/^{1}D)]. Then similar to (e_p, e_a) and in view of the discussion contained in the Sect. 5.2.1 on pages 122-129, the two photoelectrons $(e_p, e_{p'})$ will also be in the pure, maximally entangled, singlet spin-state for the total spin is conserved even in a 1-DPI experiment in L-S coupling. Thus, in the present examples, the spin conservation, rather than the mechanisms for 1-DPI and 2-DPI, dominates the entanglement properties of the spin-state of two emitted electrons. Consequently, the spin state of $(e_p, e_{p'})$ and of (e_p, e_a) in the present examples have identical entanglement properties.

$S_0 = S_{2+} = S_{1+*} + \frac{1}{2}$ with $S_{1+*} > 0$.

One of the possible experiments for this case is 2-DPI of N atom with the transitions N($1s^2 2s^2 2p^3\ ^4$S) $\longrightarrow$ N^{1+*}($1s^1 2s^2 2p^3\ ^3$S) $\longrightarrow$ N^{2+}($1s^2 2s^1 2p^2\ ^4$P) corresponding to $(S_0, S_{1+*}, S_{2+}) = (3/2, 1, 3/2)$. Here, mixing parameter, calculated from (6.9a), is $p = \frac{1}{6}$ and (e_p, e_a) are now represented by the mixed separable entangled state (6.10a) with negativity and concurrence both equal to zero as obtained from first of (6.14). The entanglement of formation, calculated using (2.47), is obviously zero as well.

Let us now consider the transition N($1s^2 2s^2 2p^3\ ^4$S) $\longrightarrow$ N^{2+}($1s^2 2s^1 2p^2\ ^4$P) taking place in the 1-DPI process (1.2) caused due to the simultaneous emission of two photoelectrons $(e_p, e_{p'})$ following the absorption of a single photon γ_r in the $E1$ approximation. According to the discussion presented in the Sect. 5.2.3 contained on pages 131–133, the spin state of $(e_p, e_{p'})$ is now given by (5.32a). Although similar to the state (6.10a) of (e_p, e_a), (5.32a) too is a mixed and separable Werner state of $(e_p, e_{p'})$; however, the value of the mixing parameter in this case is $p = \frac{1}{3}$, which is twice of the value calculated in the immediately preceding paragraph for the same initial and final states of N atom and of its dication N^{2+} participating in 2-DPI.

$S_0 = S_{2+} = S_{1+*} - \frac{1}{2}$ with $S_{1+*} > \frac{1}{2}$.

Here, transitions N($1s^2 2s^2 2p^3\ ^2$P/^{2}D) $\longrightarrow$ N^{1+*}($1s^1 2s^2 2p^3\ ^3$P/^{3}D) $\longrightarrow$ N^{2+}($1s^2 2s^2$ $2p^1\ ^2$P)] generate (e_p, e_a) in the Werner state (6.10b) which is mixed, but entangled. For, one has, in this case, $(S_0, S_{1+*}, S_{2+}) = (1/2, 1, 1/2)$. This, according to (6.9b), gives $p = \frac{2}{3}$. Both of the negativity and concurrence are obtained to be equal to one-half from (6.14) as well as from Fig. 6.2; whereas the entanglement of formation, calculated using (2.47) for the density matrix (6.10b) with $(S_0, S_{1+^*}, S_{2+}) = (1/2, 1, 1/2)$, is shown to be in this figure to be less than $\frac{1}{2}$.

In the present case as well, (5.32a) describes the properties of the spin state of photoelectrons $(e_p, e_{p'})$ emitted in the 1-DPI process (1.2) with $N(^2P/^2D) \longrightarrow N^{2+}(^2P)$. Thus, a change in mechanism of ejection of two electrons from 2-DPI to 1-DPI has again significantly altered the nature of the non-local correlation present in each of the pairs of electrons (e_p, e_a) and $(e_p, e_{p'})$, respectively: (e_p, e_a) in 2-DPI form a mixed entangled state but $(e_p, e_{p'})$ are in a mixed product state, although, in both of these processes, the electronic state of the initial atom N is $^2P/^2D$ and that of the residual dication N^{2+} is ^2P in L-S coupling in the absence of SOI.

6.2.2 $\left|S_0 \left(= \left|S_{1+^*} \pm \frac{1}{2}\right|\right) - S_{2+} \left(= \left|S_{1+^*} \mp \frac{1}{2}\right|\right)\right| = 1$

On substituting $S_0 = \left|S_{1+^*} \pm \frac{1}{2}\right|$ and $S_{2+} = \left|S_{1+^*} \mp \frac{1}{2}\right|$ in (6.5), spin state (6.6) representing electrons (e_p, e_a) in the present case becomes [188, 262]

$$\begin{aligned}\sigma^{(2-\mathrm{DPI})}(S_0 &= |S_{1+*} \pm \frac{1}{2}|; S_{2+} = |S_{1+*} \mp \frac{1}{2}|; S_{1+*}; \hat{\boldsymbol{u}}_p; \hat{\boldsymbol{u}}_a \\ &= -\frac{1}{3}\,\sigma_1(\hat{\boldsymbol{u}}_p;\, \hat{\boldsymbol{u}}_a) + \frac{4}{3}\sigma_0 \end{aligned} \tag{6.15}$$

Although in this calculation no value of S_{1+^*} for the excited photoion was specified, the state (6.15) is, however, independent of the spin quantum numbers of each of the species $(\mathfrak{T}, \mathfrak{T}^{1+^*}, \mathfrak{T}^{2+})$ participating in the 2-DPI process (1.3) as long as (S_0, S_{1+^*}, S_{2+}) meet the present requirements. This result is very unlike the one discussed in the Sect. 6.2.1 wherein mixing parameter p depends upon spins (S_0, S_{1+^*}, S_{2+}). According to (6.15), the sequentially emitted electrons (e_p, e_a) are in a mixed state with no non-local correlation between them. For illustration, let us consider a 2-DPI experiment in a carbon atom in L-S coupling without SOI. Transitions (1.3) involving electronic states $[C(1s^2 2s^2 2p^2\,{}^1S/{}^1D)$, $C^{1+^*}(1s^1 2s^2 2p^2\,{}^2S/{}^2D)$, $C^{2+}(1s^2 2s^0 2p^2\,{}^3P)]$ will generate a photo-Auger electron pair in the state (6.15).

States of the kind (6.15) of two electronic qubits are generated also in the 1-DPI process (1.2) and were discussed in the Sect. 5.2.2 on pages 129-131. The physical conditions needed for generating the spin-state (5.29) of two photoelectrons $(e_p, e_{p'})$ are almost identical to those required for producing (6.15) in the present case except that, unlike in 2-DPI (1.3), both $(e_p, e_{p'})$ in 1-DPI (1.2) come out from a single electronic state $|0\rangle$ of $\mathfrak{T}$.

Chapter 7
Fine-Structure Entanglement: Bipartite States of Flying Particles with Rest Mass Different from Zero

In this chapter, we relax the condition that electrons inside the atomic target $\mathfrak{T}$, or after leaving it, do not experience any SDIs. Hence, the spin-states of a photoelectron plus a photoion, or of two electrons, are now generated in the presence of both the C+SOIs. A proper description of the physical situation arising from the presence of both of these two important interactions demands [10, 60] that the required density matrices be now calculated in the j-j coupling scheme of angular momenta. This chapter, therefore, first presents a reformulation in j-j coupling of the density matrices for each of the processes of 1-SPI (1.1), 1-DPI (1.2), and 2-DPI (1.3), hitherto calculated in the L-S coupling in the respective Chaps. 4–6. These density matrices are then analyzed to learn about the entanglement properties of the bipartite spin-states of flying particles consisting of (photoelectron + photoion), two photoelectrons, or (photoelectron + Auger electron). The first Sect. 7.1 in this chapter describes entanglement properties of $(e_p, \mathfrak{T}^{1+})$ generated in the 1-SPI process (1.1), the second Sect. 7.2 considers non-local correlation between photoelectrons $(e_p, e_{p'})$ liberated in 1-DPI (1.2); whereas the last Sect. 7.3 contains a detailed discussion of these properties for a pair of (e_p, e_a) electrons produced in the 2-DPI process (1.3).

On the inclusion of SOI, the entanglement properties of each of the pairs $(e_p, \mathfrak{T}^{1+})$, $(e_p, e_{p'})$, and (e_p, e_a), generated in the respective processes (1.1)–(1.3), are considerably different in many ways from those when only Coulomb interaction is taken into account. For example, the density matrices obtained in this chapter are no where comparable to those already derived in Chaps. 4–6. Now, the entanglement properties very much depend upon the kinematics as well as dynamics of the process—in addition to the characteristics (e.g., frequency, polarization) of the absorbed photon γ_r—which generates a $(e_p, \mathfrak{T}^{1+})$, $(e_p, e_{p'})$, or (e_p, e_a) pair. Consequently, unlike in the case when Coulomb interaction is only considered, it is no longer possible to learn, *a priori*, any thing about the entanglement of $(e_p, \mathfrak{T}^{1+})$, $(e_p, e_{p'})$, or of (e_p, e_a) when one of the respective processes 1-SPI, 1-DPI, or 2-DPI takes place in an atomic target $\mathfrak{T}$ in the presence C+SOIs. Moreover, also the properties of entanglement between two particles in a pair in the presence of SOI are

N. Chandra and R. Ghosh, *Quantum Entanglement in Electron Optics*, Springer Series on Atomic, Optical, and Plasma Physics 67, DOI 10.1007/978-3-642-24070-6_7,

not necessarily the same as they are in the absence of any SDIs, produced purely by the Coulomb forces experienced by them (i.e., the particles) in an atomic target $\mathfrak{T}$.

7.1 Entanglement Between Spin of a Photoelectron e_p and a Fine-Structure Level of a Photoion $\mathfrak{T}^{1+}$ Generated in 1-SPI

This section presents a study of the properties of the non-local correlation, which may exist in a (e_p, $\mathfrak{T}^{1+}$) pair when both Coulomb and SOIs are taken into account. The present section is, therefore, a counter part of a study of Coulombic entanglement performed in Chap. 4 between the spins of (e_p, $\mathfrak{T}^{1+}$) in the absence of SOI.

7.1.1 Density Operator

It has already been explained on pages 66 and 67 in Sect. 3.2.2 that a proper description of the physical situation arising from the presence of SOI requires [10, 60] an application of the framework of j-j coupling. This, in other words, means that (E_0, E_{1^+}) in the conservation condition (4.1)—which is valid in the present case as well—are the energies of the (atom $\mathfrak{T}$, photoion $\mathfrak{T}^{1+}$) in their electronic states ($|0\rangle \equiv |J_0\, M_{J_0}\rangle$, $|1^+\rangle \equiv |J_{1^+}\, M_{J_{1^+}}\rangle$) participating in the process (1.1) in j-j coupling. However, the two conditions (4.2a), (4.2b), needed for total orbital and total spin angular momenta to conserve separately in L-S coupling, are now replaced by the following single requirement

$$\boldsymbol{\ell}_r + \boldsymbol{J}_0 = \boldsymbol{J}_{1^+} + \boldsymbol{j}_p \left\{ = \boldsymbol{\ell}_p + \boldsymbol{s}_p \left[= \left(\frac{\mathbf{1}}{\mathbf{2}}\right)_p \right] \right\} \tag{7.1}$$

for the conservation of the total angular momentum ($\boldsymbol{J}_0 = \boldsymbol{L}_0 + \boldsymbol{S}_0$, $\boldsymbol{J}_{1^+} = \boldsymbol{L}_{1^+} + \boldsymbol{S}_{1^+}$) in the j-j coupling. This condition is similar to that already given in (3.34a).

Equation (3.5b) gives a generic expression for the density operator for any one-step process taking place in an unpolarized target due to the absorption in $E1$ approximation of a single photon in a definite state of polarization. Its specialized forms applicable to each of the two processes of 1-SPI and 1-DPI in L-S coupling are given in (4.3) and (5.2), respectively. On specializing (3.5b) to any one-step process taking place in j-j coupling, one has

$$\rho^{(1)} = \frac{\mathcal{K}_p}{2J_0 + 1} \sum_{M_0=-J_0}^{J_0} F_p \left(|0; 1m_r\rangle\langle 0; 1m_r| \right) F_p^\dagger. \tag{7.2}$$

This expression contains averaging over all the unobserved Zeeman components [10, 60] of the unpolarized target $\mathfrak{T}$ in its electronic state $|J_0\, M_{J_0}\rangle$ in j-j coupling.

7.1.2 Density Matrix

The required density matrix for studying (e_p, $\mathfrak{T}^{1+}$) entanglement in the 1-SPI process (1.1) in j-j coupling can now be written as [220]

$$\begin{aligned}&\langle J_{1^+} M_{1^+};\; \mu_p\, \hat{\boldsymbol{u}}_p\, \boldsymbol{k}_p|\, \rho^{\text{(1-SPI)}}\, |J_{1^+} M'_{1^+};\; \mu'_p\, \hat{\boldsymbol{u}}_p\, \boldsymbol{k}_p\rangle \\ &\quad \equiv \langle J_{1^+} M_{1^+};\; \mu_p\, \hat{\boldsymbol{u}}_p\, \boldsymbol{k}_p|\, \rho^{(1)}\, |J_{1^+} M'_{1^+};\; \mu'_p\, \hat{\boldsymbol{u}}_p\, \boldsymbol{k}_p\rangle. \end{aligned} \tag{7.3}$$

This, on substituting (7.2), finally becomes

$$\begin{aligned}&\langle J_{1^+} M_{1^+};\; \mu_p\, \hat{\boldsymbol{u}}_p\, \boldsymbol{k}_p|\, \rho^{\text{(1-SPI)}}\, |J_{1^+} M'_{1^+};\; \mu'_p\, \hat{\boldsymbol{u}}_p\, \boldsymbol{k}_p\rangle = \frac{\mathcal{K}_p}{2J_0+1} \sum_{M_0} \\ &\times \langle 1^+;\; \mu_p\, \hat{\boldsymbol{u}}_p\, \boldsymbol{k}_p|\, F_p\, |0;\; 1m_r\rangle \langle 0;\; 1m_r|\, F_p^\dagger\, |J_{1^+} M'_{1^+};\; \mu'_p\, \hat{\boldsymbol{u}}_p\, \boldsymbol{k}_p\rangle. \end{aligned} \tag{7.4}$$

The sum over M_0, present on the right-hand side of this last expression has already been evaluated in (3.42) for the transition (1.3a) when the atomic target $\mathfrak{T}$, after the departure of an inner-shell electron on absorbing a photon γ_r, is left in its excited electronic state $|1^{+^*}\rangle \equiv |J_{1^{+*}}\, M_{J_{1^{+*}}}\rangle$. In the present case, on the other hand, the photoelectron e_p in (1.1) belonged to the outer-most shell of $\mathfrak{T}$. Consequently, the photoion $\mathfrak{T}^{1+}$ formed in (1.1) is in its ground electronic state $|1^+\rangle \equiv |J_{1^+}\, M_{J_{1^+}}\rangle$. Hence, expressions derived in (3.42) can directly be used also in (7.4) provided the quantum numbers ($J_{1^{+*}}$, $M_{1^{+*}}$) belonging to the excited electronic state $|1^{+^*}\rangle$ of the photoion $\mathfrak{T}^{1+^*}$ therein are replaced by those [i.e., (J_{1^+}, M_{1^+})], which specify the ground electronic state $|1^+\rangle$ of the residual photoion $\mathfrak{T}^{1+}$ formed in the 1-SPI process (1.1) taking place in the j-j coupling.

Furthermore, expression (3.42b) is in terms of the angular momentum transferred[1] from the absorbed photon to the observed photoelectron e_p, wherein the residual photoion $\mathfrak{T}^{1+^*}$ is not detected. In studying (e_p, $\mathfrak{T}^{1+}$) entanglement, one will simultaneously be observing, on the other hand, both the photoelectron and the photoion. An expression for the density matrix (7.4) in terms of the total angular momentum [$\boldsymbol{J} \equiv \boldsymbol{\ell}_r + \boldsymbol{J}_0 = \boldsymbol{J}_{1^+} + \boldsymbol{j}_p$, see (7.1)] will, therefore, be more appropriate in the present case. This desired expression for the density matrix can, however, be obtained in two alternative ways: According to one of these procedures, we need to substitute expression (3.40b) for Dirac's bracket $\langle J_{1^{+*}}\, j_p|F(j_t)|J_0\, 1\rangle$ present

[1] See the discussion on page 70 related to (3.40).

in (3.42b). The consequent (3.42b) will have, among other things, a coherent sum over (j_t, j_t') containing a product of three 6-j and one 9-j symbols. This coherent sum can analytically be evaluated using identity (36) given on page 471 in [67] and the subsequent expression simplified to the form given hereunder in (7.5).

The other is an *ab-initio* approach requiring substitutions of (3.39a), along with (3.39b), in (7.4). For simplifying the consequent (7.4), one requires, among several identities from Racah algebra, also the one [i.e., (5)] given by Varshlovich et al. [67] on page 453 of their book. The use of this identity transforms the sum over M_0 of a product of two 3-j symbols, present in the consequent (7.4), into a double sum containing two 3-j and one 6-j symbols. The finally simplified expression for (7.4) can be written as [220]

$$\begin{aligned}
\langle J_{1+} M_{1+};\, \mu_p\, \hat{\boldsymbol{u}}_p\, \boldsymbol{k}_p |\, \rho^{(1\text{-SPI})}\, | J_{1+} M_{1+}';\, \mu_p'\, \hat{\boldsymbol{u}}_p\, \boldsymbol{k}_p \rangle &= (-1)^{\mu_p' + M_{1+}} \sum_{J_T\, M_{J_p}} \\
\times \sum_{S_p\, M_{S_p}\, N_{S_p}} \begin{pmatrix} \frac{1}{2} & \frac{1}{2} & S_p \\ \mu_p & -\mu_p' & M_{S_p} \end{pmatrix} \begin{pmatrix} J_{1+} & J_{1+} & J_T \\ M_{1+} & -M_{1+}' & M_{J_p} \end{pmatrix} \left[\mathcal{D}^{S_p}_{M_{S_p}\, N_{S_p}}(\omega_p) \right]^* \\
\times A^{(1\text{-SPI})}\Big(S_p\, N_{S_p};\; J_T\, M_{J_p};\; m_r;\; \boldsymbol{k}_p \Big),
\end{aligned} \tag{7.5a}$$

where

$$\begin{aligned}
A^{(1\text{-SPI})}\Big(S_p\, N_{S_p};\; J_T\, M_{J_p};\; m_r;\; \boldsymbol{k}_p \Big) &= (-1)^{\frac{1}{2} + m_r + J_0 - J_{1+} + J_T} (2J_T + 1) \\
\times (2S_p + 1) \frac{\mathcal{K}_p}{(2J_0 + 1)\sqrt{4\pi}} & \sum_{\substack{\ell_p\, j_p\, J \\ \ell_p'\, j_p'\, J'}} \sum_{\substack{L_p\, M_{L_p} \\ J_p\, L_r}} (-1)^{\ell_p' + j_p + j_p' + J + L_r} \\
\times (2L_r + 1)(2J_p + 1)\sqrt{2L_p + 1} & \\
\times \begin{pmatrix} \ell_p & \ell_p' & L_p \\ 0 & 0 & 0 \end{pmatrix} \begin{pmatrix} 1 & 1 & L_r \\ m_r & -m_r & 0 \end{pmatrix} & \begin{pmatrix} L_p & S_p & J_p \\ M_{L_p} & N_{S_p} & M_{J_p} \end{pmatrix} \\
\times \begin{pmatrix} J_p & L_r & J_T \\ M_{J_p} & 0 & -M_{J_p} \end{pmatrix} \begin{Bmatrix} 1 & 1 & L_r \\ J & J' & J_0 \end{Bmatrix} & \begin{Bmatrix} \ell_p & \ell_p' & L_p \\ \frac{1}{2} & \frac{1}{2} & S_p \\ j_p & j_p' & J_p \end{Bmatrix} \begin{Bmatrix} j_p & j_p' & J_p \\ J & J' & L_r \\ J_{1+} & J_{1+} & J_T \end{Bmatrix} \left[Y^{L_p}_{M_{L_p}}(\hat{\boldsymbol{k}}_p) \right]^* \\
\times \Big[(-i)^{\ell_p} (-1)^{\ell_p + j_p}\, \mathrm{e}^{\mathrm{i}\sigma_{\ell_p j_p}} & (2J + 1)\sqrt{(2\ell_p + 1)(2j_p + 1)}\, \langle J_{1+}\, j_p | F(J) | J_0 1 \rangle \Big] \\
\times \Big[(-i)^{\ell_p'} (-1)^{\ell_p' + j_p'}\, \mathrm{e}^{\mathrm{i}\sigma_{\ell_p' j_p'}} & (2J' + 1)\sqrt{(2\ell_p' + 1)(2j_p' + 1)}\, \langle J_{1+}\, j_p' | F(J') | J_0 1 \rangle \Big]^*
\end{aligned} \tag{7.5b}$$

with amplitudes $\langle J_{1+}\, j_p | F(J) | J_0 1 \rangle$ already defined in (3.39b).

7.1.3 Properties of the Density Matrix

Density matrix (7.5) simultaneously represents both photoion $\mathfrak{T}^{1+}$ in its state $|J_{1+}M_{1+}\rangle$ plus angle- and spin-resolved photoelectron e_p generated in the 1-SPI process (1.1) taking place in the atomic target $\mathfrak{T}$ in $E1$ approximation in j-j coupling. This expression is completely general and independent of all dynamical models. It takes fully into account both C+SOIs present in an atom. Here, the angular momenta ($\boldsymbol{J}_0$, $\boldsymbol{J}_{1+}$) of ($\mathfrak{T}$, $\mathfrak{T}^{1+}$) are quantized along the polar axis of the SF shown in Fig. 4.1 (on page 95); whereas the directions ($\hat{\boldsymbol{k}}_p$, $\hat{\boldsymbol{u}}_p$) of (motion, spin quantization) of the photoelectron e_p emitted in (1.1) are completely arbitrary in the frame of reference shown in this figure. Radtke et al. [264, 265] too have derived expressions for the density matrices for the process (1.1) taking relativistic and nondipolar effects [264] plus electron–electron interaction [265] into account. Unlike the presently derived density matrices (4.7) and (7.5), the expressions obtained in [264, 265] are for an experimental geometry wherein polar axis is along the direction of propagation of the photoelectron e_p (i.e., OZ-axis in Fig. 4.1 is taken along $\hat{\boldsymbol{k}}_p$ by Radtke et al.) and the photoelectron therein is [264, 265] longitudinally polarized (i.e., $\hat{\boldsymbol{u}}_p \parallel \hat{\boldsymbol{k}}_p$) with SOI always taken into account.

It is obvious that any property of a (e_p, $\mathfrak{T}^{1+}$)-system, calculated from (7.5), will depend, among other things, on the total angular momenta ($\boldsymbol{J}_0$, $\boldsymbol{J}_{1+}$) of ($\mathfrak{T}$, $\mathfrak{T}^{1+}$) which determine fine-structure levels of these two respective species. Expression (7.5) can, therefore, appropriately be called as a fine-structure density matrix and the non-local correlation determined from it in a pair of (e_p, $\mathfrak{T}^{1+}$) as fine-structure entanglement of these two particles. This entanglement will be between the spin angular momentum of photoelectron and the total angular momentum $\boldsymbol{J}_{1+}$ of $\mathfrak{T}^{1+}$. In other words, while e_p will always be a qubit; the photoion, on the other hand, will be a qubit for $J_{1+} = \frac{1}{2}$, a qutrit for $J_{1+} = 1$, and a qudit for $J_{1+} \geq \frac{3}{2}$. Thus, the simple 1-SPI process (1.1), both in the absence [viz., (4.7)] as well in the presence [viz., (7.5)] of SOI, provides a readily available technique of realistically generating in any laboratory qubit–qudit pairs of flying particles possessing rest masses different from zero.

The diagonal terms in (7.5) are the cross-sections for angle- and spin-resolved photoelectron spectroscopy in $E1$ approximation with the photoion being observed in one of its $|J_{1+}\,M_{J_{1+}}\rangle$ electronic state in j-j coupling; whereas its non-diagonal elements ($\mu_p \neq \mu_p'$ and/or $M_{1+} \neq M_{1+}'$) represent coherence effects in the 1-SPI process (1.1) in the presence of C+SOIs. The diagonal part of the (2×2) matrix,[2] obtained after reducing (7.5) over the photoion $\mathfrak{T}^{1+}$, describes, in the SF of Fig. 4.1, angular distribution of photoelectrons going in direction $\hat{\boldsymbol{k}}_p$ with their spin

[2] One can readily write, using either of the definitions (A.26) and (A.27a),

$$\rho^{(e_p)} = \mathrm{Tr}_{\mathfrak{T}^{1+}}\left(\rho^{(1\text{-SPI})}\right) \tag{7.6a}$$

and, hence,

quantized along $\hat{\boldsymbol{u}}_p$. The consequent expression, obtained after simplifying (7.6b), is identical to that already given in [266] in terms of the total angular momentum $\boldsymbol{J}$, or equivalent to that derived by Chandra [190] using the angular momentum transfer $\boldsymbol{j}_t$.[3]

On the other hand, the $[(2J_{1^+}+1)\times(2J_{1^+}+1)]$ matrix[4] (7.7b), obtained after reducing (7.5) over the photoelectron e_p, represents only the residual photoion $\mathfrak{T}^{1+}$ formed after the departure of e_p from the target $\mathfrak{T}$ in the 1-SPI process (1.1). This matrix can be used to calculate *state multipoles* (also called *statistical tensors*) [61, 68, 267] for $\mathfrak{T}^{1+}$ in j-j coupling. These state multipoles describe [61, 68, 267], among other things, orientation [see footnote (1) on page 191] and alignment [see footnote (2) on page 191] of the photoion $\mathfrak{T}^{1+}$.

Let us now compare the Coulombic and fine-structure density matrices (4.7) and (7.5), respectively, both derived for the 1-SPI process (1.1). These two expressions for the same process are obviously completely different from each other. For example, unlike (4.7), the present density matrix (7.5) cannot be written as a product of two independent terms. Consequently, even those factors [e.g., experimental geometry, dynamics of the 1-SPI process (1.1), characteristics of the ionizing photon γ_r], which did not matter at all in the Coulombic entanglement of the spin state (4.7) of $(e_p,\ \mathfrak{T}^{1+})$, will now play important roles in determining properties of the fine-structure entanglement of these two particles in the presence of both C+SOIs. Thus, unlike the Coulombic entanglement, the fine-structure entanglement between e_p and $\mathfrak{T}^{1+}$ cannot be studied without a complete prior knowledge of the

---

$$\langle \mu_p\, \hat{\boldsymbol{u}}_p\, \boldsymbol{k}_p |\, \rho^{(e_p)}\, | \mu_p'\, \hat{\boldsymbol{u}}_p\, \boldsymbol{k}_p \rangle$$

$$= \sum_{M_{1^+}=-J_{1^+}}^{J_{1^+}} \langle J_{1^+} M_{1^+};\ \mu_p\, \hat{\boldsymbol{u}}_p\, \boldsymbol{k}_p |\, \rho^{(1\text{-SPI})}\, | J_{1^+} M_{1^+};\ \mu_p'\, \hat{\boldsymbol{u}}_p\, \boldsymbol{k}_p \rangle \qquad (7.6b)$$

in view of (A.27b). Here, superscript e_p means that the density operator and matrix in (7.6a) and (7.6b), respectively, describe only the observed photoelectron.

[3]For definition of $\boldsymbol{j}_t$, see, for example, footnote (1) on page 151.

[4]In the present case, one has

$$\rho^{(\mathfrak{T}^{1+})} = \int \left[\mathrm{Tr}_{e_p}\left(\rho^{(1\text{-SPI})}\right)\right] \mathrm{d}\hat{\boldsymbol{k}}_p. \qquad (7.7a)$$

The corresponding density matrix is, therefore, given by

$$\langle J_{1^+} M_{1^+} |\, \rho^{(\mathfrak{T}^{1+})}\, | J_{1^+} M_{1^+}' \rangle$$

$$= \int \left(\sum_{\mu_p=-\frac{1}{2}}^{+\frac{1}{2}} \langle J_{1^+} M_{1^+};\ \mu_p\, \hat{\boldsymbol{u}}_p\, \boldsymbol{k}_p |\, \rho^{(1\text{-SPI})}\, | J_{1^+} M_{1^+}';\ \mu_p'\, \hat{\boldsymbol{u}}_p\, \boldsymbol{k}_p \rangle \right) \mathrm{d}\hat{\boldsymbol{k}}_p. \qquad (7.7b)$$

The superscript $\mathfrak{T}^{1+}$, in (7.7), is used to indicate physical quantities related to only the photoion in the process (1.1).

photoionization dynamics of (1.1); the inclusion of SOI has coupled the $(e_p, \mathfrak{T}^{1+})$ entanglement with both the external environment (i.e., experimental geometry as well as characteristics of γ_r) and the internal forces which come into play during photoionization of an atom in j-j coupling.

7.1.4 Example

Parida and Chandra [220] have studied (e_p, Xe^{1+}) fine-structure entanglement for the 1-SPI process (1.1) in the outer-most 5p sub-shell of a Xe atom, i.e., for the transition $Xe(4d^{10}\ 5s^2\ 5p^6\ {}^1S_0) \rightarrow Xe^{1+}\ (4d^{10}\ 5s^2\ 5p^5\ {}^2P_{1/2})$. Thus, in the present example, residual photoion Xe^{1+} (with its total angular momentum $J_{1+} = \frac{1}{2}$) too is a qubit and one is looking at the entanglement in a (e_p, Xe^{1+})-state generated in the 1-SPI process (1.1) in $E1$ approximation in the presence of SOI. In [220], the fine-structure density matrix (7.5) was specialized to an experimental geometry wherein the spin of e_p is longitudinally polarized to its direction of propagation, i.e., $\hat{\boldsymbol{u}}_p\ (\vartheta_p, \varphi_p) || \ \hat{\boldsymbol{k}}_p\ (\theta_p, \phi_p)$. (The angular momenta of Xe, as well as of its photoion Xe^{1+}, are quantized along the OZ-axis in Fig. 4.1.) All the three measures of entanglement, i.e., negativity, concurrence, and entanglement of formation, were calculated [220] for several frequencies ν_r of the ionizing radiation γ_r in (7.5). For each of these ν_r, electromagnetic waves of four different polarizations [i.e., LP ($m_r = 0$), RCP ($m_r = +1$), LCP ($m_r = -1$), and UP (an even mixture of RCP and LCP)] were considered in the density matrix (7.5).

It was found in [220] that qubits (e_p, Xe^{1+}) were always in a pure state if the ionizing photon in (7.5) is LP, RCP or LCP; but, for an UP γ_r in (7.5), these two particles formed a mixed state. Consequently, in agreement with the theoretical prediction [142], in all those cases [220] wherein (e_p, Xe^{1+}) are in a pure entangled state (i.e., when Xe is ionized by a LP. RCP, or LCP γ_r), negativity and concurrence always have identical values; but, these two measures of entanglement differ [220] for ionization by UP light wherein (e_p, Xe^{1+}) are in a mixed state. Furthermore [220], the properties of the (e_p, Xe^{1+}) fine-structure entanglement did not change at all with a change in the helicity of the ionizing radiation, i.e., there is no circular dichroism[5] in the present example of fine-structure entanglement of (e_p, Xe^{1+}).

Figures 7.1a–c and 7.2a–c contain variations with the polar angle θ_p of the above-mentioned three measures of entanglement, calculated in [220] for ionization in the 5p sub-shell of a Xe atom by photons of energies $E_r = 1.0$ a.u. and 3.0 a.u., respectively. It is obvious from the figures that entanglement of formation for most of θ_p is usually less than concurrence/negativity. The pure (e_p, Xe^{1+})-state, according to Fig. 7.1a, is always maximally entangled for all directions of emission of a photoelectron; whereas, in Fig. 7.2a, it is so if photoelectron is moving either

[5]For a brief description of circular dichroism, see, for example, footnote (3) in Chap. 11.

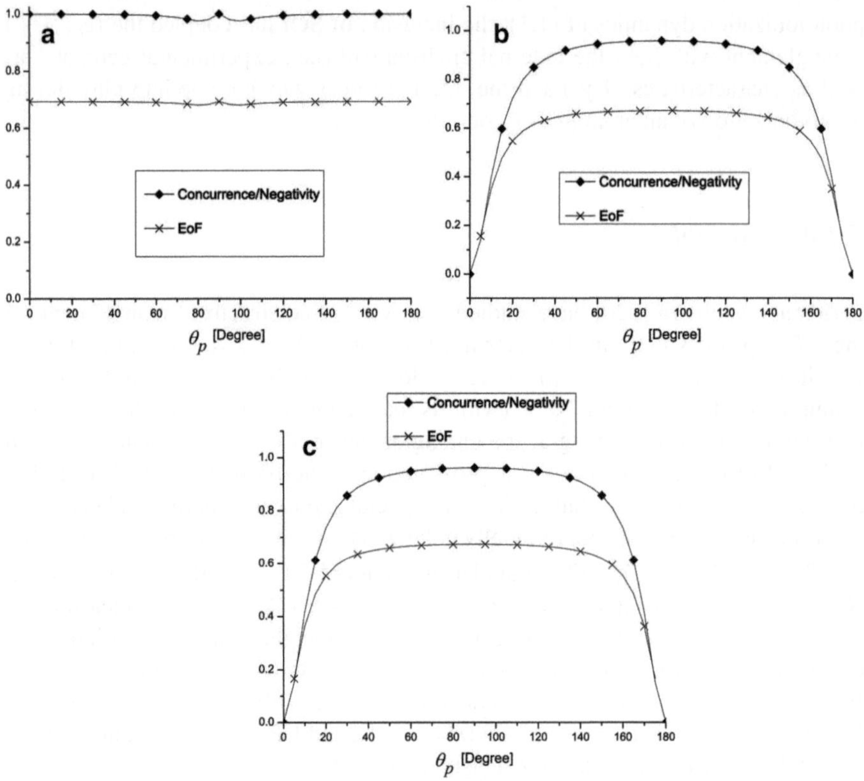

Fig. 7.1 Variations with the polar angle θ_p (of the photoelectron's direction of motion) of the concurrence/negativity and of the entanglement of formation for fine-structure entanglement between spin (longitudinally polarized to the direction of propagation) of e_p and angular momentum J_{1+} (quantized along the OZ-axis in Fig. 4.1) of the residual photoion Xe^{1+} generated in the 1-SPI process (1.1) taking place in 5p sub-shell of a Xe atom in the presence of SOI. These measures have been calculated for energy $E_r = 1.0$ a.u. of the ionizing radiation γ_r. In the present example, the use of LCP (i.e., $m_r = -1$) γ_r in (7.5) gave the values of the entanglement measures identical to those when a Xe atom is, instead, ionized by a RCP ($m_r = +1$) γ_r. This, in other words, means that there is no circular dichroism [see footnote (5)] in the fine-structure (e_p, Xe^{1+}) entanglement. Moreover, on account of the pure nature of a (e_p, Xe^{1+})-state, generated in photoionization by a LP or RCP/LCP γ_r in the present example, negativity and concurrence have identical values in parts (**a**) and (**b**) of this figure. Although, these two measures of entanglement for ionization by an UP radiation are not equal because (e_p, Xe^{1+}) are now in a mixed fine-structure state; their differences are, however, not visible on the scale of part (**c**) of the present figure. (**a**) Ionizing radiation in (7.5) LP; (**b**) ionizing radiation in (7.5) RCP/LCP; (**c**) ionizing radiation in (7.5) UP. (First published in [220].)

parallel (i.e., $\theta_p \cong 0°, 180°$), or perpendicular (i.e., $\theta_p \cong 90°$), to the electric field vector of LP ionizing radiation. In Figs. 7.1b and 7.2b, on the other hand, (e_p, Xe^{1+}) form a pure product state whenever photoelectron is ejected either along (i.e., $\theta_p \cong 0°$), or opposite (i.e., $\theta_p \cong 180°$), to the direction of incidence of circularly

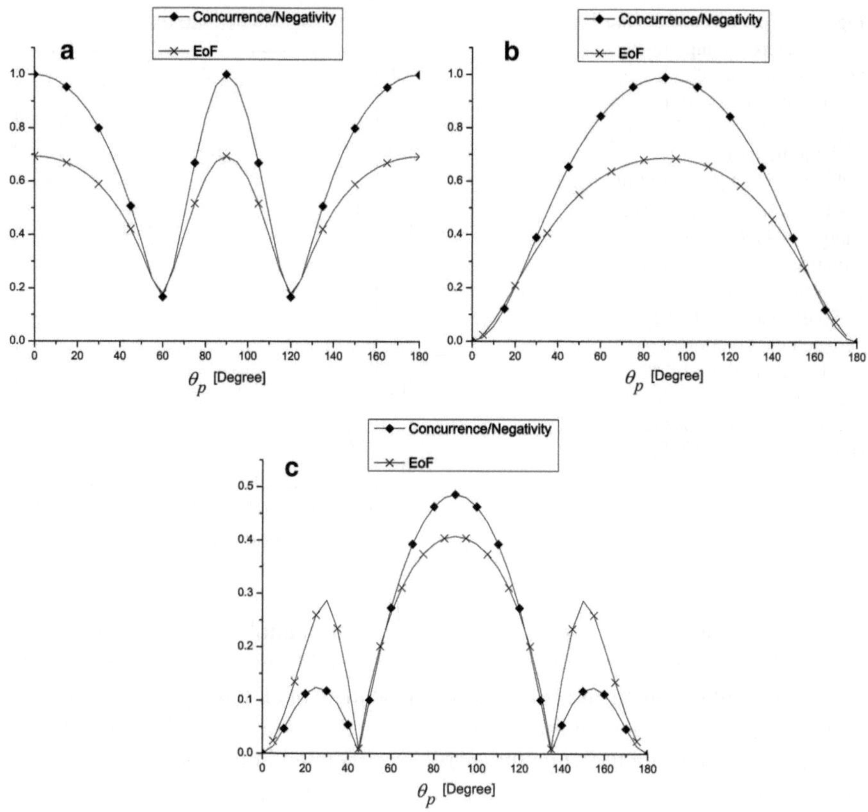

Fig. 7.2 Same as Fig. 7.1, but for $E_r = 3.00$ a.u. for energy of the ionizing radiation γ_r in the fine-structure density matrix (7.5). (First published in [220].)

polarized ionizing γ_r; for some of the other values of θ_p in these two figures, photoelectron-photoion may be maximally entangled. As far as ionization by UP radiation is concerned, the entanglement measures for $E_r = 1.0$ a.u. (Fig. 7.1c) are almost twice of those calculated for $E_r = 3.0$ a.u. (Fig. 7.2c). Moreover, in both of these last two cases, (e_p, Xe^{1+}) are not entangled at $\theta_p \cong 0°$ and 180°; in addition, there is no entanglement also at $\theta_p \cong 45°$ and 135° in Fig. 7.2c. Furthermore, the three measures of entanglement are almost same in Fig. 7.1b, c for ionization by circularly polarized (i.e., RCP/LCP) and UP γ_r, respectively; for $E_r = 3.0$ a.u., on the other hand, the entanglement measures in Fig. 7.2c are very different—in both magnitude and shape—from those seen in Fig. 7.2a, b.

These figures clearly show that the presence of the SOI results in strongly coupling, in a very complicated manner, the fine-structure (e_p, Xe^{1+}) entanglement with the kinematics as well as dynamics of the 1-SPI process (1.1), in addition to the characteristics (i.e., energy and polarization) of the ionizing radiation γ_r. In the Coulombic entanglement (discussed on pages 100–102 in the Sect. 4.2.1), on the

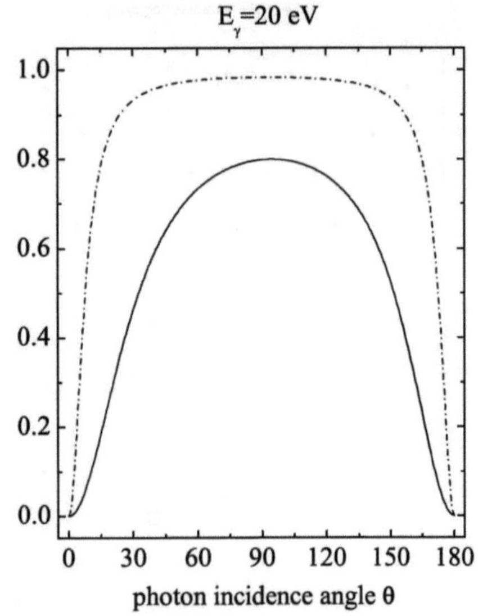

Fig. 7.3 Concurrence for the fine-structure entanglement between the spin of photoelectron e_p and the total angular momentum $J_{1+} = \frac{1}{2}$ of the ionic qubit of $Sr^{1+}(5s^1\ ^2S_{\frac{1}{2}})$ generated in the 1-SPI process (1.1). The energy of the CP ionizing radiation γ_r is E_r (shown as E_γ in the present figure) $= 20$ eV and θ is the angle between its direction of incidence and that of the ejection of e_p. ——: length gauge; ......: velocity gauge. (First published in [265].)

other hand, spins of a (e_p, Xe^{1+}) pair, in the present example, are always in the Bell state (2.17b) which is pure and maximally entangled. Moreover, the properties of the Coulombic entanglement are completely independent of all those physical attributes which are playing important roles in determining the properties of the fine-structure entanglement between the same two particles generated in the same 1-SPI process (1.1), but in the presence of SOI and discussed in the two preceding paragraphs.

Earlier, Radtke et al. [265] have considered the fine-structure entanglement for the 1-SPI process (1.1) taking place in an Sr atom. They [265] looked into the photoionizing transition $Sr(4p^6\,5s^2\ ^1S) \longrightarrow Sr^{1+}(4p^6\,5s^1\ ^2S_{\frac{1}{2}})$. Thus, the photoion in this example [265] is a qubit. Consequently, the fine-structure entanglement studied by Radtke et al. [265] is between two qubits. The entanglement measure calculated in [265] is concurrence. They [265] have calculated concurrence as a function of the angle θ ($0^o \le \theta \le 180^o$) for three different frequencies ν_r of the RCP ionizing radiation γ_r. Here, θ is the angle between the direction of incidence of γ_r with that of departure of e_p in the 1-SPI process (1.1). The variation of concurrence with the energy E_r (from 5 to 30 eV), but for θ ($= 30^o$, 60°, 90°), has also been given in [265] both for RCP and UP ionizing radiation γ_r. Figures 7.3 and 7.4 herein are taken from [265]. These [265] results show, among other things, that for certain values of θ and/or ν_r, (e_p, $\mathfrak{T}^{1+}$) are in a state which has maximum possible fine-structure entanglement; whereas for some other values of these two physical quantities, concurrence becomes zero, i.e., the entanglement between (e_p, Sr^{1+}) totally disappears and, hence, these two qubits are in a separable state. Further, a change from RCP to UP ionizing radiation in (1.1) has negligible effects on the properties of the (e_p, γ_d) fine-structure entanglement.

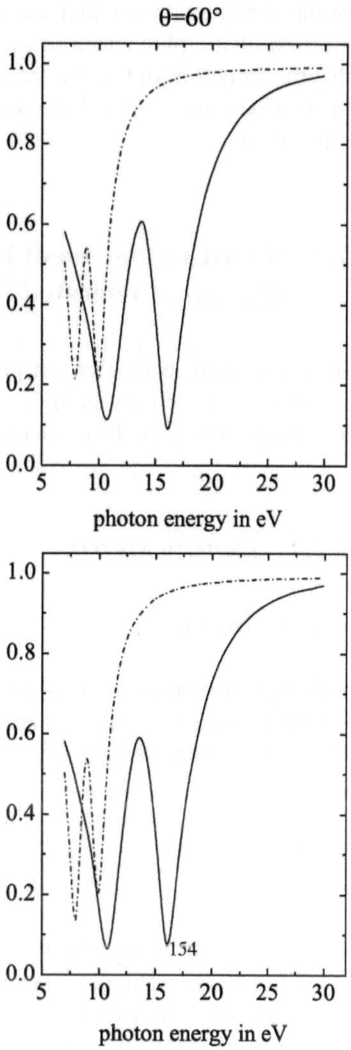

Fig. 7.4 Same as Fig. 7.3, but for fixed $\theta = 60°$ and varying energy (5 eV $\leq E_r \leq$ 30 eV) of the incident photon γ_r. The upper and lower figures are for ionization of Sr in the process (1.1) by RCP and UP γ_r, respectively. (First published in [265].)

This dependence of the properties of the fine-structure entanglement on the characteristics of the photon γ_r as well as on the direction of observation of e_p is a consequence of the fact that density matrix used by Radtke et al. [265] also involves both the geometrical and the dynamical factors determining the 1-SPI process (1.1). This again is completely different from the Coulombic entanglement discussed in the Chap. 4 for the same 1-SPI process (1.1). According to the discussion given on pages 100–102 in Sect. 4.2.1 for Coulombic entanglement, the transition $\mathrm{Sr}(4p^6\,5s^2\;{}^1S) \longrightarrow \mathrm{Sr}^{1+}(4p^6\,5s^1\;{}^2S_{\frac{1}{2}})$ will always generate a pure and maximally entangled spin state of $(e_p,\,\mathrm{Sr}^{1+})$ in the absence of SOI, whatever may be the frequency ν_r

of the absorbed photon and the angle θ between its direction of incidence and of ejection of the photoelectron e_p. Both above examples show, on the other hand, that the properties of the fine-structure entanglement are totally different from those of the Coulombic entanglement even for the same photoionizing transition in the same atom.

7.2 Spin-Entanglement Between Two Photoelectrons (e_p, $e_{p'}$) Generated in 1-DPI

In this section, entanglement properties of a state of two photoelectrons, who simultaneously left an atom $\mathfrak{T}$ following the absorption in $E1$ approximation of a single photon γ_r in the presence of C+SOIs, are discussed.

7.2.1 *Density Matrix*

7.2.1.1 Derivation

Although the condition (5.1a) for the conservation of energy holds even in the j-j coupling; however, (E_0, E_{2+}) are now the energies of the electronic states ($|J_0, M_0\rangle$, $|J_{2+}, M_{2+}\rangle$) representing ($\mathfrak{T}$, $\mathfrak{T}^{2+}$) in the presence of C+SOIs. The respective conditions (5.1b) and (5.1c) for the conservation of the orbital and spin angular momenta in L-S coupling in the absence of SOI, are now to be replaced by the single requirement

$$\boldsymbol{\ell}_r + \boldsymbol{J}_0 = \boldsymbol{J}_{2+} + \boldsymbol{j}\Big[= \boldsymbol{j}_p + \boldsymbol{j}_{p'}\Big]. \tag{7.8}$$

Here, $\boldsymbol{j}_p$ [$= \boldsymbol{\ell}_p + (\frac{1}{2})_p$] and $\boldsymbol{j}_{p'}$ [$= \boldsymbol{\ell}_{p'} + (\frac{1}{2})_{p'}$] are the total (i.e., sum of the orbital and spin) angular momenta of the respective photoelectrons e_p and $e_{p'}$ simultaneously emitted in the 1-DPI process (1.2).

The required density matrix

$$\begin{aligned}
&\langle J_{2+};\ \mu_p\,\hat{\boldsymbol{u}}_p\,\boldsymbol{k}_p;\ \mu_{p'}\,\hat{\boldsymbol{u}}_{p'}\,\boldsymbol{k}_{p'}|\ \rho^{\text{(1-DPI)}}\ |J_{2+};\ \mu'_p\,\hat{\boldsymbol{u}}_p\,\boldsymbol{k}_p;\ \mu'_{p'}\,\hat{\boldsymbol{u}}_{p'}\,\boldsymbol{k}_{p'}\rangle \\
&= \frac{\mathcal{K}_p}{(2J_0+1)} \sum_{M_{J_0}\,M_{J_{2+}}} \langle J_{2+}\ M_{J_{2+}};\ \mu_p\,\hat{\boldsymbol{u}}_p\,\boldsymbol{k}_p;\ \mu_{p'}\,\hat{\boldsymbol{u}}_{p'}\,\boldsymbol{k}_{p'}|\ F_p\ |J_0\ M_{J_0};\ 1m_r\rangle \\
&\quad\times\langle J_0\ M_{J_0};\ 1m_r|\ F_p^\dagger\ |J_{2+}\ M_{J_{2+}};\ \mu'_p\,\hat{\boldsymbol{u}}_p\,\boldsymbol{k}_p;\ \mu'_{p'}\,\hat{\boldsymbol{u}}_{p'}\,\boldsymbol{k}_{p'}\rangle.
\end{aligned} \tag{7.9}$$

in the presence of SOI, corresponding to that given in (5.4) when no SOI is taken into account, is obtained by averaging and summing over the Zeeman components [10, 60] of the state $|J_0 M_{J_0}\rangle$ of the atomic target $\mathfrak{T}$ and state $|J_{2+} M_{J_{2+}}\rangle$ of the residual photodication $\mathfrak{T}^{2+}$, respectively. As before, the averaging and summing operations have been performed assuming that atom $\mathfrak{T}$ is not polarized and dication $\mathfrak{T}^{2+}$ has not been observed. The density matrix (7.9) too satisfies the requirements (3.16) and (3.17); its diagonal elements have the same interpretation as given on page 116, but unlike in (5.4), this time 1-DPI is taking place in the presence of SOI in j-j coupling.

In order to evaluate the amplitudes present in (7.9), one needs an expression for a state of the ($\mathfrak{T}^{1+^*} + e_p + e_{p'}$)-system wherein angular momenta be coupled according to the scheme suggested on the right-hand side of the conservation condition (7.8) applicable in j-j coupling. Using the continuum orbital (3.38a) for the photoelectron e_p, and a similar expression for its companion $e_{p'}$, one can readily write the following expression for the desired state [252, 258].

$$|J_{2^+}\, M_{J_{2^+}};\, \mu_p\, \hat{\boldsymbol{u}}_p\, \boldsymbol{k}_p;\, \mu_{p'}\, \hat{\boldsymbol{u}}_{p'}\, \boldsymbol{k}_{p'}\rangle \equiv \mathcal{A}\Big(|2^+\rangle\, |\mu_p\, \hat{\boldsymbol{u}}_p\, \boldsymbol{k}_p\rangle\, |\mu_{p'}\, \hat{\boldsymbol{u}}_{p'}\, \boldsymbol{k}_{p'}\rangle\Big) \quad (7.10a)$$

$$= \sum_{\substack{\ell_p\, m_{\ell_p}\, \nu_p\, j_p\, m_{j_p}\, j\, m_j \\ \ell_{p'}\, m_{\ell_{p'}}\, \nu_{p'}\, j_{p'}\, m_{j_{p'}}}} i^{\ell_p+\ell_{p'}}\, e^{-i\,(\sigma_{\ell_p j_p}+\sigma_{\ell_{p'} j_{p'}})}$$

$$\times (-1)^{1+\ell_p+\ell_{p'}-j_p+j_{p'}-m_{j_p}-m_{j_{p'}}-m_j} \sqrt{(2j_p+1)(2j_{p'}+1)(2j+1)}$$

$$\times \begin{pmatrix} \ell_p & \frac{1}{2} & j_p \\ m_{\ell_p} & \nu_p & -m_{j_p} \end{pmatrix} \begin{pmatrix} \ell_{p'} & \frac{1}{2} & j_{p'} \\ m_{\ell_{p'}} & \nu_{p'} & -m_{j_{p'}} \end{pmatrix} \begin{pmatrix} j_p & j_{p'} & j \\ m_{j_p} & m_{j_{p'}} & -m_j \end{pmatrix} |2^+; j m_j\rangle$$

$$\times \left[Y_{\ell_p}^{m_{\ell_p}}(\hat{\boldsymbol{k}}_p)\right]^* \left[Y_{\ell_{p'}}^{m_{\ell_{p'}}}(\hat{\boldsymbol{k}}_{p'})\right]^* \left[\mathcal{D}^{\frac{1}{2}}_{\mu_p\, \nu_p}(\omega_p)\right]^* \left[\mathcal{D}^{\frac{1}{2}}_{\mu_{p'}\, \nu_{p'}}(\omega_{p'})\right]^*. \quad (7.10b)$$

Here, in (7.10a) and in the following expression (7.10c), operator $\mathcal{A}$ antisymmetrizes photoelectrons e_p and $e_{p'}$ with the bound electrons of the dication $\mathfrak{T}^{2+}$; while, in (7.10b)

$$|2^+;\ j m_j\rangle = \mathcal{A}\Bigg[|2^+\rangle\, (-1)^{-j_p+j_{p'}-m_j}\, \sqrt{2j+1}$$

$$\times\, R^-_{j_p}(r_p; k_p) R^-_{j_{p'}}(r_{p'}; k_{p'}) \sum_{m_p\, m_{j'}} \begin{pmatrix} j_p & j_{p'} & j \\ m_{j_p} & m_{j_{p'}} & -m_j \end{pmatrix} |j_p m_{j_p}\rangle\, |j_{p'} m_{j_{p'}}\rangle\Bigg]. \quad (7.10c)$$

The two kets, present on the right-hand side in (7.10c), are defined by the relation (3.36b) for each of the two photoelectrons e_p and $e_{p'}$, respectively. Angular momentum coupling, required by the right-hand side of the conservation condition (7.8) is, on the other hand, introduced as follows

$$|(J_{2^+} j) J M_J\rangle = (-1)^{-J_{2^+}+j-M_J}\sqrt{2J+1} \times \sum_{M_{J_{2^+}}\, m_j} \begin{pmatrix} J_{2^+} & j & J \\ M_{J_{2^+}} & m_j & -M_J \end{pmatrix} |2^+;\ j m_j\rangle. \tag{7.10d}$$

With the information given in (3.35) and (7.10), one can immediately obtain an expression for the amplitude present in the density matrix (7.9) for the 1-DPI process (1.2) in the j-j coupling. The resulting expression will contain, among other things, the Dirac bracket [252, 258]

$$\langle 2^+;\ j m_j | F_p | 0,\ 1 m_r\rangle = (-1)^{1-j-J_0+J_{2^+}} \sum_J (2J+1) \times \begin{pmatrix} J_0 & 1 & J \\ M_{J_0} & m_r & -M_J \end{pmatrix} \begin{pmatrix} J_{2^+} & j & J \\ M_{J_{2^+}} & m_j & -M_J \end{pmatrix} \langle (J_{2^+} j) J | F_p | (J_0 1) J\rangle, \tag{7.11a}$$

which satisfies the conservation condition (7.8). A recoupling of the angular momenta according to the procedure suggested elsewhere [190], simplifies (7.11a) to [252, 258]

$$\langle 2^+;\ j m_j | F_p | 0,\ 1 m_r\rangle = (-1)^{-J_0-J_{2^+}-M_{J_0}-m_j} \sum_{j_t\, m_t} (2j_t+1) \times \begin{pmatrix} 1 & j & j_t \\ m_r & -m_j & m_t \end{pmatrix} \begin{pmatrix} J_{2^+} & J_0 & j_t \\ M_{J_{2^+}} & -M_{J_0} & m_t \end{pmatrix} \langle J_{2^+} j | F(j_t) | J_0 1\rangle \tag{7.11b}$$

with

$$\langle J_{2^+} j | F(j_t) | J_0 1\rangle = \sum_J (-1)^J (2J+1) \times \begin{Bmatrix} 1 & J_0 & J \\ J_{2^+} & j & j_t \end{Bmatrix} \langle (J_{2^+} j) J | F_p | (J_0 1) J\rangle. \tag{7.11c}$$

In (7.11), $\boldsymbol{j}_t = \boldsymbol{\ell}_r - \boldsymbol{j} = \boldsymbol{J}_{2^+} - \boldsymbol{J}_0$ is the angular momentum transferred from the absorbed photon γ_r to the observed spin-resolved photoelectrons $(e_p,\ e_{p'})$ simultaneously emitted in the 1-DPI process (1.2) [252, 258].

After these, and some additional simplifications, the amplitude for 1-DPI finally becomes [252]

$$\langle J_{2^+}\, M_{J_{2^+}};\ \mu_p\, \hat{\boldsymbol{u}}_p\, \boldsymbol{k}_p;\ \mu_{p'}\, \hat{\boldsymbol{u}}_{p'}\, \boldsymbol{k}_{p'} |\, F_p\, | J_0\, M_{J_0};\ 1 m_r\rangle = (-1)^{1-J_0-J_{2^+}-M_{J_0}} \times \sum_{\substack{\ell_p\, m_{\ell_p}\, \nu_p\, j_p\, m_{j_p}\, j\, m_j \\ \ell_{p'}\, m_{\ell_{p'}}\, \nu_{p'}\, j_{p'}\, m_{j_{p'}}\, j_t\, m_t}} (-i)^{\ell_p+\ell_{p'}}\, \mathrm{e}^{\mathrm{i}\,(\sigma_{\ell_p j_p}+\sigma_{\ell_{p'} j_{p'}})}\, (-1)^{\ell_p+\ell_{p'}+j_p-j_{p'}+m_j}$$

$$\times (2j_t+1)\sqrt{(2j_p+1)(2j_{p'}+1)(2j+1)}\begin{pmatrix} \ell_p & \frac{1}{2} & j_p \\ m_{\ell_p} & \nu_p & -m_{j_p} \end{pmatrix}\begin{pmatrix} \ell_{p'} & \frac{1}{2} & j_{p'} \\ m_{\ell_{p'}} & \nu_{p'} & -m_{j_{p'}} \end{pmatrix}$$

$$\times \begin{pmatrix} j_p & j_{p'} & j \\ m_{j_p} & m_{j_{p'}} & -m_j \end{pmatrix}\begin{pmatrix} 1 & j & j_t \\ m_r & -m_j & m_t \end{pmatrix}\begin{pmatrix} J_{2+} & J_0 & j_t \\ M_{J_{2+}} & -M_{J_0} & m_t \end{pmatrix}$$

$$\times Y_{\ell_p}^{m_{\ell_p}}(\hat{\boldsymbol{k}}_p)\, Y_{\ell_{p'}}^{m_{\ell_{p'}}}(\hat{\boldsymbol{k}}_{p'})\, \mathcal{D}^{\frac{1}{2}}_{\mu_p \nu_p}(\omega_p)\, \mathcal{D}^{\frac{1}{2}}_{\mu_{p'} \nu_{p'}}(\omega_{p'})\langle J_{2+}\, j\,|F(j_t)|J_0 1\rangle. \tag{7.12}$$

Here, $\langle J_{2+}\, j\,|F(j_t)|J_0 1\rangle$ are already defined in (7.11c).

For calculating the density matrix, required in the present case, we now substitute amplitude (7.12) and its Hermitian conjugate in (7.9). The consequent expression needs several simplifications. Some of these are [252] applications of identities (3.7.8), (6.2.8), (4.3.2), and (4.6.5) from Edmonds [67]. The first two of these four identities are used only once, while each of the remaining needs to be applied twice. In addition, one twice needs to convert—with the help of the identity (14.42) given in [64]—a quadruple sum of the product of four 3-j symbols to a double sum of the product of two 3-j and one 9-j symbols. After all these simplifications, the density matrix (7.9) in its final form can be written as [252]

$$\langle J_{2+};\, \mu_p\, \hat{\boldsymbol{u}}_p\, \boldsymbol{k}_p;\, \mu_{p'}\, \hat{\boldsymbol{u}}_{p'}\, \boldsymbol{k}_{p'}|\, \rho^{\text{(1-DPI)}}\, |J_{2+};\, \mu_p'\, \hat{\boldsymbol{u}}_p\, \boldsymbol{k}_p;\, \mu_{p'}'\, \hat{\boldsymbol{u}}_{p'}\, \boldsymbol{k}_{p'}\rangle$$

$$= (-1)^{\mu'_p+\mu'_{p'}} \sum_{\substack{S_1\, M_1\, M_{S_1} \\ S_2\, M_2\, M_{S_2}}} (-1)^{M_1+M_2}\begin{pmatrix} \frac{1}{2} & \frac{1}{2} & S_1 \\ \mu_p & -\mu_p' & M_1 \end{pmatrix}\begin{pmatrix} \frac{1}{2} & \frac{1}{2} & S_2 \\ \mu_{p'} & -\mu_{p'}' & M_2 \end{pmatrix}$$

$$\times \mathcal{D}^{S_1}_{-M_1\, M_{S_1}}(\omega_p)\, \mathcal{D}^{S_2}_{-M_2\, M_{S_2}}(\omega_{p'}) A^{S_1\, S_2}_{M_{S_1},\, M_{S_2}}(m_r;\, \boldsymbol{k}_p,\, \boldsymbol{k}_{p'}) \tag{7.13a}$$

with

$$A^{S_1\, S_2}_{M_{S_1},\, M_{S_2}}(m_r;\, \boldsymbol{k}_p,\, \boldsymbol{k}_{p'}) = -1^{1+m_r+S_1+S_2}\,(2S_1+1)(2S_2+1)\frac{\mathcal{K}_p}{4\pi(2J_0+1)}$$

$$\times \sum_{\substack{\ell_p\, \ell_p'\, j_p\, j_p'\, L_1\, M_{L_1}\, J_1\, J_2\, M_J \\ \ell_{p'}\, \ell_{p'}'\, j_{p'}\, j_{p'}'\, L_2\, M_{L_2}\, j\, j'\, j_t\, L_r}} (-1)^{\ell_p'+\ell_{p'}'+j+j_t}(2j_t+1)(2L_r+1)$$

$$\times (2J_1+1)(2J_2+1)\sqrt{(2L_1+1)(2L_2+1)}\begin{pmatrix} 1 & 1 & L_r \\ m_r & -m_r & 0 \end{pmatrix}$$

$$\times \begin{pmatrix} \ell_p & \ell_p' & L_1 \\ 0 & 0 & 0 \end{pmatrix}\begin{pmatrix} \ell_{p'} & \ell_{p'}' & L_2 \\ 0 & 0 & 0 \end{pmatrix}\begin{pmatrix} L_1 & S_1 & J_1 \\ M_{L_1} & M_{S_1} & M_J \end{pmatrix}\begin{pmatrix} L_2 & S_2 & J_2 \\ M_{L_2} & M_{S_2} & -M_J \end{pmatrix}$$

$$\times \begin{pmatrix} J_1 & J_2 & L_r \\ M_J & -M_J & 0 \end{pmatrix} \begin{Bmatrix} 1 & 1 & L_r \\ j & j' & j_t \end{Bmatrix}$$

$$\times \begin{Bmatrix} \ell_p & \frac{1}{2} & j_p \\ \ell'_p & \frac{1}{2} & j'_p \\ L_1 & S_1 & J_1 \end{Bmatrix} \begin{Bmatrix} \ell_{p'} & \frac{1}{2} & j_{p'} \\ \ell'_{p'} & \frac{1}{2} & j'_{p'} \\ L_2 & S_2 & J_2 \end{Bmatrix} \begin{Bmatrix} j_p & j_{p'} & j \\ j'_p & j'_{p'} & j' \\ J_1 & J_2 & L_r \end{Bmatrix} Y_{L_1}^{M_{L_1}}(\hat{\boldsymbol{k}}_p) Y_{L_2}^{M_{L_2}}(\hat{\boldsymbol{k}}_{p'})$$

$$\times\, d_{\ell_p \ell_{p'}}\Big(J_{2^+}\big(j_p j_{p'}\big)j;\ J_0 1;\ j_t\Big) \Big[d_{\ell'_p \ell'_{p'}}\Big(J_{2^+}\big(j'_p j'_{p'}\big)j';\ J_0 1;\ j_t\Big)\Big]^* \tag{7.13b}$$

and

$$d_{\ell_p \ell_{p'}}\Big(J_{2^+}\big(j_p j_{p'}\big)j;\ J_0 1;\ j_t\Big) = (-1)^{\ell_p + \ell_{p'}}\, \mathrm{e}^{\mathrm{i}\,(\sigma_{\ell_p j_p} + \sigma_{\ell_{p'} j_{p'}})}$$
$$\times \sqrt{(2\ell_p + 1)(2\ell_{p'} + 1)(2j_p + 1)(2j_{p'} + 1)(2j + 1)}\ \ \langle J_{2^+}\, j \,|F(j_t)|\, J_0 1\rangle. \tag{7.13c}$$

7.2.1.2 Properties

One can readily show that the density matrix (7.13) is Hermitian, i.e., satisfies a condition equivalent to (3.16) applicable in the present case. The diagonal terms $\langle J_{2^+};\ \mu_p \hat{\boldsymbol{u}}_p \boldsymbol{k}_p;\ \mu_{p'} \hat{\boldsymbol{u}}_{p'} \boldsymbol{k}_{p'} |\, \rho^{(1\text{-DPI})}\, | J_{2^+};\ \mu_p \hat{\boldsymbol{u}}_p \boldsymbol{k}_p;\ \mu_{p'} \hat{\boldsymbol{u}}_{p'} \boldsymbol{k}_{p'}\rangle$ in (7.13b) are the angular distribution [258] of spin-resolved photoelectrons (e_p, $e_{p'}$) simultaneously emitted in the 1-DPI process (1.2) in the presence of C+SOIs in j-j coupling. These terms are, therefore, never less than zero. The non-diagonal elements (i.e., terms with $\mu_p \neq \mu'_p$ and/or $\mu_{p'} \neq \mu'_{p'}$) in (7.13b) represent, on the other hand, coherence effects in 1-DPI. Moreover, as usual, this density matrix too is always positive. [6]

Let us now compare the present density matrix (7.13) with that given in (5.9). Both of these matrices include in full the correlation effects present in an interacting many-body quantum system. The later one is, however, a Coulombic density matrix as it was obtained in the presence of only the Coulomb forces experienced by electrons inside an atom—SOI was not taken into account while deriving (5.9). But, in arriving at (7.13), both C+SOIs have been included.

These two density matrices are, obviously, totally different from each other. For example, unlike the Coulombic density matrix, (7.13) cannot be written as a product of two independent terms representing purely angular and purely spin correlations between two simultaneously emitted photoelectrons (e_p, $e_{p'}$). Consequently, while it was only the spin-correlation part (5.11) of the density matrix (5.9) which

[6] See footnote (6) on page 39.

completely determined the properties of the Coulombic entanglement in Chap. 5, now the whole of the density matrix (7.13) needs to be considered for investigating the entanglement in the 1-DPI process (3.2) in the presence of C+SOI. It, in other words, means that entanglement between the spin angular momenta of $(e_p, e_{p'})$ now depends upon the properties (e.g., polarization m_r, frequency ν_r) of the ionizing radiation γ_r in (3.2), kinematics {i.e., linear momenta $[\boldsymbol{k}_p\ (k_p, \hat{\boldsymbol{k}}_p\ (\theta_p, \phi_p)), \boldsymbol{k}_{p'}\ (k_{p'}, \hat{\boldsymbol{k}}_{p'}\ (\theta_{p'}, \phi_{p'}))]$ or, equivalently, on the energies $[\epsilon_p, \epsilon_{p'}]$ and directions $[\hat{\boldsymbol{k}}_p, \hat{\boldsymbol{k}}_{p'}]$ of ejection/propagation} of photoelectrons $[e_p, e_{p'}]$, experimental geometry {i.e., directions $[\hat{\boldsymbol{u}}_p\ (\vartheta_p, \varphi_p), \hat{\boldsymbol{u}}_{p'}\ (\vartheta_{p'}, \varphi_{p'})]$ of quantization of the spins of $[e_p, e_{p'}]$}, angular momenta $[J_0, J_{2+}]$ of the electronic states $[|0\rangle, |2^+\rangle]$ of $[\mathfrak{T}, \mathfrak{T}^{2+}]$, as well as the dynamics [i.e., phases and amplitudes (7.13c)] of the process of 1-DPI. But the Coulombic entanglement (5.11) depends on none of these physical variables; it, on the other hand, is completely characterized merely by a knowledge of the spin multiplicities of the electronic states of $(\mathfrak{T}, \mathfrak{T}^{2+})$ in L-S coupling.

Thus, the total angular momenta (J_0, J_{2+}) of the electronic states $\left(|J_0\, M_{J_0}\rangle, |J_{2+}\, M_{J_{2+}}\rangle\right)$ of $(\mathfrak{T}, \mathfrak{T}^{2+})$ participating in the 1-DPI process (1.2) play important roles in the density matrix (7.13). Fine-structure levels of the atomic target $\mathfrak{T}$ and of its residual dication $\mathfrak{T}^{2+}$ are specified [10, 60] by J_0 and J_{2+}, respectively. Expression (7.13) can, therefore, be appropriately termed a fine-structure density matrix for 1-DPI. Consequently, an investigation of non-local correlation in the spin-state (7.13) will, henceforth, be called as a study of fine-structure entanglement of a $(e_p, e_{p'})$-pair.

It is obvious from the above discussion that a study of the fine-structure entanglement between the spins of $(e_p, e_{p'})$, unlike that of their Coulombic entanglement, necessarily requires a knowledge of the dynamical parameters d's present in the density matrix (7.13). These can be calculated in a hierarchy of approximations, starting from semi-empirical phenomenological to sophisticated *ab-initio* methods. Thus, the presence of SOI has coupled the entanglement of the spins of $(e_p, e_{p'})$ with both the kinematics and dynamics of the 1-DPI process (1.2).

7.2.2 A Generic Example

It is extremely difficult to calculate the sixteen elements of the (4×4) matrix (7.13a) as each of them contains a large number of complicated coefficients A's, defined in (7.13b), which requires a knowledge of the dynamical parameters (7.13c) [i.e., $d_{\ell_p \ell_{p'}}$], among other things. However, one can qualitatively learn about the important properties of the fine-structure entanglement of the spin angular momenta of $(e_p, e_{p'})$ in the presence of C+SOIs even without using any $d_{\ell_p \ell_{p'}}$ present in this density matrix.

In order to simplify the analysis, let us assume, following Chandra and Chakraborty [252], that the photoelectrons e_p and $e_{p'}$ are observed parallel and anti parallel to the OZ-axis in Fig. 5.1. This, in other words, means that $(e_p, e_{p'})$ are moving in a collinear geometry in opposite directions parallel to the OZ-axis. Obviously, then, $\boldsymbol{k}_p = (k_p, \hat{\boldsymbol{e}}_z)$ and $\boldsymbol{k}_{p'} = (k_{p'}, -\hat{\boldsymbol{e}}_z)$.

The entanglement of a system of particles is always independent of the basis set used and of the directions of quantization. In view of this, Chandra and Chakraborty [252] further simplified their density matrix (7.13) by assuming that the spins of both of the photoelectrons (e_p, $e_{p'}$) are quantized in the $X-Y$ plane of the SF shown in Fig. 5.1. This corresponds to taking the Euler angles ω_p (φ_p, $\vartheta_p = \pi/2, 0$) and $\omega_{p'}$ ($\varphi_{p'}, \vartheta_{p'} = \pi/2, 0$) in (7.13a).

This reduces expression (7.13a) to the following form [252]

$$\begin{aligned}
&\langle J_{2+};\ \mu_p\,\hat{\boldsymbol{u}}_p\,\boldsymbol{k}_p;\ \mu_{p'}\,\hat{\boldsymbol{u}}_{p'}\,\boldsymbol{k}_{p'}|\,\rho^{(\text{1-DPI})}\,|J_{2+};\ \mu'_p\,\hat{\boldsymbol{u}}_p\,\boldsymbol{k}_p;\ \mu'_{p'}\,\hat{\boldsymbol{u}}_{p'}\,\boldsymbol{k}_{p'}\rangle \\
&\quad\longrightarrow \rho(\varphi_p;\,\varphi_{p'})_{\mu_p,\mu_{p'};\mu'_p,\mu'_{p'}} = (-1)^{\mu'_p+\mu'_{p'}} \\
&\quad\times \sum_{\substack{S_1\,S_2\,S\\ M_1\,M_2\,M}} (-1)^{M_1+M_2} \begin{pmatrix} \frac{1}{2} & \frac{1}{2} & S_1 \\ \mu_p & -\mu'_p & M_1 \end{pmatrix} \begin{pmatrix} \frac{1}{2} & \frac{1}{2} & S_2 \\ \mu_{p'} & -\mu'_{p'} & M_2 \end{pmatrix} \begin{pmatrix} S_1 & S_2 & S \\ M & -M & 0 \end{pmatrix} \\
&\quad\times \mathcal{D}^{S_1}_{-M_1\,M}(\varphi_p, \pi/2, 0)\,\mathcal{D}^{S_2}_{-M_2\,-M}(\varphi_{p'}, \pi/2, 0)\ A^{S_1\,S_2}_S(m_r),
\end{aligned} \tag{7.14a}$$

where

$$\begin{aligned}
&A^{S_1\,S_2}_S(m_r) = (-1)^{1+m_r+S_1+S_2} \\
&\quad\times (2S_1+1)(2S_2+1)(2S+1)\frac{\mathcal{K}_p}{16\pi^2(2J_0+1)} \\
&\quad\times \sum_{\substack{\ell_p\,\ell'_p\,j_p\,j'_p\,L_1\,J_1\,J_2\,L\\ \ell_{p'}\,\ell'_{p'}\,j_{p'}\,j'_{p'}\,L_2\,j\,j'\,j_t\,L_r}} (-1)^{\ell'_p+\ell_{p'}+j+j_t} \\
&\quad\times (2j_t+1)(2L_r+1)(2L_1+1)(2L_2+1)(2L+1)(2J_1+1)(2J_2+1) \\
&\quad\times \begin{pmatrix} 1 & 1 & L_r \\ m_r & -m_r & 0 \end{pmatrix} \begin{pmatrix} \ell_p & \ell'_p & L_1 \\ 0 & 0 & 0 \end{pmatrix} \begin{pmatrix} \ell_{p'} & \ell'_{p'} & L_2 \\ 0 & 0 & 0 \end{pmatrix} \begin{pmatrix} L_1 & L_2 & L_r \\ 0 & 0 & 0 \end{pmatrix} \begin{pmatrix} L & S & L_r \\ 0 & 0 & 0 \end{pmatrix} \\
&\quad\times \begin{Bmatrix} 1 & 1 & L_r \\ j & j' & j_t \end{Bmatrix} \begin{Bmatrix} \ell_p & \frac{1}{2} & j_p \\ \ell'_p & \frac{1}{2} & j'_p \\ L_1 & S_1 & J_1 \end{Bmatrix} \begin{Bmatrix} \ell_{p'} & \frac{1}{2} & j_{p'} \\ \ell'_{p'} & \frac{1}{2} & j'_{p'} \\ L_2 & S_2 & J_2 \end{Bmatrix} \begin{Bmatrix} j_p & j_{p'} & j \\ j'_p & j'_{p'} & j' \\ J_1 & J_2 & L_r \end{Bmatrix} \begin{Bmatrix} L_1 & S_1 & J_1 \\ L_2 & S_2 & J_2 \\ L & S & L_r \end{Bmatrix} \\
&\quad\times d_{\ell_p\ell_{p'}}\big(J_{2+}\big(j_p j_{p'}\big)j;\ J_0 1;\ j_t\big)\Big[d_{\ell'_p\ell'_{p'}}\big(J_{2+}\big(j'_p j'_{p'}\big)j';\ J_0 1;\ j_t\big)\Big]^*.
\end{aligned} \tag{7.14b}$$

Thus, each element of the density matrix can now contain up to a maximum of six A's (namely, A^{00}_0, A^{01}_1, A^{10}_1, A^{11}_0, A^{11}_1, A^{11}_2). Expression (7.14a) can now be written as [252]

$$
\rho(\varphi_p; \varphi_{p'}) =
\begin{array}{ccccc}
\mu_p\mu_{p'}/\mu'_p\mu'_{p'} & \frac{1}{2}\frac{1}{2} & \frac{1}{2}-\frac{1}{2} & -\frac{1}{2}\frac{1}{2} & -\frac{1}{2}-\frac{1}{2} \\
\frac{1}{2}\frac{1}{2} & A_0 + cA_p & A_{01} + cA_{11} & A_{10} - cA_{11} & A_m - cA_p \\
 & +isA_{11} & +isA_p & -isA_p & -isA_{11} \\
\frac{1}{2}-\frac{1}{2} & A_{01} - cA_{11} & A_0 - cA_p & A_m + cA_p & A_{10} + cA_{11} \\
 & -isA_p & -isA_{11} & +isA_{11} & +isA_p \\
-\frac{1}{2}\frac{1}{2} & A_{10} + cA_{11} & A_m + cA_p & A_0 - cA_p & A_{01} - cA_{11} \\
 & +isA_p & +isA_{11} & -isA_{11} & -isA_p \\
-\frac{1}{2}-\frac{1}{2} & A_m - cA_p & A_{10} - cA_{11} & A_{01} + cA_{11} & A_0 + cA_p \\
 & -isA_{11} & -isA_p & +isA_p & +isA_{11}.
\end{array}
\tag{7.15a}
$$

The partial transpose of this matrix with respect to the photoelectron $e_{p'}$ is [252]

$$
\rho^{T_{p'}}(\varphi_p; \varphi_{p'}) =
\begin{array}{ccccc}
\mu_p\mu_{p'}/\mu'_p\mu'_{p'} & \frac{1}{2}\frac{1}{2} & \frac{1}{2}-\frac{1}{2} & -\frac{1}{2}\frac{1}{2} & -\frac{1}{2}-\frac{1}{2} \\
\frac{1}{2}\frac{1}{2} & A_0 + cA_p & A_{01} - cA_{11} & A_{10} - cA_{11} & A_m + cA_p \\
 & +isA_{11} & -isA_p & -isA_p & +isA_{11} \\
\frac{1}{2}-\frac{1}{2} & A_{01} + cA_{11} & A_0 - cA_p & A_m - cA_p & A_{10} + cA_{11} \\
 & +isA_p & -isA_{11} & -isA_{11} & +isA_p \\
-\frac{1}{2}\frac{1}{2} & A_{10} + cA_{11} & A_m - cA_p & A_0 - cA_p & A_{01} + cA_{11} \\
 & +isA_p & -isA_{11} & -isA_{11} & +isA_p \\
-\frac{1}{2}-\frac{1}{2} & A_m + cA_p & A_{10} - cA_{11} & A_{01} - cA_{11} & A_0 + cA_p \\
 & +isA_{11} & -isA_p & -isA_p & +isA_{11}.
\end{array}
\tag{7.15b}
$$

Here, [252]

$$
\begin{aligned}
c &\equiv \cos\varphi, \quad s \equiv \sin\varphi, \quad \varphi \equiv \varphi_p - \varphi_{p'}, \quad i \equiv \sqrt{-1}, \\
A_0 &\equiv -\frac{1}{2}A_0^{00}, A_{01} \equiv \frac{1}{6}A_1^{01}, A_{10} \equiv \frac{1}{6}A_1^{10}, A_{11} \equiv \frac{1}{6\sqrt{6}}A_1^{11}, \\
A_p &= \frac{1}{6\sqrt{3}}\Big(A_0^{11} + \frac{1}{\sqrt{10}}A_2^{11}\Big), A_m = \frac{1}{6\sqrt{3}}\Big(A_0^{11} - \sqrt{\frac{2}{3}}A_2^{11}\Big)
\end{aligned}
\tag{7.16a}
$$

with

$$
\begin{aligned}
(A_0)^* = A_0, \quad (A_{01})^* = A_{01}, \quad (A_{10})^* = A_{10}, \\
(A_m)^* = A_m, \quad A_p + A_{11} = (A_p)^* - (A_{11})^*.
\end{aligned}
\tag{7.16b}
$$

The fact that each of A_0, A_{01}, A_{10}, A_m, and $(A_p)^2 - (A_{11})^2$ is pure real, suggested by (7.16b), is followed [252] from the Hermiticity of the density matrix (7.15a).

On diagonalizing matrices (7.15), one finds that [252]

$$\begin{aligned}
E_1 &= A_0 + A_{01} + A_{10} + A_m, \quad E_2 = A_0 - A_{01} - A_{10} + A_m, \\
E_3 &= A_0 - A_m + \sqrt{\left(A_{01} - A_{10}\right)^2 - 4\left((A_{11})^2 - (A_p)^2\right)} \\
E_4 &= A_0 - A_m - \sqrt{\left(A_{01} - A_{10}\right)^2 - 4\left((A_{11})^2 - (A_p)^2\right)}
\end{aligned} \tag{7.17a}$$

are the eigenvalues of the matrix (7.15a); while, those of its partial transpose (7.15b) are given by

$$\begin{aligned}
E_1' &= A_0 + A_{01} - A_{10} - A_m, \quad E_2' = A_0 - A_{01} + A_{10} - A_m, \\
E_3' &= A_0 + A_m + \sqrt{\left(A_{01} + A_{10}\right)^2 - 4\left((A_{11})^2 - (A_p)^2\right)} \\
E_4' &= A_0 + A_m - \sqrt{\left(A_{01} + A_{10}\right)^2 - 4\left((A_{11})^2 - (A_p)^2\right)}
\end{aligned} \tag{7.17b}$$

1. In order for the fine-structure density matrix (7.15a) to represent a physically acceptable state of the photoelectrons $(e_p, e_{p'})$ ejected in the 1-DPI process (1.2) in the presence of C+SOIs, none of the four eigenvalues (7.17a) should be less than zero. This requires, from E_1, that $A_0 + A_{01} \geq -(A_{10} + A_m)$ which, in turn, suggests that E_1' is never negative. Eigenvalue E_1 further suggests that $A_0 + A_{10} \geq -(A_{01} + A_m)$. This immediately shows that the eigenvalue E_2' too cannot be less than zero either. One further finds, $A_0 + A_m \geq 0$ due to $E_1 + E_2 \geq 0$; on the other hand, $E_3 + E_4 \geq 0$ implies $A_0 - A_m \geq 0$.
2. Eigenvalues of a Hermitian matrix are always pure real.[7] Then the eigenvalues E_3, E_4, E_3', and E_4' lead to one or the other of the two conditions $\left(A_{01} \pm A_{10}\right)^2 \geq 4\left((A_{11})^2 - (A_p)^2\right)$. On combining this condition with $A_0 + A_m \geq 0$, one concludes that the eigenvalue E_3' of the partial transpose (7.15b) can never be negative. On the other hand, E_4' can be less than zero if $A_0 + A_m$ is less than $\left[\left(A_{01} + A_{10}\right)^2 - 4\left((A_{11})^2 - (A_p)^2\right)\right]^{1/2}$. Similarly, signs of E_1' and E_2' will depend upon the value of $A_0 - A_m$ relative to the values of $A_{01} - A_{10}$.

Not more than one eigenvalues of the partial transpose of an entangled bipartite state of qubits can be negative. The above analysis suggests that if density matrix (7.15a) is to represent an entangled state of photoelectrons $(e_p, e_{p'})$ then one of E_1', E_2', E_4' must be less than zero. This result is applicable to 1-DPI (1.2)

[7] Partial transpose of a Hermitian matrix is also Hermitian.

of all atoms—as well as to all electronic states of atom A and of its residual photodication A^{2+}—in the presence of C+SOIs in j-j coupling, whatever may be the energy or polarization of the absorbed photon γ_r. Here, photoelectrons $(e_p,\ e_{p'})$ are observed in opposite directions parallel to the OZ-axis in Fig. 5.1. In the present discussion, although spins of both e_p and $e_{p'}$ are taken to be polarized in the $X - Y$ plane; however, choice of any directions of spin quantization should not affect the entanglement properties of the two photoelectrons.

Purely Coulomb interaction, i.e., the absence of SOI—in an atom was found to entangle spins of $(e_p,\ e_{p'})$ in Chap. 5 when both atom $\mathfrak{T}$ and photodication $\mathfrak{T}^{2+}$ in the 1-DPI process (1.2) are only in their singlet electronic states. This Coulombic entanglement between $(e_p,\ e_{p'})$ is completely lost whenever either or both of the species $(\mathfrak{T},\ \mathfrak{T}^{2+})$ have spins (S_0, S_{2^+}) of their respective electronic states $(|0 > \equiv |L_0 S_0 M_{L_0} M_{S_0}\rangle,\ |2^+\rangle \equiv |L_{2^+} S_{2^+} M_{L_{2^+}} M_{S_{2^+}}\rangle)$ different from zero. Although the two photoelectrons may not be entangled, but they are certainly correlated. The presence of both C+SOIs, on the other hand, has changed the properties of the non-local correlation between e_p and $e_{p'}$, although the correlation responsible for their simultaneous emission in the process (1.2) remains the same. This is reflected in the possibility of $(e_p,\ e_{p'})$ having fine-structure entanglement even for the spin multiplicities of the electronic states $\big(|J_0\, M_{J_0}\rangle,\ |J_{1^+}\, M_{J_{1^+}}\rangle\big)$ of either or both of $(\mathfrak{T},\ \mathfrak{T}^{2+})$ being different from one.

The dependence of the fine-structure entanglement on both the kinematic and dynamics of the 1-DPI process also means that, unlike the Coulombic entanglement, it is now no longer possible to either *a-priori* or analytically determine the properties of the non-local correlation between the spin angular momenta of $(e_p,\ e_{p'})$. Without numerically calculating the density matrix (7.13) using dynamical amplitudes (7.13c), one will not be able to definitively conclude any thing about the fine-structure entanglement between the spins of photoelectrons' $(e_p,\ e_{p'})$.

7.3 Spin-Entanglement in a Photo-Auger Electron Pair $(e_p,\ e_a)$ Generated in 2-DPI

Let us now study the effects of the presence of SOI, in addition to the Coulomb forces, on the properties of spin-state of a photoelectron plus Auger electron emitted sequentially in the 2-DPI process (1.3) following the absorption of photon γ_r by the atomic target $\mathfrak{T}$ in $E1$ approximation.

7.3.1 Density Matrix

7.3.1.1 Derivation

Mathematical expressions for most of the physical quantities needed to obtain density matrix for studying entanglement between the spin angular momenta of photoelectron e_p and Auger electron e_a, emitted in two sequential steps in the 2-DPI process (1.3) in the presence of C+SOIs in j-j coupling, have already been given in Sect. 3.2.2 on pages 66–73. Now the energies (E_0, $E_{1^{+*}}$, E_{2^+}) to be used in the conservation conditions (6.1) are for the electronic states ($|0\rangle = |J_0 M_0\rangle$, $|1^{+^*}\rangle = |J_{1^{+*}} M_{1^{+*}}\rangle$, $|2^+\rangle = |J_{2^+} M_{2^+}\rangle$) of the species ($\mathfrak{T}$, $\mathfrak{T}^{1+^*}$, $\mathfrak{T}^{2+}$) in j-j coupling. Expression (3.15) for the density matrix for 2-DPI in j-j coupling becomes

$$
\begin{aligned}
&\langle J_{2^+};\, \mu_p\, \hat{\boldsymbol{u}}_p\, \boldsymbol{k}_p;\, \mu_a\, \hat{\boldsymbol{u}}_a\, \boldsymbol{k}_a |\, \rho^{\text{(2-DPI)}}\, | J_{2^+};\, \mu_p'\, \hat{\boldsymbol{u}}_p\, \boldsymbol{k}_p;\, \mu_a'\, \hat{\boldsymbol{u}}_a\, \boldsymbol{k}_a \rangle \\
&= \frac{\mathcal{K}_{ps}}{2J_0+1} \sum_{M_{1^{+*}}\, M'_{1^{+*}}} \Bigg[\sum_{M_0} \langle J_{1^{+*}}\, M_{1^{+*}};\, \mu_p\, \hat{\boldsymbol{u}}_p\, \boldsymbol{k}_p |\, F_p\, | 0;\, 1 m_r \rangle \\
&\quad \times \langle 0;\, 1 m_r |\, F_p^\dagger\, | J_{1^{+*}}\, M'_{1^{+*}};\, \mu_p'\, \hat{\boldsymbol{u}}_p\, \boldsymbol{k}_p \rangle \Bigg] \\
&\quad \times \Bigg[\sum_{M_{2^+}} \langle J_{2^+}\, M_{2^+};\, \mu_a\, \hat{\boldsymbol{u}}_a\, \boldsymbol{k}_a |\, F_s\, | J_{1^{+*}}\, M_{1^{+*}} \rangle \\
&\quad \times \langle J_{1^{+*}}\, M'_{1^{+*}} |\, F_s^\dagger\, | J_{2^+}\, M_{2^+};\, \mu_a'\, \hat{\boldsymbol{u}}_a\, \boldsymbol{k}_a \rangle \Bigg] .
\end{aligned}
\tag{7.18}
$$

Here, $\mathcal{K}_{ps}$ is already defined immediately after (6.2) on page 137; Zeeman components [10, 60] of the state $|0\rangle$ have been averaged, and those of $|2^+\rangle$ summed, over assuming that atom $\mathfrak{T}$ before its photoionization in the step (1.3a) is unpolarized and dication $\mathfrak{T}^{2+}$ formed in the second step (1.3b) remains unobserved.

Sums over M_0 and M_{2^+}, present on the right-hand sides of the density matrix (7.18), have already been evaluated in (3.42) and (3.45), respectively. After replacing these two sums in (7.18) by the right-hand sides of (3.42b) and (3.45b), the resulting expression is simplified using the procedures similar to those already explained in arriving at the density matrices (5.9), (6.3), (7.5), and (7.13). Finally, the density matrix for the 2-DPI process (1.3) in the presence of both C+SOIs in j-j coupling can be written as [188]

$$
\begin{aligned}
&\langle J_{2^+};\, \mu_p\, \hat{\boldsymbol{u}}_p\, \boldsymbol{k}_p;\, \mu_a\, \hat{\boldsymbol{u}}_a\, \boldsymbol{k}_a |\, \rho^{\text{(2-DPI)}}\, | J_{2^+};\, \mu_p'\, \hat{\boldsymbol{u}}_p\, \boldsymbol{k}_p;\, \mu_a'\, \hat{\boldsymbol{u}}_a\, \boldsymbol{k}_a \rangle \\
&= (-1)^{\mu_p'+\mu_a'} \sum_{\substack{S_p\, M_{S_p}\, N_{S_p} \\ S_a\, M_{S_a}\, N_{S_{ap}}}} \begin{pmatrix} \frac{1}{2} & \frac{1}{2} & S_p \\ \mu_p & -\mu_p' & M_{S_p} \end{pmatrix} \begin{pmatrix} \frac{1}{2} & \frac{1}{2} & S_a \\ \mu_a & -\mu_a' & M_{S_a} \end{pmatrix} \\
&\quad \times \left[\mathcal{D}^{S_p}_{M_{S_p}\, N_{S_p}}(\omega_p) \right]^* \left[\mathcal{D}^{S_a}_{M_{S_a}\, N_{S_a}}(\omega_a) \right]^* A^{S_p\, S_a}_{N_{S_p}\, N_{S_a}}(m_r;\, \hat{\boldsymbol{k}}_p\, \hat{\boldsymbol{k}}_a)
\end{aligned}
\tag{7.19a}
$$

with

$$A^{S_p\,S_a}_{N_{S_p}\,N_{S_a}}(m_r;\,\hat{\boldsymbol{k}}_p\,\hat{\boldsymbol{k}}_a) = (-1)^{1+m_r+J_0+2J_{1^+*}+J_{2^+}}\,\frac{(2J_{1^+*}+1)\,\mathcal{K}_{sa}}{4\pi(2J_0+1)}$$

$$\times \sum_{\substack{\ell_p\,\ell_p'\,j_p\,j_p'\,L_p\,M_{L_p}\,J_p\,J_a\,j_t\\ \ell_a\,\ell_a'\,j_a\,j_a'\,L_a\,M_{L_a}\,M\,L_r\,j_t'}} (-1)^{\ell_p'+\ell_a-j_a'+j_t}\,(2S_p+1)(2S_a+1)(2J_p+1)$$

$$\times (2J_a+1)(2L_r+1)\sqrt{(2L_p+1)(2L_a+1)}\begin{pmatrix}\ell_p & \ell_p' & L_p\\ 0 & 0 & 0\end{pmatrix}\begin{pmatrix}\ell_a & \ell_a' & L_a\\ 0 & 0 & 0\end{pmatrix}$$

$$\times \begin{pmatrix}L_p & S_p & J_p\\ M_{L_p} & N_{S_p} & -M\end{pmatrix}\begin{pmatrix}L_a & S_a & J_a\\ M_{L_a} & N_{S_a} & M\end{pmatrix}\begin{pmatrix}1 & 1 & L_r\\ m_r & -m_r & 0\end{pmatrix}\begin{pmatrix}J_p & J_a & L_r\\ -M & M & 0\end{pmatrix}$$

$$\times \begin{Bmatrix}J_{1^+*} & J_{1^+*} & J_a\\ j_t & j_t' & J_0\end{Bmatrix}\begin{Bmatrix}J_{1^+*} & J_{1^+*} & J_a\\ j_a & j_a' & J_{2^+}\end{Bmatrix}$$

$$\times \begin{Bmatrix}\ell_p & \ell_p' & L_p\\ \frac{1}{2} & \frac{1}{2} & S_p\\ j_p & j_p' & J_p\end{Bmatrix}\begin{Bmatrix}\ell_a & \ell_a' & L_a\\ \frac{1}{2} & \frac{1}{2} & S_a\\ j_a & j_a' & J_a\end{Bmatrix}\begin{Bmatrix}1 & 1 & L_r\\ j_p & j_p' & J_p\\ j_t & j_t' & J_a\end{Bmatrix}$$

$$\times \left[Y^{M_{L_p}}_{L_p}(\hat{\boldsymbol{k}}_p)\right]^* P_e(J_{1^+*}\,j_p;\;J_0\,1;\;j_t;\;\ell_p)\left[P_e(J_{1^+*}\,j_p';\;J_0\,1;\;j_t';\;\ell_p')\right]^*$$

$$\times \left[Y^{M_{L_a}}_{L_a}(\hat{\boldsymbol{k}}_a)\right]^* A_e(J_{2^+}\,j_a;\;J_{1^+*};\;\ell_a)\left[A_e(J_{2^+}\,j_a';\;J_{1^+*};\;\ell_a')\right]^*. \qquad (7.19b)$$

Here,

$$P_e(J_{1^+*}\,j_p;\;J_0\,1;\;j_t;\;\ell_p) = (-i)^{\ell_p}\,e^{i\sigma_{\ell_p j_p}}\,(-1)^{\ell_p+j_t}\,(2j_t+1)$$
$$\times\sqrt{(2\ell_p+1)(2j_p+1)}\,\langle J_{1^+*}\,j_p|F(j_t)|J_0 1\rangle \qquad (7.20a)$$

and

$$A_e(J_{2^+}\,j_a;\;J_{1^+*};\;\ell_a) = (-i)^{\ell_a}\,e^{i\sigma_{\ell_a j_a}}$$
$$\times\sqrt{(2\ell_a+1)(2j_a+1)}\,\langle (J_{2^+}\,j_a)J_{1^+*}|F_a|J_0 1\rangle \qquad (7.20b)$$

are the dynamical terms, which contain amplitudes defined in (3.40b) and (3.45b) for the respective photoionization and Auger decay parts of the 2-DPI process (1.3).

7.3.1.2 Properties

Expression (7.19) for the density matrix for 2-DPI process (1.3) in the presence of C+SOIs in j-j coupling is completely general and independent of all models, which one may like to use for calculating the dynamical terms (7.20) present therein. The diagonal terms $\langle J_{2^+};\ \mu_p\,\hat{\boldsymbol{u}}_p\,\boldsymbol{k}_p;\ \mu_a\,\hat{\boldsymbol{u}}_a\,\boldsymbol{k}_a|\,\rho^{(2\text{-DPI})}\,|J_{2^+};\ \mu_p\,\hat{\boldsymbol{u}}_p\,\boldsymbol{k}_p;\mu_a\,\hat{\boldsymbol{u}}_a\,\boldsymbol{k}_a\rangle$ in (7.19) describe angular distributions of a photo-Auger electron pair $[e_p,\ e_a]$ moving in the directions $[\hat{\boldsymbol{k}}_p\ (\theta_p,\ \phi_p),\ \hat{\boldsymbol{k}}_a\ (\theta_a,\ \phi_a)]$ with energies $[\epsilon_p,\ \epsilon_a]$ and their spins polarized along $[\hat{\boldsymbol{u}}_p\ (\vartheta_p,\ \varphi_p),\ \hat{\boldsymbol{u}}_a\ (\vartheta_a,\ \varphi_a)]$; whereas, non-diagonal (i.e., $\mu_p \neq \mu_p'$ and/or $\mu_a \neq \mu_a'$) terms in (7.19) represent coherence effects in a 2-DPI process. Similar to other density matrices, hitherto calculated in this monograph, (7.19) too is positive and Hermitian.

Unlike the Coulombic density matrix (6.3), which describes entanglement between the spins of $(e_p,\ e_a)$ in the absence of SOI, the present density matrix (7.19) cannot be written as a product of two parts separately representing angular and spin correlations between photoelectron and Auger electron. Thus, the presence of SOI, in addition to the Coulomb forces, in an atom has coupled the angular and spin correlation between e_p and e_a. This, in turn, makes the spin-entanglement properties of a $(e_p,\ e_a)$ pair to depend upon both the kinematics as well as dynamics of the 2-DPI process (1.3). This is very similar to what was found in the previous two Sects. 7.1 and 7.2 of this chapter describing spin-entanglement between an electronic qubit (e_p) plus ionic qudit $(\mathfrak{T}^{1+})$ generated in the 1-SPI process (1.1), or between two photoelectrons $(e_p,\ e_{p'})$ simultaneously emitted in the 1-DPI process (1.2), in the presence of C+SOIs.

Consequently, unlike in Chap. 6, it now becomes absolutely necessary to know the dynamical amplitudes, present in (7.20), for analyzing the properties of the spin-state (7.19) of a photo-Auger electron pair $(e_p,\ e_a)$. This implies that 2-DPI in each atom has to be studied individually with specific photoionization and Auger transitions [(1.3a) and (1.3b)] one is interested in to learn about the spin-entanglement properties of a bipartite state of $(e_p,\ e_a)$ formed in that process. In addition, one needs to know also the frequency ν_r and polarization [specified by the parameter m_r in (7.19)] of the photon γ_r absorbed for ionization in (1.3a). Moreover, the entanglement properties of $(e_p,\ e_a)$, in the present case, will depend also upon the directions $(\hat{\boldsymbol{k}}_p,\ \hat{\boldsymbol{k}}_a)$ these two electrons are receding from the dication residue $\mathfrak{T}^{2+}$ formed in (1.3b) and from each other.

Furthermore, similar to the density matrices (7.5) and (7.13) for $(e_p,\ \mathfrak{T}^{1+})$- and $(e_p,\ e_{p'})$-systems, expressions (7.19) and (7.20) also involve fine-structure levels $(J_0,\ J_{1^{+*}},\ J_{2^+})$ of the atomic species $(\mathfrak{T},\ \mathfrak{T}^{1+^*},\ \mathfrak{T}^{2+})$ participating in the 2-DPI process (1.3) in the presence of C+SOIs in j-j coupling. The properties of a $(e_p,\ e_a)$ pair will, therefore, be different for different fine structure levels specifying the electronic states $(|0\rangle = |J_0\,M_{J_0}\rangle,\ |1^{+^*}\rangle = |J_{1^{+*}}\,M_{J_{1^{+*}}}\rangle,\ |2^+\rangle = |J_{2^+}\,M_{J_{2^+}}\rangle)$ of $(\mathfrak{T},\ \mathfrak{T}^{1+^*},\ \mathfrak{T}^{2+})$ in the process (1.3). Expression (7.19) too can, therefore, be called as a fine-structure density matrix, which will determine properties of the fine-structure entanglement between the spins of a photo-Auger electron pair

(e_p, e_a) generated on the inclusion of C+SOIs in the 2-DPI process (1.3) in j-j coupling.

7.3.2 *Example: Fine-Structure Spin-Entanglement in a (e_p, e_a) Pair*

Kämmerling and Schmidt [268–270] have experimentally measured triply differential cross-sections in 2-DPI of a Xe atom. The inner-shell photoionizing transition

$$\gamma_r + Xe(4d^{10}\,5s^2\,5p^6\,{}^1S_0) \longrightarrow Xe^{1+^*}(4d^9\,5s^2\,5p^6\,{}^2D_{5/2}) + e_p(p_{3/2};\, f_{5/2,7/2}) \tag{7.21a}$$

in their [268–270] experiment was followed by the Auger emission

$$Xe^{1+^*}(4d^9\,5s^2\,5p^6\,{}^2D_{5/2}) \longrightarrow Xe^{2+}(4d^{10}\,5s^0\,5p^6\,{}^1S_0) + e_a(d_{5/2}). \tag{7.21b}$$

Kämmerling et al. [269], in addition, found that the probabilities for the non-radiative, spontaneous decay of $[Xe^{1+^*}(4d^9\,5s^2\,5p^6\,{}^2D_{3/2}),\ Xe^{1+^*}(4d^9\,5s^2\,5p^6\,{}^2D_{5/2})]$ are not only very high [0.783, 0.825], but also have little dependence on the energy of absorbed radiation γ_r which creates these excited photoions. Moreover, a recent study [271] has shown that non-relativistic description of photoionization of as heavy an atom as Xe is sufficient as well as satisfactory. Further, Kämmerling and Schmidt [268, 269] from their measurements of angular distributions in (7.21a) have extracted magnitudes and phases of the complex amplitudes for the photoionization channels $4d_{5/2} \longrightarrow p_{3/2}$, $4d_{5/2} \longrightarrow f_{5/2}$, $4d_{5/2} \longrightarrow f_{7/2}$. These experimental values of the photoionization amplitudes agree with the theoretical calculations reported by Johnson and Cheng [272] in relativistic random-phase approximation [273, 274].

In view of these both experimental and theoretical studies on Xe, Chandra and Ghosh [188] performed a detailed investigation of spin-entanglement of a (e_p, e_a) pair generated sequentially in the 2-DPI process (7.21) in the presence of C+SOIs in j-j coupling. In this [188] very first investigation of 2-DPI of any atom in the presence C+SOIs using the tools from quantum information science, photoionization was considered by linearly polarized (LP), left circularly polarized (LCP), right circularly polarized (RCP), as well as by an unpolarized (UP) photon γ_r absorbed in (7.21a). It was done in order to assess the effects of different kinds of polarization of the ionizing radiation on the entanglement of (e_p, e_a).

For simplicity, Chandra and Ghosh [188] considered an experiment for 2-DPI (7.21) in a geometry of diametric emission of e_p and e_a in the coordinate system shown in Fig. 6.1. This means, if

$$\hat{\boldsymbol{k}}_p(\theta_p,\, \phi_p) \parallel \hat{\boldsymbol{k}}(\theta,\, \phi), \tag{7.22a}$$

then

$$\hat{\boldsymbol{k}}_a(\theta_a,\, \phi_a) \parallel -\hat{\boldsymbol{k}},\; ie\; (\theta_a = \pi - \theta,\; \phi_a = \pi + \phi) \tag{7.22b}$$

Entanglement properties of a state of qubits are well known [17,24,34,43,99,132] to be totally independent of the local choice of the bases and of the coordinate system used to describe them. Therefore, for simplicity, spins of each of the photoelectron e_p—emitted in (7.21a)—and of the Auger electron e_a—ejected in (7.21b)—were taken [188] to be longitudinally polarized to these electrons' directions of departure from Xe^{1+^*} and Xe^{2+}, respectively. That is

$$\begin{aligned} \hat{\boldsymbol{u}}_p &\parallel \hat{\boldsymbol{k}}_p,\; i.e.,\; (\vartheta_p = \theta,\; \varphi_p = \phi), \\ \hat{\boldsymbol{u}}_a &\parallel \hat{\boldsymbol{k}}_a,\; i.e.,\; (\vartheta_a = \pi - \theta,\; \varphi_a = \pi + \phi). \end{aligned} \tag{7.22c}$$

Choices (7.22a) and (7.22b) for the propagation directions means that the polar axis of the space-frame $OXYZ$ in Fig. 6.1 and the line joining oppositely moving e_p and e_a are in the same plane, hereafter called [188], unless specified otherwise, emission plane. Experimental geometries specified by (7.22) also reduce the number of independent angles present in the density matrix (7.19) from eight to the two spherical angles of $\hat{\boldsymbol{k}}$ (θ, ϕ) defined in (7.22a).

For the 2-DPI process (7.21), [268, 269] contain, in the notation of this monograph, magnitude and phases of the complex amplitudes $\langle J_{1^{+*}}\, j_p | F(J) | J_0\, 1\rangle^-$, present on the right-hand side of (3.40b), for J_0, $J_{2^+} = 0$, and $(J_{1^{+*}}, j_p) = (5/2, 3/2), (5/2, 5/2), (5/2, 7/2)$. Meanings of these quantum numbers are obvious from the angular momentum conservation conditions (3.34). The total angular momentum $J = 1$ (for $\boldsymbol{J} = \boldsymbol{\ell}_r + \boldsymbol{J}_0 = \boldsymbol{J}_{1^{+*}} + \boldsymbol{j}_p$ and $J = |\boldsymbol{J}|$) is obtained form the left-hand side of (3.34a) in the $E1$ approximation.

In the present case, the complex amplitude for the Auger electron e_a is, however, not needed because a single d-partial wave contributes to the Auger transition (7.21b). Consequently, during the normalization to unit trace of the density matrix (7.19), the contribution of this amplitude is cancelled. Although in the 2-DPI process (7.21), SOI does not contribute[8] to the bound state $|0\rangle$ of the atomic target Xe and the state $|2^+\rangle$ of its residual dication Xe^{2+}; nevertheless, it will certainly be important for the electronic state $|1^{+^*}\rangle$ of the excited photoion Xe^{1+^*}, as well as in the continuum orbitals (3.38a) of the photoelectron e_p and (3.43a) of the Auger electron e_a because both of these electrons are represented in (7.21) by other than s-partial wave.

[8] In the present example of (7.21), each of both Xe and Xe^{2+} is in $^1S_0$ electronic state.

On the other hand, if one disregards SOI in the 2-DPI process (7.21), these two transitions simply become

$$\gamma_r + Xe(4d^{10}\,5s^2\,5p^6\,{}^1S) \longrightarrow Xe^{1+^*}(4d^9\,5s^2\,5p^6\,{}^2D) + e_p(p, f), \quad (7.23a)$$

$$Xe^{1+^*}(4d^9\,5s^2\,5p^6\,{}^2D) \longrightarrow Xe^{2+}(4d^{10}\,5s^0\,5p^6\,{}^1S) + e_a(f). \quad (7.23b)$$

The properties of the Coulombic entanglement, generated in (7.23) have been discussed in detail in the Chap. 6. According to the discussion given therein (see, for example, pages 146 and 147), (e_p, e_a) in (7.23) are always in a singlet spin state, which is pure as well as maximally entangled, whatever may be the properties of the ionizing radiation γ_r and directions of observations of the two sequentially emitted electrons.

Appendix E describes a generic computer program for calculating the coefficients $A^{S_p\,S_a}_{N_{S_p}\,N_{S_a}}(m_r;\,\hat{\boldsymbol{k}}_p\,\hat{\boldsymbol{k}}_a)$ defined in (7.19b). This programme was readily adapted [188] for calculating A's in the experimental geometry (7.22). These coefficients were then used in a small programme—run on MATHMATICA [275]—for calculating each of the sixteen elements of the density matrix (7.19a). The results reported in [188] on the fine-structure entanglement of the spins of (e_p, e_a) in the presence of C+SOIs in j-j coupling for the experimental geometry specified by (7.22) can be summarized as follows:

1. According to the presently used experimental geometry, each element of the density matrix (7.19) contains both of the angles (θ, ϕ); however, none of the properties of the spin-entanglement discussed in [188] was found to depend on the azimuthal angle ϕ. Thus, the fine-structure entanglement between the spins of (e_p, e_a) is quite independent of the orientation of the emission plane in the SF of Fig. 6.1 for diametric emission and longitudinal polarization of these two electrons.
2. The spin-entanglement between (e_p, e_a) is significantly affected by a change in the polar angle θ (specifying orientation with respect to the OZ-axis in Fig. 6.1 of the line joining oppositely moving e_p and e_a in the emission plane). But the Coulombic entanglement in (7.23) never depends upon the geometry of an experiment.
3. For LP, LCP, or RCP photon γ_r, absorbed in the 2-DPI process (7.21), (e_p, e_a) were always in a pure spin state. However, these two electrons formed a mixed spin-state whenever an UP radiation was used for ionization in (7.21). But, (e_p, e_a) in the Coulombic entanglement (7.23) always formed the Bell state (2.17b)—a singlet state of two spin-$\frac{1}{2}$ particles—for any polarization of the photon absorbed in the 2-DPI process (1.3) taking place in the absence of SOI.
4. Figure 7.5 shows variation of each of the three measures (i.e., negativity, concurrence, and entanglement of formation) of entanglement with the angle θ—defined in (7.22)—when the γ_r absorbed in (7.22) is LP. On account of the purity of the state of (e_p, e_a) in the present case, negativity and concurrence

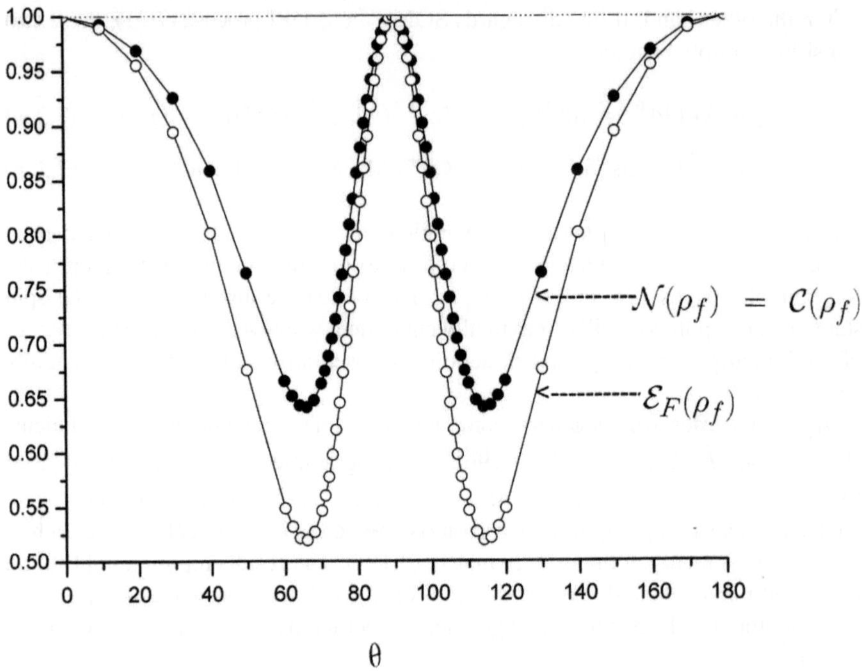

Fig. 7.5 Variations of the three measures (i.e., negativity, concurrence, and entanglement of formation) of fine-structure entanglement for the bipartite state (7.19) of electronic qubits (e_p, e_a) generated in the 2-DPI process (7.21) in Xe when the absorbed photon γ_r is LP. The broken horizontal line at the top of this figure represents value of each of these three measures for Coulombic entanglement for the 2-DPI process (7.23) also in Xe, but in the absence of SOI. For the Coulombic entanglement, all the three measures considered herein, always have unit value corresponding to a maximally entangled state of (e_p, e_a) in L-S coupling. The negativity, concurrence, and entanglement of formation are calculated using the respective (2.45c), (2.46), and (2.47). On account of the pure nature of the bipartite state (7.19) of (e_p, e_a) emitted during the transitions (7.21) on the inclusion of C+SOIs in j-j coupling, negativity and concurrence in this figure have identical values. Results shown herein for the fine-structure entanglement are calculated for the experimental geometry specified by (7.22). (First published in [188].)

have identical values. Thus, unlike in the Coulombic case (7.23) wherein the spin-state is always pure and entanglement (shown by the upper horizontal broken line in each of the three Figs. 7.5, 7.6, and 7.7) maximum for any experimental geometry (i.e., values of $\boldsymbol{k}_p$, $\boldsymbol{k}_a$, $\hat{\boldsymbol{u}}_p$, $\hat{\boldsymbol{u}}_a$), the presence of C+SOIs dilutes the degree of non-separability of these two electrons in all directions except when they are moving either parallel or perpendicular to the polar OZ-axis [i.e., the electric field vector of the LP ionizing radiation in (7.21a)] in Fig. 6.1, although (e_p, e_a) continue to remain in a pure spin-state.

5. The variations of the negativity, concurrence, and entanglement of formation with the direction of observation of oppositely moving, but longitudinally polarized, (e_p, e_a) is shown in Fig. 7.6 while γ_r in (7.21) is taken to have circular polarization. The present variations of the three measures of bipartite

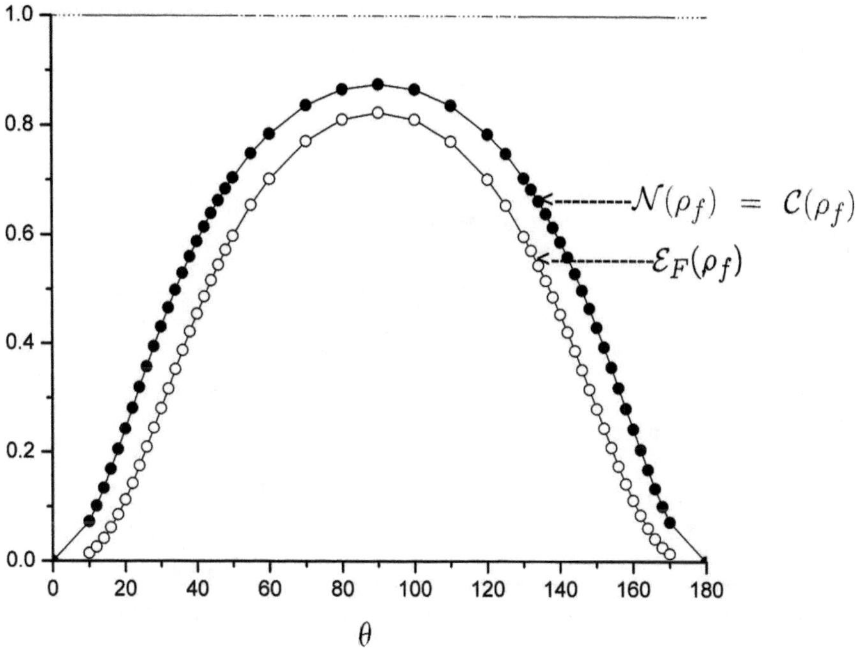

Fig. 7.6 This is same as Fig. 7.5, but now ionization in (7.21) takes place due to the absorption of a right circularly polarized photon γ_r. Here again, both negativity and concurrence have equal values. According to the discussion given in point 5 in the present Sect. 7.3.2, this figure represents also the values of all the three measures of entanglement when Xe atom in (7.21a) is ionized by absorbing a left circularly polarized photon γ_r as well. (First published in [188].)

entanglement are very different from those seen in Fig. 7.5 for an LP γ_r. Moreover [188], the behaviours in Fig. 7.6 do not change whether the radiation used for ionization in (7.21) is LCP or RCP.[9] Unlike in Fig. 7.5, none of the three measures of entanglement in Fig. 7.6 ever attains its maximum allowed value, i.e., unity. In the present case, the fine-structure entanglement is maximum (but less than one by a non-negligible amount) when the line joining (e_p, e_a) lies in the $X-Y$ plane of the SF Fig. 6.1 [*ie*, perpendicular to the direction of incidence of the CP ionizing radiation in (7.21a)]; the fine-structure spin state in Fig. 7.6, on the other hand, becomes completely separable when the two electrons are moving in opposite directions parallel to the polar OZ-axis of Fig. 6.1. The (e_p, e_a)-pair is always in a pure state in this case as well.

6. Finally, Fig. 7.7 contains negativity, concurrence, and entanglement of formation for a fine-structure state of electronic qubits (e_p, e_a), while Xe in (7.21a) is

[9]It, in other words, means that there is no circular dichroism [see footnote (3) in Chapter 11] in the fine-structure entanglement between the spins of (e_p, e_a) in the presently being considered experimental geometry (7.22).

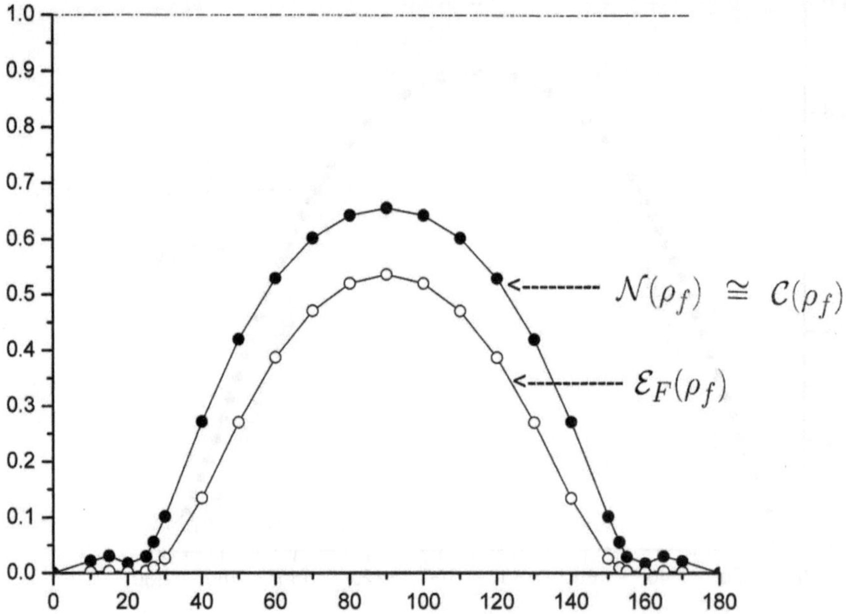

Fig. 7.7 This is again same as Figs. 7.5 and 7.6, but for unpolarized ionizing radiation γ_r in the 2-DPI process (7.21). In this case, $(e_p,\, e_a)$ were found to be in a mixed bipartite state. Consequently, negativity and concurrence do not have same values which are, however, indistinguishable on the scale of this figure. (First published in [188].)

ionized by an UP electromagnetic wave. (An UP radiation is an even mixture of RCP and LCP electromagnetic waves.) The bipartite spin-state of $(e_p,\, e_a)$ in this case is not pure, negativity and concurrence, therefore, do not have same values; however, on the scale of the Fig. 7.7, these two measures of entanglement cannot be distinguished from each other. The maximum entanglement (at $\theta = \frac{\pi}{2}$) between $(e_p,\, e_a)$, in the present case, is even smaller than that found in Fig. 7.6 for a circularly polarized ionizing radiation. In Fig. 7.7, electrons $(e_p,\, e_a)$ are in a spin state which is almost separable for $\theta = 0^\circ$ - 30°, 150° - 160° with respect to the direction of incidence of the UP ionizing radiation in (7.21a).

Thus, the presence of SOI, in addition to the Coulomb forces, in an atom couples the entanglement between the spin angular momenta of photoelectron e_p and Auger electron e_a to the kinematic variables and dynamical parameters of a 2-DPI process (1.3), as well as to the polarization of the ionizing radiation γ_r. As a result of this, the nature of the entanglement between $(e_p,\, e_a)$ in the presence of C+SOIs—herein called fine-structure entanglement—is quite different from that of the Coulombic entanglement.

Chapter 8
Bipartite States of Photonic and Flying Electronic Qubits

Chapter 6 and (pages 169–178 in Chap. 7 containing) Sect. 7.3 have been concerned with the non-radiative, spontaneous decay of an inner-shell vacancy created in an atom in its photoionization. Consequently, at both of these places, properties of the Coulombic and of fine-structure entanglement of bipartite spin-states of a pair of (photoelectron e_p, Auger electron e_a) have, respectively, been investigated. [On the other hand, a single photoelectron e_p in 1-SPI, or two simultaneous photoelectrons $(e_p, e_{p'})$ in 1-DPI, are emitted from the outermost shell(s) of a target $\mathfrak{T}$ so that the residual photoion $\mathfrak{T}^{1+}$, or photo-dication $\mathfrak{T}^{2+}$, formed in the process (1.1), or (1.2), are always in their ground electronic states $|1^+\rangle$, or $|2^+\rangle$, respectively.] The other mode of spontaneous decay of an inner-shell vacancy in an atom is by emitting a photon, i.e., radiative. While an Auger electron is more likely to be emitted in the decay of a vacancy in one of the outer (inner) shells of a heavy (light) element; radiative decay is, however, predominant for a vacancy in the inner most shell of a heavy element. Thus, as far as heavy elements are concerned, radiative and non-radiative decay of a vacancy are two competitive processes. Hence, inner-shell photoionization (1.4a), followed by spontaneous emission of a photon γ_d in (1.4b), is a very common and important 2-step process both in pure and applied sciences. But, why should one study entanglement between the spin of photoelectron e_p and polarization of photon γ_d, emitted in two sequential steps of the process (1.4)?

Several studies have recently been performed both on the discrete variable entanglement [276–280], and on continuous variable entanglement [281–283] of a single atom and a single photon. A photonic qubit and a qubit in the form of a trapped $^{111}Cd^{1+}$ [277,279], or trapped ^{87}Rb [280]; or, probabilistic entanglement of a photon and collective atomic excitation [278] are some of the recent examples of atom–photon entangled states in discrete variables. On the other hand, continuous kinetic variables experiments on photon–atom entanglement consists of spontaneous emission [281], Raman or Rayleigh scattering [282], and resonance scattering [283] of a photon from an atom. This matter-light hybrid entanglement has been suggested [276–283] to be useful in many ways. In addition to testing [279] violation of Bell inequality in a hybrid system, non-separable states of light and matter have

N. Chandra and R. Ghosh, *Quantum Entanglement in Electron Optics*, Springer Series on Atomic, Optical, and Plasma Physics 67, DOI 10.1007/978-3-642-24070-6_8,

been suggested to be crucial in long range quantum communication for these make it possible to entangle quantum memory (in the form of matter) with communication channel (of a photon). Following these [276–283] developments on atom–photon entanglement, Flindt et al. [284] have generated entanglement between the spin state of an electron and frequency of a photon in a double quantum dot experiment.

The 2-step process (1.4) too provides [263] one of the possible schemes for interfacing stable quantum memory (in the form of a photoelectron e_p) with carriers of quantum information (i.e., an optical photon γ_d). There are, however, many important differences between the scheme originally suggested by Chandra and Ghosh [263] and those studied in [276–284] for matter–light entanglement. For example, unlike a flying electronic qubit in [263], matter in each of the experiments [276–283] is an ion/atom—free or trapped. Similar to [263], Flindt et al. [284] too have generated entangled states of electron–photon; but, in the later study [284], electron is in a double quantum dot in a semiconductor device and its spin is being entangled with the frequency of the photon; whereas in [263], spin of the free electron e_p (which can readily be made stationary, if desired) is entangled with the polarization of the photon γ_d. Moreover, unlike in [284], noise does not play any significant role in the scheme (1.4).

The present Chap. 8 of this monograph, therefore, analyses the properties of entanglement between the spin of photoelectron e_p and polarization of the photon γ_d emitted sequentially in two different steps of the process (1.4). Both, for the absorbed and detected photons γ_r and γ_d, respectively, linear, left circular, and right circular polarizations, as well as unpolarized, cases are considered. The discrete variable (e_p, γ_d) entanglement described in this chapter is both without and with SOI taken into account.

The energy conservation conditions applicable to the inner-shell photoioniza-tion (1.4a) and spontaneous radiative decay (1.4b) in the 2-step process (1.4) are

$$h\,\nu_r + E_0 = E_{1^{+*}} + \epsilon_p \tag{8.1a}$$

and

$$E_{1^{+*}} = E_{1^+} + h\,\nu_d, \tag{8.1b}$$

respectively. Here, condition (8.1a) is identical to (6.1a) applicable to the first step of the 2-DPI process (1.3a); whereas in (8.1b), E_{1^+} is the energy of the singly charged positive ion $\mathfrak{T}^{1+}$ formed after the emission, in the second step of the process (1.4), of the photon γ_d of frequency ν_d in the spontaneous radiative decay of the excited photoion $\mathfrak{T}^{1+^*}$ possessing energy $E_{1^{+*}}$. Obviously, similar to the energy ϵ_a of the Auger electron in (6.1b), frequency ν_d of γ_d is independent of the energy $E_r (= h\nu_r)$

of the ionizing photon γ_r in (8.1a) and does not change until either or both of the energies $E_{1^{+*}}$ and E_{1^+} change.

Figure 8.1 shows geometry of a possible experiment for studying (e_p, γ_d) entanglement generated in the 2-step process (1.4). In this figure, $OX_dY_dZ_d$ is detector's frame (DF), i.e., the frame of reference of the detector used to observe the emitted photon γ_d. Its orientation with respect to the SF $OXYZ$ is specified by the Euler's angles ω_d $(\phi_d, \theta_d, \xi_d)$ [i.e., rotations [65] by ω_d will bring SF in coincidence with the DF]. On taking the OZ_d-axis to be the direction of emission of the photon γ_d in SF, its polarization in the DF is then represented [68] by the vector

$$\hat{\boldsymbol{u}}_d = \hat{\boldsymbol{e}}_{x_d} \cos\beta_d + i\,\hat{\boldsymbol{e}}_{y_d} \sin\beta_d, \tag{8.2}$$

where $\hat{\boldsymbol{e}}_{x_d}$ and $\hat{\boldsymbol{e}}_{y_d}$ are the unit vectors along the OX_d and OY_d axes of the DF. Here, $\beta_d = 0$, or $\frac{\pi}{2}$, means that γ_d is LP with its electric field vector along the OX_d-, or OY_d-, axis, respectively. On the other hand, $\beta_d = \frac{\pi}{4}$, or $\frac{3\pi}{4}$, corresponds, respectively, to a γ_d with positive helicity (i.e., RCP), or negative helicity (i.e., LCP). The quantity β_d is, therefore, called [68] polarization parameter for the photon γ_d emitted in the spontaneous radiative decay of the excited photoion $\mathfrak{T}^{1+^*}$ in the second step of the process (1.4).

In the $E1$ photoemission operator (3.9), the polarization vector $\hat{\boldsymbol{u}}_d$ for the emitted photon γ_d is referred to the DF $OX_dY_dZ_d$; whereas the vector $\boldsymbol{r}_i$, representing the position of the i-th electron in excited photoion $\mathfrak{T}^{1+^*}$, is with respect to the SF $OXYZ$ in Fig. 8.1. In order to refer $\hat{\boldsymbol{u}}_d$ to the SF, we first expand it in its spherical components [65]

$$\left(\hat{\boldsymbol{u}}_d\right)^* = \sum_q (-1)^q \left[\left(u_d\right)^*\right]_q \hat{\boldsymbol{e}}_{-q} \tag{8.3a}$$

and then rotate the spherical unit vectors $\hat{\boldsymbol{e}}_{-q}$ from DF to SF. This operation takes (8.3a) to the following form

$$\left(\hat{\boldsymbol{u}}_d\right)^* = \sum_{q\,q_1} (-1)^q \left[\left(u_d\right)^*\right]_q \left[\mathcal{D}^1_{-q\,q_1}(\omega_d)\right]^* \hat{\boldsymbol{\zeta}}_{q_1}. \tag{8.3b}$$

Here, $\hat{\boldsymbol{\zeta}}_{q_1}$ and $\hat{\boldsymbol{e}}_q$ (with $q_1, q = -1, 0, +1$) are the spherical unit vectors referred to the SF and DF, respectively. Consequently [65],

$$\begin{aligned}
\hat{\boldsymbol{\zeta}}_{-1} &= \frac{1}{\sqrt{2}}\left(\hat{\boldsymbol{e}}_x - i\hat{\boldsymbol{e}}_y\right), & \hat{\boldsymbol{e}}_{-1} &= \frac{1}{\sqrt{2}}\left(\hat{\boldsymbol{e}}_{x_d} - i\hat{\boldsymbol{e}}_{y_d}\right);\\
\hat{\boldsymbol{\zeta}}_0 &= \hat{\boldsymbol{e}}_z, & \hat{\boldsymbol{e}}_0 &= \hat{\boldsymbol{e}}_{z_d};\\
\hat{\boldsymbol{\zeta}}_{+1} &= -\frac{1}{\sqrt{2}}\left(\hat{\boldsymbol{e}}_x + i\hat{\boldsymbol{e}}_y\right), & \hat{\boldsymbol{e}}_{+1} &= -\frac{1}{\sqrt{2}}\left(\hat{\boldsymbol{e}}_{x_d} + i\hat{\boldsymbol{e}}_{y_d}\right).
\end{aligned} \tag{8.4}$$

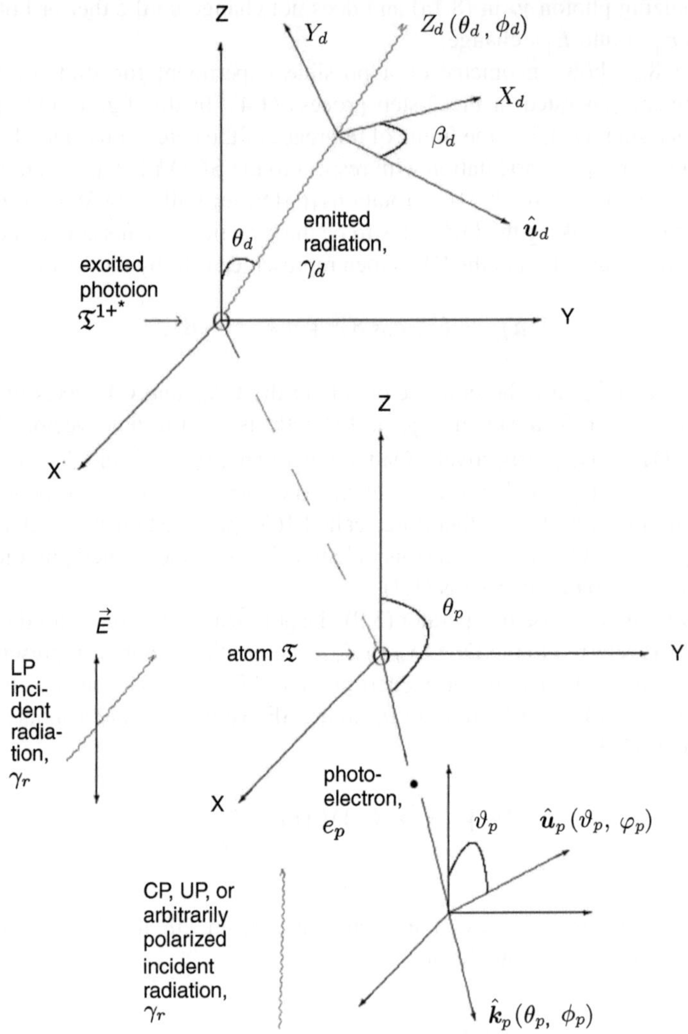

Fig. 8.1 Coordinate system showing the space-fixed frame of reference $OXYZ$, photon spectrometer's frame of reference $OX_dY_dZ_d$, and propagation direction $\hat{\boldsymbol{k}}_p(\theta_p,\ \phi_p)$ as well as the spin quantization direction $\hat{\boldsymbol{u}}_p(\vartheta_p,\ \varphi_p)$ of the photoelectron e_p. The orientation of the photon spectrometer's frame with respect to the SF is defined by the Euler angles $\omega_d\ (\phi_d,\ \theta_d,\ \xi_d)$. Following the absorption of a single photon γ_r, the photoelectron e_p and the detected photon γ_d are sequentially ejected from the atom $\mathfrak{T}$ and from its excited photoion $\mathfrak{T}^{1+^*}$, placed at the origin "O" of our coordinate system. The OZ-axis is the quantization direction for the electronic states of the atom $\mathfrak{T}$, excited photoion $\mathfrak{T}^{1+^*}$, and of the residual ion $\mathfrak{T}^{1+}$ participating in the 2-step process (1.4). Photoelectron enters the Mott detector [285], which records whether e_p's spin is up or down with respect to its own quantization direction $\hat{\boldsymbol{u}}_p$; whereas γ_d enters the photon spectrometer measuring its polarization with respect to the $OX_dY_dZ_d$ frame. The quantities referred to $OXYZ$ and $OX_dY_dZ_d$ frames of reference, shown in this figure, are related by (8.3). Parameter β_d, also shown in the figure and explained in the discussion related to (8.2), specifies polarization of the emitted photon γ_d

Similarly, the spherical components [65] of the polarization vectors

$$\hat{\boldsymbol{u}}_d = \hat{\boldsymbol{e}}_{x_d} \cos\beta_d + i\,\hat{\boldsymbol{e}}_{y_d} \sin\beta_d \quad \text{and} \qquad (8.5a)$$
$$(\hat{\boldsymbol{u}}_d)^* = \hat{\boldsymbol{e}}_{x_d} \cos\beta_d - i\,\hat{\boldsymbol{e}}_{y_d} \sin\beta_d$$

in DF are [263]

$$\begin{aligned} (u_d)_{-1} &= \tfrac{1}{\sqrt{2}}(x_d - i\,y_d) = \tfrac{1}{\sqrt{2}}(\cos\beta_d + \sin\beta_d), \\ (u_d)_0 &= z_d = 0, \\ (u_d)_{+1} &= -\tfrac{1}{\sqrt{2}}(x_d + i\,y_d) = -\tfrac{1}{\sqrt{2}}(\cos\beta_d - \sin\beta_d); \end{aligned} \qquad (8.5b)$$

and

$$\begin{aligned} [(u_d)^*]_{-1} &= \tfrac{1}{\sqrt{2}}(x_d - i\,y_d) = \tfrac{1}{\sqrt{2}}(\cos\beta_d - \sin\beta_d), \\ [(u_d)^*]_0 &= z_d = 0, \\ [(u_d)^*]_{+1} &= -\tfrac{1}{\sqrt{2}}(x_d + i\,y_d) = -\tfrac{1}{\sqrt{2}}(\cos\beta_d + \sin\beta_d); \end{aligned} \qquad (8.5c)$$

respectively.

8.1 Coulombic Entanglement Between (Spin, Polarization) of (e_p, γ_d)

This section contains a discussion of the bipartite entanglement between spin of the photoelectron e_p and polarization of the photon γ_d, generated sequentially in the 2-step process (1.4), by taking only the Coulomb forces inside an atom into account. The following is, therefore, a counter part of the bipartite Coulombic entanglement discussed in Chaps. 4–6 between two particles both of which have, unlike herein, rest mass different from zero.

8.1.1 Density Operator and Density Matrix

When the SOI present in an atom is either negligible or discarded, then

$$\boldsymbol{\ell}_r + \boldsymbol{L}_0 = \boldsymbol{L}_{1^{+*}} + \boldsymbol{\ell}_p, \qquad \boldsymbol{L}_{1^{+*}} = \boldsymbol{L}_{1^+} + \boldsymbol{\ell}_d \qquad (8.6a)$$

and

$$\boldsymbol{S}_0 = \boldsymbol{S}_{1^{+*}} + \boldsymbol{s}_p\left[= \left(\frac{1}{2}\right)_p\right], \qquad \boldsymbol{S}_{1^{+*}} = \boldsymbol{S}_{1^+} \qquad (8.6b)$$

are the conditions for the conservation of the orbital (8.6a) and of the spin (8.6b) angular momenta for the process (1.4). Here, $\boldsymbol{\ell}_d$ (with $|\boldsymbol{\ell}_d| = 1$ in $E1$ approximation) is the angular momentum of the detected photon γ_d. Further, $(\boldsymbol{L}_{1^+}, \boldsymbol{S}_{1^+})$ in (8.6) are (orbital, spin) angular momenta of the residual positive ion $\mathfrak{T}^{1+}$ left in its electronic state $|1^+\rangle = |L_{1^+}\ S_{1^+}\ M_{L_{1^+}}\ M_{S_{1^+}}\rangle$ after the radiative decay of the state $|1^{+^*}\rangle = |L_{1^{+*}}\ S_{1^{+*}}\ M_{L_{1^{+*}}}\ M_{S_{1^{+*}}}\rangle$ of the excited photoion $\mathfrak{T}^{1+^*}$ in the absence of SOI in L-S coupling.

The density operator for the present 2-step process is obtained from the expression (3.12) with the operator F_s for spontaneous radiative decay given by (3.9). The corresponding density matrix now becomes

$$
\begin{aligned}
&\langle L_{1^+} S_{1^+};\ \mu_p\, \hat{\boldsymbol{u}}_p\, \boldsymbol{k}_p;\ \omega_d\ \beta_d |\, \rho^{(2)}\, | L_{1^+} S_{1^+};\ \mu_p'\, \hat{\boldsymbol{u}}_p\, \boldsymbol{k}_p;\ \omega_d\ \beta_d' \rangle \\
&= \frac{\mathcal{K}}{(2L_0+1)(2S_0+1)} \sum_{\substack{M_{L_0}\, M_{L_{1^{+*}}}\, M'_{L_{1^{+*}}}\, M_{L_{1^+}} \\ M_{S_0}\, M_{S_{1^{+*}}}\, M'_{S_{1^{+*}}}\, M_{S_{1^+}}}} \\
&\quad \times \langle 1^+;\ \omega_d\ \beta_d |\, F_\gamma\, | 1^{+^*}\rangle \langle 1^{+^*};\ \mu_p\, \hat{\boldsymbol{u}}_p\, \boldsymbol{k}_p |\, F_p\, |0;\ 1m_r\rangle \\
&\quad \times \langle 0;\ 1m_r | (F_p)^\dagger | 1'^{+^*};\ \mu_p'\, \hat{\boldsymbol{u}}_p\, \boldsymbol{k}_p\rangle \langle 1'^{+^*} | (F_\gamma)^\dagger | 1^+;\ \omega_d\ \beta_d' \rangle. \qquad (8.7)
\end{aligned}
$$

As usual, for a bipartite state of qubits, (8.7) is a (4×4) matrix whose diagonal (i.e., $\mu_p = \mu_p'$, $\beta_d = \beta_d'$) elements are the angular distribution of spin-resolved photoelectron e_p and polarized photon γ_d; whereas non-diagonal (i.e., $\mu_p \neq \mu_p'$ and/or $\beta_d \neq \beta_d'$) elements represent coherence effects. The diagonal terms in the matrix (8.7) can be shown to simplify to those given elsewhere [177, 286, 287].

The product of the photoionization amplitudes, present in the density matrix (8.7) with sum over M_{S_0} and division by $(2\ S_0 + 1)$, has already been evaluated in (3.27) and (3.28). The photoemission amplitude [263]

$$
\begin{aligned}
\mathcal{P}_\gamma &\equiv \langle 1^+;\ \omega_d\ \beta_d |\, F_\gamma\, | 1^{+^*}\rangle \\
&= \langle L_{1^+}\ S_{1^+}\ M_{L_{1^+}}\ M_{S_{1^+}};\ \omega_d\ \beta_d |\, F_\gamma\, | L_{1^{+*}}\ S_{1^{+*}}\ M_{L_{1^{+*}}}\ M_{S_{1^{+*}}}\rangle, \qquad (8.8a) \\
&= \delta_{S_{1^+} S_{1^{+*}}}\, \delta_{M_{S_{1^+}} M_{S_{1^{+*}}}}\, (-1)^{L_{1^+} - M_{L_{1^+}}}\ \langle L_{1^+} \|\, \boldsymbol{r}\, \| L_{1^{+*}}\rangle \\
&\quad \times \sum_{q\, q_1} (-1)^q \begin{pmatrix} L_{1^+} & 1 & L_{1^{+*}} \\ -M_{L_{1^+}} & q_1 & M_{L_{1^{+*}}} \end{pmatrix} \left[(u_d)^* \right]_q \left[\mathcal{D}^1_{-q\, q_1}(\omega_d) \right]^*, \qquad (8.8b)
\end{aligned}
$$

on the other hand, is readily obtained on using the transformation (8.3b) and Wigner–Eckart theorem [65, 68]. Further in (8.8b), $\langle L_{1^+} \|\, \boldsymbol{r}\, \| L_{1^{+*}}\rangle$ is the reduced amplitude [65, 68]. Expression (8.8b) uses the facts that the $E1$ operator (3.9) for spontaneous radiative decay does not depend on any spin variables and, hence, the conservation of angular momenta expressed in (8.6). With the help of (8.8b), one readily calculates the following expression needed in the present density matrix (8.7).

$$\sum_{M_{L_{1^+}}\, M_{S_{1^+}}} \mathcal{P}_\gamma\, (\mathcal{P}'_\gamma)^\dagger \equiv \sum_{M_{L_{1^+}}\, M_{S_{1^+}}} \langle 1^+;\, \omega_d\, \beta_d |\, F_\gamma\, | 1^{+^*} \rangle \langle 1^{+^*} |\, (F_\gamma)^\dagger\, | 1^+;\, \omega_d\, \beta'_d \rangle$$

$$= \delta_{S_{1^+} S_{1^{+*}}}\, \delta_{M_{S_{1^+}} M_{S_{1^{+*}}}}\, (-1)^{L_{1^+} + M_{L'_{1^{+*}}}} \left| \langle L_{1^+} \|\, \boldsymbol{r}\, \| L_{1^{+*}} \rangle \right|^2$$

$$\times \sum_{K\,Q\,Q_1} (-1)^{Q_1} \sqrt{2K+1} \begin{pmatrix} L_{1^{+*}} & K & L_{1^{+*}} \\ -M'_{L_{1^{+*}}} & -Q_1 & M_{L_{1^{+*}}} \end{pmatrix} \begin{Bmatrix} L_{1^{+*}} & L_{1^{+*}} & K \\ 1 & 1 & L_{1^+} \end{Bmatrix}$$

$$\times \left[\hat{\boldsymbol{u}}'_d \otimes (\hat{\boldsymbol{u}}_d)^* \right]^{(K)}_Q \mathcal{D}^K_{Q\, Q_1}(\omega_d). \tag{8.9}$$

Let us now substitute (3.28) and (8.9) in (8.7). The consequent expression is simplified using some of the identities—already mentioned elsewhere in this monograph—from Racah algebra. Finally, one obtains

$$\langle L_{1^+} S_{1^+};\, \mu_p\, \hat{\boldsymbol{u}}_p\, \boldsymbol{k}_p;\, \omega_d\, \beta_d |\, \rho^{(2)}\, | L_{1^+} S_{1^+};\, \mu'_p\, \hat{\boldsymbol{u}}_p\, \boldsymbol{k}_p;\, \omega_d\, \beta'_d \rangle$$

$$= \delta_{S_{1^+} S_{1^{+*}}}\, (-1)^{m_r + L_0 + L_{1^{+*}} + L_{1^+}} \left| \langle L_{1^+} \|\, \boldsymbol{r}\, \| L_{1^{+*}} \rangle \right|^2 \frac{\mathcal{K}}{2(2L_0+1)}$$

$$\times \sum_{\ell_p\, \ell'_p\, L_p\, L_r L\, L'\, K\, Q\, Q_1} (-1)^{\ell'_p + L + Q_1} (2L_r+1) \sqrt{(2L_p+1)(2K+1)}$$

$$\times \begin{pmatrix} \ell_p & \ell'_p & L_p \\ 0 & 0 & 0 \end{pmatrix} \begin{pmatrix} 1 & 1 & L_r \\ m_r & -m_r & 0 \end{pmatrix} \begin{pmatrix} L_p & L_r & K \\ Q_1 & 0 & -Q_1 \end{pmatrix}$$

$$\times \begin{Bmatrix} 1 & 1 & L_r \\ L & L' & L_0 \end{Bmatrix} \begin{Bmatrix} 1 & 1 & K \\ L_{1^{+*}} & L_{1^{+*}} & L_{1^+} \end{Bmatrix} \begin{Bmatrix} \ell_p & \ell'_p & L_p \\ L & L' & L_r \\ L_{1^{+*}} & L_{1^{+*}} & K \end{Bmatrix}$$

$$\times \left[\hat{\boldsymbol{u}}'_d \otimes (\hat{\boldsymbol{u}}_d)^* \right]^{(K)}_Q \left[Y^{Q_1}_{L_p}(\hat{\boldsymbol{k}}_p) \right]^* \mathcal{D}^K_{Q\, Q_1}(\omega_d)$$

$$\times P_e(L_{1^{+*}}\, \ell_p;\, L_0\, 1;\, L) \left[P_e(L_{1^{+*}}\, \ell'_p;\, L_0\, 1;\, L') \right]^* \tag{8.10}$$

for the density matrix needed for studying entanglement between spin of photoelectron e_p and polarization of photon γ_d emitted in two sequential steps of the process (1.4) in the absence of the SOI in L-S coupling. Present in (8.10) is the photoionization amplitude $P_e(L_{1^{+*}}\, \ell_p;\, L_0\, 1;\, \mathrm{L})$ already defined in (6.4b).

8.1.2 Properties of the Density Matrix

It is obvious from its derivation that the expression (8.10) is completely general and independent of all dynamical models, which can be used in the evaluation of the photoionization amplitude (6.4b) and the reduced matrix element $\langle L_{1+} \| \boldsymbol{r} \| L_{1+^*} \rangle$ in the absence of SOI in L-S coupling. It is easy to show that (8.10) satisfies the Hermiticity and positivity conditions (3.16) and (3.17), respectively. In order for (8.10) to meet the normalization requirement (3.18), one may need to divide each term in this density matrix by its trace.

8.1.3 Are Spin of Photoelectron e_p and Polarization of Photon γ_d Entangled?

Looking carefully, one finds that the right-hand side of (8.10) does not contain any of the variables (i.e., $S_0, S_{1+*}, S_{1+}, \mu_p, \mu_p', \hat{\boldsymbol{u}}_p/\omega_p$) related to the spin of the atom $\mathfrak{T}$, its excited photoion $\mathfrak{T}^{1+^*}$, residual ion $\mathfrak{T}_{1+}$, or photoelectron e_p, but were originally present in the photoionization amplitude (3.26) [or in the expression (3.28), for that matter], used in the derivation of the above density matrix; although, (8.10) does contain variables $(\hat{\boldsymbol{u}}_d, \hat{\boldsymbol{u}}_d')$ which, in turn, depend upon the parameters (β_d, β_d') [in (8.2) and (8.5)] needed for specifying polarization of the detected photon γ_d. The total absence of $(\mu_p, \mu_p', \hat{\boldsymbol{u}}_p/\omega_p)$ from the density matrix (8.10) simply means that spin of photoelectron e_p is not at all correlated with the polarization of the detected photon γ_d. Thus, purely Coulomb forces inside an atom are incapable of generating (spin, polarization) entanglement in a (e_p, γ_d) pair. This is a completely general result applicable to the 2-step process (1.4) involving any of the electronic states of an unpolarized atom $\mathfrak{T}$ in the absence of SOI in L-S coupling. Consequently, (8.10) represents a photoelectron-photon state, which is always separable (i.e., not entangled) with respect to the spin angular momentum of e_p and polarization of γ_d.

We have seen in Chap. 6, on the other hand, that if the same excited photoion $\mathfrak{T}^{1+^*}$, whose radiative decay in the 2-step process (1.4) forms a separable state of the (e_p, γ_d)-system, decays non-radiatively in the 2-DPI process (1.3) then the spin of the emitted Auger electron e_a may be entangled with that of the photoelectron e_p. The circumstances in which 2-DPI in Chap. 6 takes place are identical to those present in the current discussion of the 2-step process (1.4): That is, in both cases, inner-shell ionization is caused by the absorption of a single photon γ_r in $E1$ approximation, only the Coulomb interaction present in an atom has been taken into account in L-S coupling, and the SOI has completely been discarded. Thus, the same Coulomb forces, which were successful in Chap. 6 in generating a spin-entangled state of (e_p, e_a) in 2-DPI (1.3), fail to entangle spin angular momentum of e_p with the polarization of γ_d—both generated sequentially in the 2-step process (1.4)—in the present Sect. 8.1.

The well-known selection rule applicable to the spin angular momentum for the emission of an electron from an atom due either to the absorption of a single photon in $E1$ approximation or in a spontaneous non-radiative decay is [10,60] $\Delta S = \pm\frac{1}{2}$. This rule is always obeyed whenever electron emission takes place in the presence of only the Coulomb forces and no SDIs are taken into account in an atom in L-S coupling [see, for example, (3.20) and first of (8.6b)]. This means, in the non-radiative decay on $\mathfrak{T}^{1+^*}$ in (1.3b), its spin $S_{1^{+*}}$ is equal to neither S_0 of $\mathfrak{T}$ nor S_{2^+} of $\mathfrak{T}^{2+}$.

This restriction on the spin $S_{1^{+*}}$ of $\mathfrak{T}^{1+^*}$ in 2-DPI process (1.3) and the presence of the coherent sums over its magnetic quantum numbers $(M_{S_{1^{+*}}}, M'_{S_{1^{+*}}})$ in the density matrix (6.2) provides a coupling between the spin angular momenta of (e_p, e_a) sequentially emitted in the 2-DPI process (1.3) even in the absence of the SOI. This coupling, in addition to the satisfaction of certain other requirements, already discussed in detail in Chap. 6, is found to entangle the spin angular momenta of photoelectron e_p and Auger electron e_a even in the absence of SOI.

Although, the coherent sums over the magnetic quantum numbers $(M_{S_{1^{+*}}}, M'_{S_{1^{+*}}})$ of the spin $S_{1^{+*}}$ are present also in the density matrix (8.7), which describes a (e_p, γ_d)-system in the absence of SOI in L-S coupling; however, unlike (3.20b) in the non-radiative decay of $\mathfrak{T}^{1+^*}$ in (1.3b), the second of the conservation conditions (8.6b) applicable to the radiative decay [10, 60] in (1.4b) is that the spin $S_{1^{+*}}$ of $\mathfrak{T}^{1+^*}$ must now be equal to the spin S_{1^+} of the residual ion $\mathfrak{T}^{1+}$. It makes $M_{S_{1^{+*}}} = M'_{S_{1^{+*}}}$ and, thus, reduces the coherent sums over $(M_{S_{1^{+*}}}, M'_{S_{1^{+*}}})$ to a single incoherent sum over, say, $M_{S_{1^{+*}}}$ which can analytically be performed. Consequently, photoelectron's spin variables $(\mu_p, \mu'_p, \hat{\mathbf{u}}_p/\omega_p)$, present in the original density matrix (8.7), disappear from its final expression (8.10) describing a state of the (e_p, γ_d)-system. Hence, purely Coulomb forces in an atom in the absence of the SOI become incapable of entangling spin angular momentum of photoelectron e_p with the polarization of the photon γ_d emitted sequentially in the 2-step process (1.4).

8.2 Fine-Structure Entanglement Between (Spin, Polarization) of (e_p, γ_d) in the Presence of Both Coulomb Plus Spin–Orbit Interactions

In the following, the condition imposed in the last Sect. 8.1—namely, the exclusion of SOI form the process (1.4)—has been relaxed. The current section, therefore, investigates [263], using the tools from quantum information science, properties of non-local correlation between the spin of photoelectron e_p and polarization of photon γ_d generated in the 2-step process (1.4) in the presence of C+SOIs.

8.2.1 Density Operator and Density Matrix

In the presence of C+SOIs, although energy in the 2-step process (1.4) is still conserved according to the (8.1); however, $(E_0,\ E_{1^{+*}},\ E_{1^+})$ now belong to the electronic states $(|0\rangle = |J_0\, M_{J_0}\rangle,\ |1^{+^*}\rangle = |J_{1^{+*}}\, M_{J_{1^{+*}}}\rangle,\ |1^+\rangle = |J_{1^+}\, M_{J_{1^+}}\rangle)$ of the species $(\mathfrak{T},\ \mathfrak{T}^{1+^*},\ \mathfrak{T}^{1+})$ in j-j coupling. Second, conditions (8.6) for the conservation of the angular momenta, suitable for studying Coulombic entanglement of $(e_p,\ \gamma_d)$, are no longer applicable on taking also the SOI into account in the process (1.4). The valid conditions expressing the conservation of angular momenta in the present case are, instead,

$$\boldsymbol{\ell}_r\,\Big(\big|\boldsymbol{\ell}_r\big| = 1\Big) + \boldsymbol{J}_0 = \boldsymbol{J}_{1^{+*}} + \boldsymbol{j}_p\Big[= \boldsymbol{\ell}_p + \boldsymbol{s}_p\Big(\big|\boldsymbol{s}_p\big| = \frac{1}{2}\Big)\Big] \tag{8.11a}$$

for the photoionization (1.4a) and

$$\boldsymbol{J}_{1^{+*}} = \boldsymbol{J}_{1^+} + \boldsymbol{\ell}_d\,\Big(\big|\boldsymbol{\ell}_d\big| = 1\Big) \tag{8.11b}$$

for the spontaneous radiative decay (1.4b) in the j-j coupling.

For the 2-step process (1.4) in j-j coupling, the density matrix (8.7) takes the following form [263]:

$$\begin{aligned}
&\langle J_{1^+};\ \mu_p\, \hat{\boldsymbol{u}}_p\, \boldsymbol{k}_p;\ \omega_d\, \beta_d \,|\, \rho^{(2)} \,|\, J_{1^+};\ \mu_p'\, \hat{\boldsymbol{u}}_p\, \boldsymbol{k}_p;\ \omega_d\, \beta_d' \rangle = \frac{\mathcal{K}}{(2J_0+1)} \\
&\times \sum_{\substack{M_{J_0}\, M_{J_{1^+}} \\ M_{J_{1^{+*}}}\, M_{J_{1^{+*}}}'}} \langle 1^+;\ \omega_d\, \beta_d \,|\, F_\gamma \,|1^{+^*}\rangle \langle 1^{+^*};\ \mu_p\, \hat{\boldsymbol{u}}_p\, \boldsymbol{k}_p \,|\, F_p \,|0;\ 1 m_r\rangle \\
&\times \langle 0;\ 1 m_r | (F_p)^\dagger | 1'^{+^*};\ \mu_p'\, \hat{\boldsymbol{u}}_p\, \boldsymbol{k}_p\rangle \langle 1'^{+^*} | (F_\gamma)^\dagger | 1^+;\ \omega_d\, \beta_d' \rangle.
\end{aligned} \tag{8.12}$$

Next, the amplitude for the radiative decay (1.4b) in j-j coupling, corresponding to that given in (8.8) in L-S coupling, is [263]

$$\mathcal{P}_\gamma \equiv \langle 1^+;\ \omega_d\, \beta_d \,|\, \mathcal{F}_s \,|1^{+^*}\rangle = \langle J_{1^+}\, M_{J_{1^+}};\ \omega_d\, \beta_d \,|\, \mathcal{F}_s \,|J_{1^{+*}}\, M_{J_{1^{+*}}}\rangle, \tag{8.13a}$$

$$\begin{aligned}
&= (-1)^{J_{1^+} - M_{J_{1^+}}} \langle J_{1^+} \|\, \boldsymbol{r} \,\| J_{1^{+*}} \rangle \\
&\times \sum_{q\, q_1} (-1)^q \begin{pmatrix} J_{1^+} & 1 & J_{1^{+*}} \\ -M_{J_{1^+}} & q_1 & M_{J_{1^{+*}}} \end{pmatrix} \Big[(u_d)^* \Big]_q \Big[\mathcal{D}^1_{-q\, q_1}(\omega_d) \Big]^*,
\end{aligned} \tag{8.13b}$$

with

$$\sum_{M_{J_{1^+}}} \mathcal{P}_\gamma\,(\mathcal{P}'_\gamma)^\dagger \;\equiv\; \sum_{M_{J_{1^+}}} \langle 1^+;\; \omega_d\,\beta_d|\,\mathcal{F}_s\,|1^{+^*}\rangle\langle 1^{+^*}|\,(\mathcal{F}_s)^\dagger\,|1^+;\; \omega_d\,\beta'_d\rangle$$

$$= (-1)^{J_{1^+}+M'_{1^{+*}}}\left|\langle J_{1^+}\|\,\boldsymbol{r}\,\|J_{1^{+*}}\rangle\right|^2 \sum_{K\,Q\,Q_1} (-1)^{Q_1}\sqrt{2K+1}$$

$$\times \begin{pmatrix} J_{1^{+*}} & K & J_{1^{+*}} \\ -M'_{J_{1^{+*}}} & -Q_1 & M_{J_{1^{+*}}} \end{pmatrix} \begin{Bmatrix} J_{1^{+*}} & J_{1^{+*}} & K \\ 1 & 1 & J_{1^+} \end{Bmatrix}$$

$$\times \left[\hat{\boldsymbol{u}}'_d \otimes (\hat{\boldsymbol{u}}_d)^*\right]^{(K)}_Q \mathcal{D}^K_{Q\,Q_1}(\omega_d). \tag{8.13c}$$

Finally, let us substitute (3.42b) and (8.13c) in the expression (8.12) and perform sums over $M_{J_{1^{+*}}}$, $M'_{J_{1^{+*}}}$, plus some additional simplifications. Finally, one obtains [263]

$$\langle J_{1^+};\; \mu_p\,\hat{\boldsymbol{u}}_p\,\boldsymbol{k}_p;\; \omega_d\,\beta_d|\,\rho^{(2)}\,|J_{1^+};\; \mu'_p\,\hat{\boldsymbol{u}}_p\,\boldsymbol{k}_p;\; \omega_d\,\beta'_d\rangle$$

$$\equiv \left(\rho^{(2)}\right)_{\mu_p\,\beta_d;\,\mu'_p\,\beta'_d}(\hat{\boldsymbol{u}}_p\,\boldsymbol{k}_p;\; \omega_d;\; m_r)$$

$$= (-1)^{\frac{1}{2}+\mu'_p} \sum_{\substack{S_p\,M_{S_p}\,N_{S_p}\\ K\,Q\,Q_1}} \begin{pmatrix} \frac{1}{2} & \frac{1}{2} & S_p \\ \mu_p & -\mu'_p & M_{S_p} \end{pmatrix} \left[\hat{\boldsymbol{u}}'_d \otimes (\hat{\boldsymbol{u}}_d)^*\right]^{(K)}_Q$$

$$\times \left[\mathcal{D}^{S_p}_{M_{S_p}\,N_{S_p}}(\omega_p)\right]^* \mathcal{D}^K_{Q,-Q_1}(\omega_d)\, A^{S_p\,K}_{N_{S_p}\,Q_1}(\boldsymbol{k}_p;\; m_r), \tag{8.14a}$$

where

$$A^{S_p\,K}_{N_{S_p}\,Q_1}(\boldsymbol{k}_p;\; m_r)$$

$$= (-1)^{m_r+J_0-J_f}\,(2S_p+1)\sqrt{2K+1}\left|\langle J_{1^+}\|\,\boldsymbol{r}\,\|J_{1^{+*}}\rangle\right|^2 \frac{\mathcal{K}}{(2J_0+1)\sqrt{4\pi}}$$

$$\times \sum_{\substack{\ell_p\,j_p\,j_t\,L_p\,M_{L_p}\\ \ell'_p\,j'_p\,j'_t\,L_r\,J_p}} (-1)^{\ell_p+j'_t}\,(2J_p+1)(2L_r+1)\sqrt{2L_p+1}$$

$$\times \begin{pmatrix} \ell_p & \ell'_p & L_p \\ 0 & 0 & 0 \end{pmatrix} \begin{pmatrix} L_p & S_p & J_p \\ M_{L_p} & N_{S_p} & Q_1 \end{pmatrix} \begin{pmatrix} 1 & 1 & L_r \\ m_r & -m_r & 0 \end{pmatrix} \begin{pmatrix} J_p & K & L_r \\ Q_1 & -Q_1 & 0 \end{pmatrix}$$

$$\times \begin{Bmatrix} J_{1+^*} & J_{1+^*} & K \\ j_t & j_t' & J_0 \end{Bmatrix} \begin{Bmatrix} J_{1+^*} & J_{1+^*} & K \\ 1 & 1 & J_{1+} \end{Bmatrix} \begin{Bmatrix} \ell_p & \ell_p' & L_p \\ \frac{1}{2} & \frac{1}{2} & S_p \\ j_p & j_p' & J_p \end{Bmatrix} \begin{Bmatrix} 1 & 1 & L_r \\ j_p & j_p' & J_p \\ j_t & j_t' & K \end{Bmatrix}$$

$$\times \left[Y_{L_p}^{M_{L_p}}(\hat{\boldsymbol{k}}_p)\right]^* \overline{P}_e(J_{1+^*}\, j_p;\ J_0\, 1;\ j_t;\ \ell_p)\left[\overline{P}_e(J_{1+^*}\, j_p';\ J_0\, 1;\ j_t';\ \ell_p')\right]^*. \tag{8.14b}$$

Here, $\overline{P}_e(J_{1+^*}\, j_p;\ J_0\, 1;\ j_t;\ \ell_p) = (-i)^{\ell_p} e^{\sigma_{\ell_p j_p}}(2j_t+1)\sqrt{(2\ell_p+1)(2j_p+1)}$ $\langle J_{1+^*} j_p | F(j_t) | J_0\, 1\rangle$ have already been introduced in (3.42b) with the matrix element $\langle J_{1+^*} j_p | F(j_t) | J_0\, 1\rangle$ defined in (3.40b).

8.2.2 Properties of the Density Matrix

Expression (8.14) for a state of the $(e_p,\ \gamma_d)$-system, generated in the 2-step process (1.4) in the presence of C+SOIs in j-j coupling, is completely general and independent of all dynamical models. On comparing the present density matrix (8.14) with (8.10) derived earlier taking only the Coulomb interaction into account in L-S coupling, one immediately notices some important differences between the two expressions. For example, unlike (8.10), the present density matrix (8.14) contains the variables $(\mu_p,\ \mu_p',\ \omega_p)$ related to the spin of the photoelectron e_p as well as $(\beta_d,\ \beta_d',\ \omega_d)$ needed to characterize polarization of the photon γ_d emitted in the second step of the process (1.4b). Thus, unlike in the Coulombic density matrix (8.10), the presence of the SOI, in addition to the Coulomb forces, in an atom has correlated spin of e_p with the polarization of γ_d. This correlation is, however, very much dependent not only upon the dynamics of photoionization of the atomic target $\mathfrak{T}$ and of the radiative decay of the excited photoion $\mathfrak{T}^{1+^*}$, but also upon the properties (i.e., energy and polarization) of the ionizing as well as detected photons γ_r and γ_d, respectively, in addition to the spin quantization direction $\hat{\boldsymbol{u}}_p$ and linear momentum $\boldsymbol{k}_p$ of the photoelectron e_p.

As usual, the diagonal elements of the density matrix (8.14) represent angular correlation between spin-resolved e_p and polarized γ_d emitted sequentially in the 2-step precess (1.4); whereas its non-diagonal elements describe coherence effects. In addition to being both Hermitian and positive, the normalized (8.14) will not contain reduced matrix elements $\langle J_{1+} \| \ \boldsymbol{r}\ \| J_{1+^*}\rangle$, i.e., will be independent of the dynamics of radiative decay of $\mathfrak{T}^{1+^*}$ in (1.4b).

Further, the final expression (8.14) very much involves the total angular momenta $(J_0,\ J_{1+^*},\ J_{1+})$ of the atomic species $(\mathfrak{T},\ \mathfrak{T}^{1+^*},\ \mathfrak{T}^{1+})$ participating in the two steps of the process (1.4). These quantum numbers determine the fine-structure levels of the species $(\mathfrak{T},\ \mathfrak{T}^{1+^*},\ \mathfrak{T}^{1+})$ among which photoionizing and radiative decay transitions in (1.4) take place. It will, therefore, be very appropriate to call, henceforth, the expression (8.14) as the fine-structure density matrix for the

2-step process (1.4) and any non-local correlation represented by it as fine-structure entanglement between the spin of e_p and polarization of γ_d.

8.2.3 *Example*

It is obvious from the discussion on pages 186–187 in Sect. 8.1.3, and from that contained in the last two Sects. 8.2.1 plus 8.2.2, that a (e_p, γ_d)-pair can only have fine-structure non-local correlation, i.e., entanglement between spin of photoelectron e_p and polarization of photon γ_d in the 2-step process (1.4) can exist when SOI, in addition to the Coulomb interaction, is also taken into account. Therefore, in order to see the effects of non-local correlation predicted by the density matrix (8.14) in a (e_p, γ_d)-pair, one should select an atom with large value of the SOI. For this and several other physical reasons (see, e.g., [288]), Chandra and Ghosh [263] in their theoretical study considered the respective transitions

$$\gamma_r + \mathrm{Cd}(4\mathrm{d}^{10}\,5\mathrm{s}^2\,{}^1\mathrm{S}) \longrightarrow$$
$$\mathrm{Cd}^{1+^*}(4\mathrm{d}^9\,5\mathrm{s}^2\,{}^2\mathrm{D}_{3/2}) + e_p(\mathrm{p}_{1/2,3/2};\ \mathrm{f}_{5/2}) \tag{8.15a}$$

and

$$\mathrm{Cd}^{1+^*}(4\mathrm{d}^9\,5\mathrm{s}^2\,{}^2\mathrm{D}_{3/2}) \longrightarrow \mathrm{Cd}^{1+}(4\mathrm{d}^{10}\,5\mathrm{s}^0\,5\mathrm{p}^1\,{}^2\mathrm{P}_{3/2}) + \gamma_d \tag{8.15b}$$

for photoionization and spontaneous radiative decay in a free and unpolarized Cd atom.

The fine structure splitting [10, 60], arising due only to the presence of SOI, between the ${}^2\mathrm{D}_{3/2}$ and ${}^2\mathrm{D}_{5/2}$ states of Cd^{1+^*} is significantly large at about 0.7eV. In order to see the effects of SOI on photoinduced processes taking place in Cd, a large number of experimental (see, for example, [289–293], *etc*) and theoretical (e.g., [294–299]) studies on photoionization of this atom have been made during the last forty years. In addition, the theoretical predictions that orientation[1], and alignment[2] of ion $\mathfrak{T}^{1+^*}$ generated [173, 300] in photoionization of an atom $\mathfrak{T}$ in, say, (1.4a) may be detected [301–307] by a polarization measurement of photons γ_d in (1.4b) had inspired many experiments to investigate radiative decay $\mathrm{Cd}^{1+^*}(4\mathrm{d}^9\,5\mathrm{s}^2\,{}^2\mathrm{D}_{5/2}) \longrightarrow \mathrm{Cd}^{1+}(4\mathrm{d}^{10}\,5\mathrm{s}^0\,5\mathrm{p}^1\,{}^2\mathrm{P}_{3/2})$ [308, 309], and/or the

[1]It, in a $|J_{1+^*}\ M_{J_{1+^*}}\rangle$ representation of the excited photoion $\mathfrak{T}^{1+^*}$, means (see, for example, [61, 68, 267]) different populations of $|J_{1+^*}\ M_{J_{1+^*}}\rangle$ and $|J_{1+^*}\ -M_{J_{1+^*}}\rangle$ fine-structure levels.

[2]Here, although levels $|J_{1+^*}\ M_{J_{1+^*}}\rangle$ and $|J_{1+^*}\ -M_{J_{1+^*}}\rangle$ have equal populations; the combined populations of these two levels is, however, different from those of $|J_{1+^*}\ M'_{J_{1+^*}}\rangle$ and $|J_{1+^*}\ -M'_{J_{1+^*}}\rangle$ (see, for example, [61, 68, 267]).

one [308] shown in (8.15b), for measuring polarization of the fluorescence γ_d. These [308, 309] experimental studies have been followed by theoretical analysis [298, 305–307] of the polarization of photon γ_d emitted in the radiative decay of $Cd^{1+^*}(4d^9\,5s^2\;{}^2D_{3/2})$ and of $Cd^{1+^*}(4d^9\,5s^2\;{}^2D_{5/2})$. Moreover, the unique and interesting properties of the photoionizing and radiative transitions involving 4d and 5s electrons in Cd have led to many new types of laser excitation schemes and processes [288].

Furthermore, Caldwell and Zare [308] counted 20 γ_d in one sec in the radiative decay (8.15b) in their experiment on Cd. Every emission of electron e_p in the photoionizing transition (8.15a) has to be followed by a radiative decay of Cd^{1+^*} in (8.15b); one can, therefore, expect about twenty entangled (e_p, γ_d) pairs in one second in the 2-step process (8.15), provided the fine-structure state (8.14) is non-separable.

For these and other such reasons, the 2-step process (8.15) in a Cd atom is, probably, quite suitable for quantitatively studying the properties of (spin, polarization) fine-structure entanglement predicted by the density matrix (8.14) in a (e_p, γ_d)-pair. A non-relativistic description of the process (8.15a) is also adequate if one follows a recent [271] experiment on photoionization of as heavy an atom as Xe.

In order to see the effects on (e_p, γ_d) entanglement of the polarization of the ionizing radiation in (8.15a), Chandra and Ghosh [263] considered in their theoretical study four of those different possible polarizations of γ_r, which are used in most of the experimental measurements on photoionization in a laboratory. Namely, γ_r to be LP ($m_r = 0$), LCP ($m_r =$ -1), RCP ($m_r = +1$), or UP (i.e., an even mixture of RCP and RCP photons). Similarly, the fluorescence γ_d in (8.15b) may have any polarization specified by the vector $\hat{\boldsymbol{u}}_d$ in (8.1). Treating the γ_d as a qubit, it was considered to have either LP with its electric field vector to be along (OX_d, OY_d)-axes of the photon detector (DF) shown in Fig. 8.1 as $OX_dY_dZ_d$ and specified by ($\beta_d = 0$, $\beta_d = \pi/2$)) in (8.2); or CP with (negative, positive) helicities corresponding to ($\beta_d = \pi/4$, $\beta_d = 3\pi/4$) for the polarization vector $\hat{\boldsymbol{u}}_d$ in (8.2). However, for both of these (i.e., linear and circular) polarizations, γ_d was always propagating along the OZ_d-axis in Fig. 8.1.

On the other hand, a choice of the quantization direction $\hat{\boldsymbol{u}}_p$ of the spin of the electronic qubit in (8.15a) should have no effects on the properties related to the non-separability of the state (8.14) of a (e_p, γ_d)-pair as the entanglement is well known [17, 34] to be independent of the local choices of the bases and of coordinate systems. For brevity, let us take photoelectron e_p in (8.15a) to be longitudinally polarized along its direction of propagation, i.e.,

$$\hat{\boldsymbol{u}}_p(\vartheta_p, \varphi_p) \parallel \hat{\boldsymbol{k}}_p(\theta_p, \phi_p) \parallel \hat{\boldsymbol{k}}(\theta, \phi) \text{ (say)},$$
$$i.e., (\vartheta_p = \theta_p \equiv \theta, \varphi_p = \phi_p \equiv \phi). \tag{8.16a}$$

In order to see the effects on (spin, polarization) fine-structure entanglement of the directions of emission $(\hat{\boldsymbol{k}}, \omega_d)$ of (e_p, γ_d) from (Cd, Cd^{1+^*}), Chandra and

Ghosh [263] considered following three of an infinite number of geometries in which the 2-step experiment (8.15) can possibly be performed in a laboratory:

$$\omega_d\,(\phi_d = \pi + \phi, \theta_d = \pi - \theta, \xi_d = 0). \tag{8.16b}$$

It, in other words, means that γ_d and e_p are in diametric motion in the ionization plane (containing the OZ-axis and e_p's propagation vector—both referred to the SF).

$$\left.\begin{array}{l}\omega_d\,(\phi_d = \phi, \quad \theta_d = \theta + \frac{\pi}{2}, \ \xi_d = 0), \text{ for } 0 \le \theta \le \frac{\pi}{2}, \\ \omega_d\,(\phi_d = \phi + \pi, \theta_d = \frac{3\pi}{2} - \theta, \xi_d = 0), \text{ for } \frac{\pi}{2} \le \theta \le \pi.\end{array}\right\} \tag{8.16c}$$

Here, γ_d is still observed in the ionization plane, but in a direction perpendicular to $\hat{\boldsymbol{k}}_p$. Finally,

$$\omega_d\left(\phi_d = \phi + \frac{\pi}{2}, \theta_d = \frac{\pi}{2}, \xi_d = 0\right) \tag{8.16d}$$

corresponding to an experimental geometry in which γ_d is moving perpendicular to the ionization plane.

Another advantage of choosing the experimental geometries (8.16) is that the density matrix (8.14) will now contain only two spherical angles $\hat{\boldsymbol{k}}\ (\theta, \phi)$, in place of the original four spherical angles $[\hat{\boldsymbol{k}}_p\ (\theta_p, \phi_p), \hat{\boldsymbol{u}}_p\ (\vartheta_p, \varphi_p)]$ plus three Euler angles $\omega_d\ (\theta_d, \phi_d, \xi_d)$.

8.2.3.1 Results

The results obtained by Chandra and Ghosh [263] on (spin, polarization) entanglement of a (e_p, γ_d) pair sequentially emitted in the two-step process (8.15) taking place in any of the experimental geometries (8.16a)–(8.16d) can be summarized as follows:

- *Absorbed photon γ_r LP*: Spin of the photoelectron e_p is never entangled with either a linearly or circularly polarized fluorescence γ_d in any of the experimental geometries (8.16). Moreover, (e_p, γ_d) in this case are always in a mixed state.

- *Absorbed photon γ_r CP*: Although (e_p, γ_d) are now in an entangled mixed state; however, this state does not change with the helicity of γ_r [263]. That is, there is no circular dichroism [see footnote (3) in Chapter 11] in the entanglement in the present case. This entanglement is, nevertheless, highly anisotropic as it significantly changes with a change in the angle θ defined in (8.16a) and used in (8.16b)–(8.16d). Moreover, (e_p, γ_d) in the present case always formed a mixed state.

– Experimental geometry (8.16b): Here, the nature of the variations of both concurrence and of entanglement of formation with θ depends upon whether one is looking at the entanglement of the spin of e_p with an LP or a CP of γ_d. Figure 8.2 is for the entanglement when γ_d is LP. For CP γ_d, the variations of both concurrence and entanglement of formation with θ in the present experimental geometry of (8.16b) were found [263] to be different than those shown in Fig. 8.2.
– Experimental geometry (8.16c): In this case, first, there are negligible differences whether an LP or a CP γ_d is entangled with the spin of the photoelectron e_p; second, variation of entanglement with θ, shown in Fig. 8.3, is now very different from that present in the geometry (8.16b).

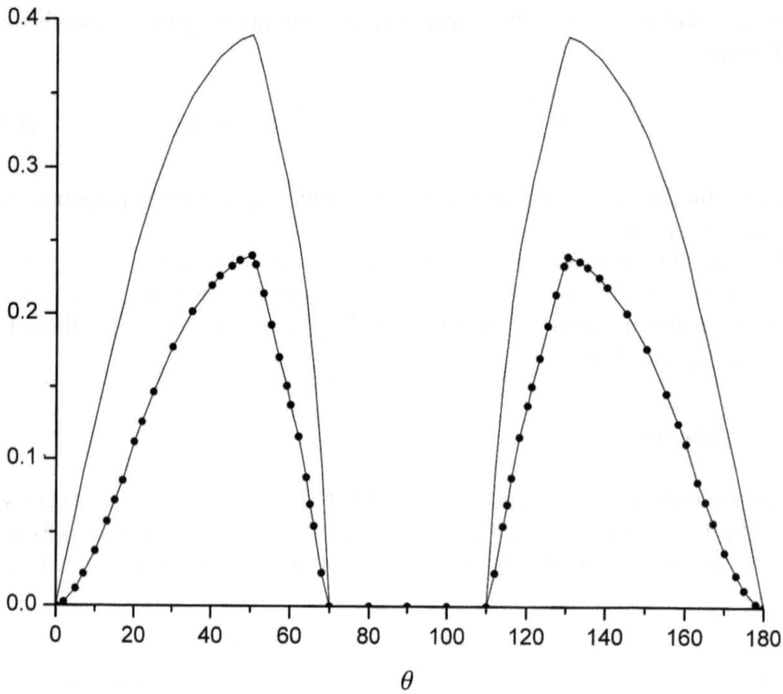

Fig. 8.2 Variation of the concurrence (continuous curve) and of the entanglement of formation (continuous curve with solid circles) with the direction of the line joining photoelectron e_p and the LP fluorescence γ_d, sequentially emitted in the process (8.15) in opposite directions according to the geometry (8.16b) in the ionization plane. Hence, the entanglement here is between the longitudinally polarized spin angular momentum of e_p and linear polarization of γ_d, while the ionizing radiation γ_r in (8.15a) is left or right circularly polarized. If, on the other hand, the detected γ_d is CP, then its entanglement with the spin of the photoelectron e_p is somewhat different from that shown in this figure (see [263]). [In this geometry, the (e_p, γ_d) entanglement is independent of the direction of the ionization plane in OXYZ frame; (e_p, γ_d) are never entangled if γ_r is either linearly polarized or unpolarized.] (First published in [263].)

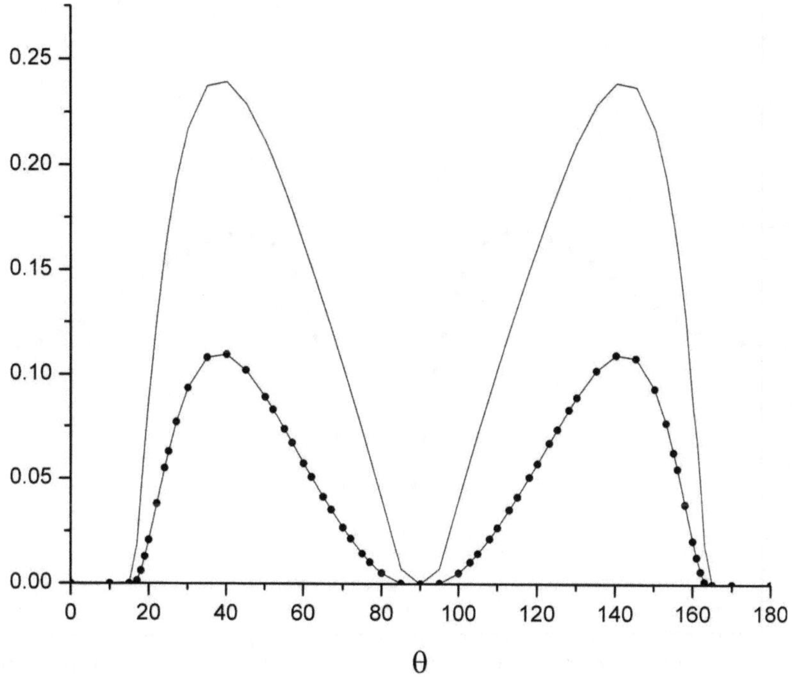

Fig. 8.3 Same as Fig. 8.2, but photoelectron e_p and fluorescence γ_d are now observed in mutually perpendicular directions in the ionization plane in the experimental geometry (8.16c). However, unlike the previous case of geometry (8.16b) shown in Fig. 8.2, there was negligible difference in the properties of the (e_p, γ_d)-state whether it is a LP or a CP γ_d entangled with the spin of e_p. [Even in the present geometry, the (e_p, γ_d) entanglement is independent of the direction of the ionization plane in OXYZ frame; (e_p, γ_d) are never entangled if γ_r is either linearly polarized or unpolarized.] (First published in [263].)

- Experimental geometry (8.16d): Although here too there are no noticeable differences whether a CP or a LP γ_d is entangled with the spin of e_p; the variation of entanglement with θ in the present experimental geometry is very different from that found in when both (e_p, γ_d) are observed in the same plane [i.e., in the set-ups represented by (8.16b) or (8.16c)]. Figure 8.4 contains concurrence and entanglement of formation calculated [263] for the present (e_p, γ_d) entanglement.

- *Absorbed photon γ_r UP*: Even now, (e_p, γ_d) are always in a mixed state. This state [263] is non-separable if these two particles are moving in perpendicular planes, i.e., in the experimental geometry represented by (8.16d); otherwise [i.e., in geometries (8.16b) or (8.16c)], spin of photoelectron e_p is never entangled with the polarization of the photon γ_d. The variation of entanglement with θ in the present case is, obviously, the same as found in the experimental set-up corresponding to (8.16d) for CP ionizing radiation γ_r and shown in Fig. 8.4.

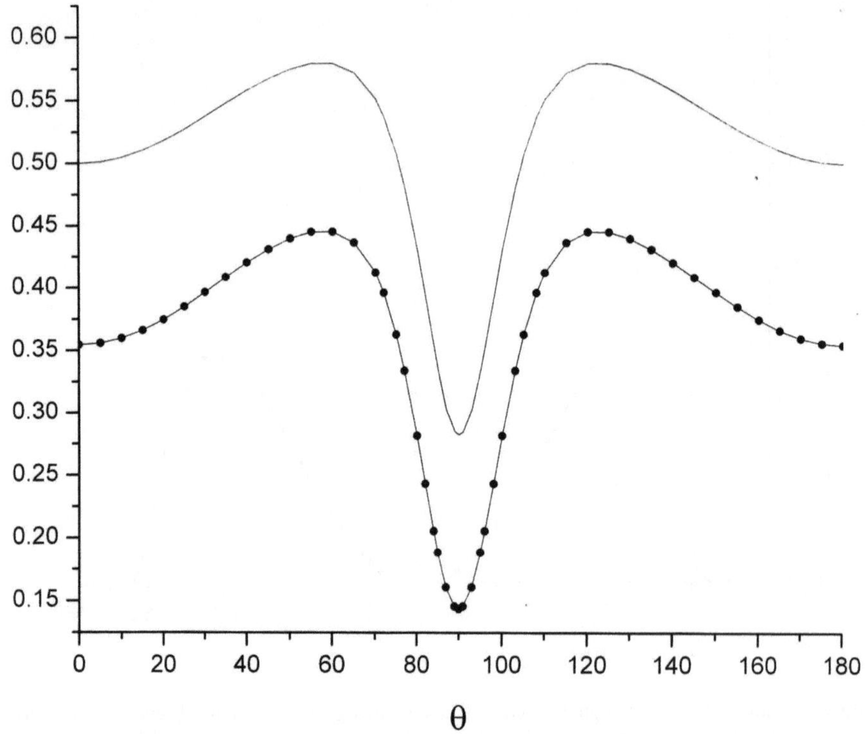

Fig. 8.4 Same as Fig. 8.2, but for the experimental geometry specified by (8.16d), i.e., fluorescence γ_d is now moving perpendicular to the ionization plane. Similar to the preceding case of geometry (8.16c) shown in the Fig. 8.3, there was negligible difference in the properties of the entanglement in the present figure whether it is a LP or CP γ_d entangled with the spin of e_p. [Even in the present geometry, the (e_p, γ_d) entanglement is independent of the direction of the ionization plane in OXYZ frame; (e_p, γ_d) are never entangled if γ_r is linearly polarized.] Thus, the present figure also shows the variation of concurrence (continuous curve) and of entanglement of formation (continuous curve with solid circles) with θ in the experimental geometry (8.16d) when the ionizing radiation γ_r in (8.15a) is UP. (First published in [263].)

These results very clearly show a very strong dependence of both nature and properties of the entanglement between the spin of the electronic qubit e_p and (linear or circular) polarization of the photonic qubit γ_d on the geometry of the experiment in which these two sequentially emitted particles are observed as well as on the polarization of the photon γ_r absorbed for ionization in (8.15a) to take place.

8.2.4 *Effects of Hyperfine Interaction and of Noise on* (e_p, γ_d) *Entanglement*

Interactions, other than the Coulomb interaction,[3] between a nucleus and the electrons of the same atom are known as hyperfine interactions (HFIs) (see, for example, [10, 60, 68], *etc*). The most important among the HFIs is the one which arises due to the interactions between the magnetic dipoles of nucleus and of electrons in a given atom. The moments of these two dipoles are related to the spin angular momentum, say, $\boldsymbol{I}$ of the nucleus and total angular momentum $\boldsymbol{J}$ of the electrons.

HFI is known [10, 60, 68] to have little effects on the energy levels of an atom. However, on account of the coupling between these two angular momenta in the electronic state $|1^{+^*}\rangle$ of the excited photoion $\mathfrak{T}^{1+^*}$, each of $\boldsymbol{J}_{1^{+*}}$ and $\boldsymbol{I}_{1^{+*}}$ precesses about their resultant $\boldsymbol{F}_{1^{+*}} = \boldsymbol{J}_{1^{+*}} + \boldsymbol{I}_{1^{+*}}$. If $\mathfrak{T}^{1+^*}$ in (1.4b) decays radiatively after the precessional motion has set in, then the reduction in the degree of polarization (called hyperfine depolarization) [61, 303, 304] of γ_d is substantial. A depolarization of an electromagnetic wave, caused by HFI or any other reasons, means [263] reduction in the difference between the relative intensities of the radiation linearly (circularly) polarized in two mutually perpendicular (opposite) directions, without affecting its total intensity. Consequently, hyperfine depolarization should have no effect on the entanglement between the spin of the photoelectron e_p and polarization of the photon γ_d emitted sequentially in the two-step process (1.4).

Similar to the entanglement discussed in other chapters of this monograph, the (spin, polarization) fine-structure entanglement in a (e_p, γ_d)-system in this chapter is generated by the mechanisms responsible for the sequential emission of these two particles and the forces present inside an atom. The environment external to the atom plays no role whatsoever. The entanglement in the present case simply depends upon the properties of the ionizing radiation γ_r, kinematics of the two-step process (1.4), and dynamics of photoionization (1.4a). Consequently, the (e_p, γ_d) entanglement generated in the present case is not coupled to the surroundings of the atomic target $\mathfrak{T}$ in which two-step process (1.4) takes place and, hence, it too, like those discussed elsewhere in this monograph, should be least prone to the harmful effects of dissipation and decoherence, which may be caused by noise external to the atomic target $\mathfrak{T}$.

[3] This is well known to be $-\sum_i Ze^2/r_i$ for an atom with atomic number Z and distance r_i of the i-th electron from its nucleus.

Part II
Molecular Processes

Chapter 9
One-Step Double Photoionization of Molecules

The process (1.2) of 1-DPI can equally take place in both atomic and molecular targets. Theoretical as well as experimental studies of 1-DPI of both linear (see, for example, [310–320]) and non-linear [321] molecules have already been performed by several workers in context other than those relevant to quantum information. Molecules are known to have both vibration and rotation of their nuclei, in addition to the motion of their electrons. It should be quite instructive—and is also, probably, important—to study the effects of these two nuclear motions, in addition to the other typical nuclear effects (e.g., non-central nature of a molecular potential, *etc*), on the spin-entanglement of electrons emitted in the 1-step and 2-step processes of (1.2) and (1.3), respectively, taking place in a molecular target. In this monograph, we limit our molecular studies only to these two processes taking place in a linear[1] system possessing either of the $C_{\infty v}$ or $D_{\infty h}$ point symmetries [68–70, 185].

Inclusion of the effects of nuclear vibration on any process taking place in a linear molecule requires, usually, averaging of the amplitude of the relevant physical quantity over the wavefunctions of the initial and final vibrational states of the molecule under consideration. Nuclear rotation in a molecule, on the other hand, has more profound effects on the molecular dynamics and, hence, on the processes taking place in it. Rotation of nuclei gives rise to both the rotational angular momentum and SRI. A proper considerations of these two and other additional effects are simultaneously possible by representing a linear molecule by parity-adapted states [68, 69, 75]. Such states for linear molecules have already been successfully prepared experimentally using multiphoton processes (see, for example, [75, 322]). Moreover, availability of electron spectrometers with high resolving powers has made it possible [322] to separate those electrons, which are produced by leaving behind molecular ions (e.g., ${H_2}^{1+}$, ${N_2}^{1+}$, ${O_2}^{1+}$, CO^{1+}, NO^{1+}) in their different rotational states. It should, therefore, be possible, in not too distant

[1] See Footnote (11) on page 74.

N. Chandra and R. Ghosh, *Quantum Entanglement in Electron Optics*, Springer Series on Atomic, Optical, and Plasma Physics 67, DOI 10.1007/978-3-642-24070-6_9,

a future, to resolve electrons according to the rotational states of molecular dications produced in processes like 1-DPI, 2-DPI, *etc.*

Chandra and coworkers [213, 215–217] have studied entanglement properties of the spin-states of $(e_p, e_{p'})$ and of (e_p, e_a) produced in 1-DPI and in 2-DPI of a RLM (rotating linear molecule) both without as well as with its SDIs taken into account using parity adapted wavefunctions for the molecular species $(\mathfrak{T}, \mathfrak{T}^{2+})$ and $(\mathfrak{T}, \mathfrak{T}^{1+^*}, \mathfrak{T}^{2+})$ participating in the two respective processes (1.2) and (1.3). Although the use of parity adapted states is not at all necessary for investigating the non-local correlation between the spins of electrons departed from a RLM in either of 1-DPI and 2-DPI; however, the use of such states facilitates the studies related to $C_{\infty v}$ and $D_{\infty h}$ in more than one ways. For example, one can investigate effects of SOI and of SRI separately, or of SDIs (*ie*, SOI + SRI jointly), on the spin-entanglement of $(e_p, e_{p'})$ and/or of (e_p, e_a) by using parity-adapted states. More importantly, applications of parity-adapted states directly give [213, 310, 311] us selection rules, which lead to 1-DPI and are applicable in $E1$ approximation. Similarly, one automatically obtains [214, 218] selection rules for non-radiative spontaneous decay of a RLM by employing parity-adapted states.

The two Sects. 9.1 and 9.2, in this Chap. 9, contain a discussion of the properties [213] of entanglement between the spins of a pair of electrons $(e_p, e_{p'})$ simultaneously emitted in the 1-DPI process (1.2) taking place in a RLM $\mathfrak{T}$ in the absence, or in the presence, of SDIs, respectively. In 1-DPI, the condition (5.1a) for the conservation of energy is always valid irrespective of the nature of the target $\mathfrak{T}$, or the interactions being taken into account, in a study of this process.

9.1 Coulombic Entanglement Between the Spins of Molecular Photoelectrons $(e_p, e_{p'})$

Unlike atoms, the constituents of a molecule experience three types of Coulomb forces. Namely, in addition to the electron–electron and electron–nucleus interactions, which are present in an atom as well, nucleus–nucleus interaction also exists in a molecule. Hence, in the absence of SDIs, entanglement between the spins of electrons liberated in a 1-DPI process (1.2) in a RLM will now be determined, among other things, by all these three kinds of Coulombic interactions.

9.1.1 Density Matrix

9.1.1.1 Derivation

In the absence of SDIs, spin angular momentum is always conserved in any photoionization process taking place in $E1$ approximation. Consequently, conservation condition (5.1c) is applicable to Coulombic 1-DPI even of a RLM. However, on account of the loss of spherical symmetry in both linear and non-linear molecular systems, orbital angular momentum is no longer a constant of the motion. Consequently, the condition (5.1b) cannot be used in the present case.

Expression (3.5b) is the density operator for any one-step process induced in any unpolarized target due to the absorption of a single photon in $E1$ approximation. This density operator has already been used for calculating, in Chaps. [4, 5] and in Sects. [7.1, 7.2], the density matrices needed for studying the Coulombic as well as fine-structure bipartite entanglements [(e_p, $\mathfrak{I}^{1+}$), ($e_p, e_{p'}$)] when the respective single-step processes [(1.1), (1.2)] take place in an atomic target. The same expression (3.5b) can be readily adapted also to the present study of non-local correlation between the spins of photoelectrons (e_p, $e_{p'}$) produced in 1-DPI of a RLM in Hund's coupling scheme (b) in the absence of SDIs. Recalling [68, 69, 75] that, in Hund's case (b), quantum numbers ($\mathfrak{p}_0$, M_{N_0}, M_{S_0}) specify states degenerate to $|0\rangle$, given by (3.48), of an unpolarized RLM, the following expression [213]

$$\rho^{(m,\,\text{1-DPI})} = \frac{\mathcal{K}_p}{(2-\delta_{0\Lambda_0})(2N_0+1)(2S_0+1)} \times \sum_{\mathfrak{p}_0\, M_{N_0}\, M_{S_0}} F_p \left(|0;\ 1m_r\rangle\langle 0;\ 1m_r|\right) F_p^{\dagger} \tag{9.1}$$

for the density operator for 1-DPI of an unpolarized molecular target $\mathfrak{T}$ without SDIs can readily be written from (3.5b). The superscript m on ρ on the left-hand side of (9.1) has been used to indicate that this density operator is meant for 1-DPI in a RLM as opposed to (5.2) used for studying the same process in an atomic target.

The density matrix for the 1-DPI process (1.2) involving the states $|0\rangle$ and $|2^+\rangle$ of the unpolarized RLM $\mathfrak{T}$ and of its residual photo-dication $\mathfrak{T}^{2+}$, respectively, plus of two photoelectrons (e_p, $e_{p'}$), is now given by [213]

$$\begin{aligned}
&\langle n_{2^+}\, N_{2^+}\, \Lambda_{2^+};\ \mu_p\, \hat{\boldsymbol{u}}_p\, \boldsymbol{k}_p;\ \mu_{p'}\, \hat{\boldsymbol{u}}_{p'}\, \boldsymbol{k}_{p'}|\rho^{(m,\,\text{1-DPI})}|n_{2^+}\, N_{2^+}\, \Lambda_{2^+};\ \mu_p'\, \hat{\boldsymbol{u}}_p\, \boldsymbol{k}_p;\\
&\mu_{p'}'\, \hat{\boldsymbol{u}}_{p'}\, \boldsymbol{k}_{p'}\rangle = \frac{\mathcal{K}_p}{(2-\delta_{0\Lambda_0})(2N_0+1)(2S_0+1)}\\
&\times \sum_{\substack{\mathfrak{p}_0\, M_{N_0}\, M_{S_0}\\ \mathfrak{p}_{2^+}\, M_{N_{2^+}}\, M_{S_{2^+}}}} \langle 2^+;\ \mu_p\, \hat{\boldsymbol{u}}_p\, \boldsymbol{k}_p;\ \mu_{p'}\, \hat{\boldsymbol{u}}_{p'}\, \boldsymbol{k}_{p'}|F_p|0;\ 1m_r\rangle\\
&\times \langle 2^+;\ \mu_p'\, \hat{\boldsymbol{u}}_p\, \boldsymbol{k}_p;\ \mu_{p'}'\, \hat{\boldsymbol{u}}_{p'}\, \boldsymbol{k}_{p'}|F_p|0;\ 1m_r\rangle^*
\end{aligned} \tag{9.2}$$

The incoherent sums ($\mathfrak{p}_{2^+}$, $M_{N_{2^+}}$, $M_{S_{2^+}}$) over the degenerate states of $|2^+\rangle$ imply that, in the present experiment on 1-DPI, the dication $\mathfrak{T}^{2+}$ remains unobserved. Further in (9.2)

$$|2^+;\ \mu_p\, \hat{\boldsymbol{u}}_p\, \boldsymbol{k}_p;\ \mu_{p'}\, \hat{\boldsymbol{u}}_{p'}\, \boldsymbol{k}_{p'}\rangle = |2^+\rangle\, |\mu_p \hat{\boldsymbol{u}}_p \boldsymbol{k}_p\rangle\, |\mu_{p'} \hat{\boldsymbol{u}}_{p'} \boldsymbol{k}_{p'}\rangle \tag{9.3}$$

represents a state of the $(\mathfrak{T}^{2+} + e_p + e_{p'})$-system. Hund's case (b) parity-adapted state $|2^+\rangle$ of $\mathfrak{T}^{2+}$ is given in (3.57) [68, 69, 75]; whereas (3.49) is the spin–orbital of photoelectron e_p in the case (b). An expression, for the spin–orbital for the companion photoelectron $e_{p'}$ is readily obtained on replacing subscript p everywhere in (3.49) by p'. After substituting expressions for $|2^+\rangle$, $|\mu_p \hat{\boldsymbol{u}}_p \boldsymbol{k}_p\rangle$, and $|\mu_{p'} \hat{\boldsymbol{u}}_{p'} \boldsymbol{k}_{p'}\rangle$ in (9.3), their respective spin parts $|S_{2+}\, M_{S_{2+}}\rangle$, $|\frac{1}{2}\, \nu_p\rangle$, and $|\frac{1}{2}\, \nu_{p'}\rangle$ are coupled according to the spin conservation condition (5.1c), which needs to be satisfied in any 1-DPI experiment performed without taking any SDIs into account. After this and some other simplifying operations, state (9.3) becomes

$$
\begin{aligned}
|2^+;\, \mu_p \hat{\boldsymbol{u}}_p \boldsymbol{k}_p;\, \mu_{p'} \hat{\boldsymbol{u}}_{p'} \boldsymbol{k}_{p'}\rangle &= \frac{1}{\sqrt{2}} (-1)^{-S_{2+}} \\
&\times \sum_{\substack{\ell_p m_p \lambda_p \ell_{p'} m_{p'} \lambda_{p'} \\ \nu_p \nu_{p'} s_{p_t} \nu_{p_t} S M_S}} (-1)^{s_{p_t} - \nu_{p_t} + M_S}\, i^{\ell_p + \ell_{p'}}\, e^{-i(\sigma_{\ell_p} + \sigma_{\ell_{p'}})} \\
&\times \sqrt{(2s_{p_t}+1)(2S+1)} \begin{pmatrix} \frac{1}{2} & \frac{1}{2} & s_{p_t} \\ \nu_p & \nu_{p'} & -\nu_{p_t} \end{pmatrix} \begin{pmatrix} S_{2+} & s_{p_t} & S \\ M_{S_{2+}} & \nu_{p_t} & -M_S \end{pmatrix} \\
&\times \left[Y_{\ell_p}^{m_p}(\hat{\boldsymbol{k}}_p)\right]^* \left[Y_{\ell_{p'}}^{m_{p'}}(\hat{\boldsymbol{k}}_{p'})\right]^* \left[\mathcal{D}_{\lambda_p m_p}^{\ell_p}(\omega)\right]^* \\
&\times \left[\mathcal{D}_{\lambda_{p'} m_{p'}}^{\ell_{p'}}(\omega)\right]^* \left[\mathcal{D}_{\mu_p \nu_p}^{\frac{1}{2}}(\omega_p)\right]^* \left[\mathcal{D}_{\mu_{p'} \nu_{p'}}^{\frac{1}{2}}(\omega_{p'})\right]^* \\
&\times |S M_S\rangle \Big[|n_{2+} \Lambda_{2+};\, \ell_p \lambda_p;\, \ell_{p'} \lambda_{p'}\rangle\, |N_{2+} \Lambda_{2+} M_{N_{2+}}\rangle \\
&\times (-1)^{\mathrm{p}_{2+}} |n_{2+} - \Lambda_{2+};\, \ell_p \lambda_p;\, \ell_{p'} \lambda_{p'}\rangle\, |N_{2+} - \Lambda_{2+} M_{N_{2+}}\rangle\Big].
\end{aligned}
\tag{9.4a}
$$

Here, we have defined

$$
|n_{2+} \Lambda_{2+};\, \ell_p \lambda_p;\, \ell_{p'} \lambda_{p'}\rangle \equiv |n_{2+} \Lambda_{2+}\rangle\, |F^-_{\ell_p \lambda_p}(k_p)\rangle\, |F^-_{\ell_{p'} \lambda_{p'}}(k_{p'})\rangle \tag{9.4b}
$$

with $|n_{2+} \Lambda_{2+}\rangle$, $|F^-_{\ell_p \lambda_p}(k_p)\rangle$, and $|F^-_{\ell_{p'} \lambda_{p'}}(k_{p'})\rangle$ occurring earlier in (3.57a), (3.49), and in the spin orbital $|\mu_{p'} \hat{\boldsymbol{u}}_{p'} \boldsymbol{k}_{p'}\rangle$ of the photoelectron $e_{p'}$.

Using (3.51), after the substitution of (3.48a) and (3.50), one readily evaluates [213, 217] the amplitude $\langle 2^+;\, \mu_p \hat{\boldsymbol{u}}_p \boldsymbol{k}_p;\, \mu_{p'} \hat{\boldsymbol{u}}_{p'} \boldsymbol{k}_{p'} | F_p | 0;\, 1 m_r\rangle$ for 1-DPI needed in the calculation of the density matrix (9.2). On substituting the consequent expression for $\langle 2^+;\, \mu_p \hat{\boldsymbol{u}}_p \boldsymbol{k}_p;\, \mu_{p'} \hat{\boldsymbol{u}}_{p'} \boldsymbol{k}_{p'} | F_p | 0;\, 1 m_r\rangle$ and for its Hermitian conjugate, one can simplify (9.2) to the following final form [213]

$$\langle n_{2^+}\; N_{2^+}\; \Lambda_{2^+};\; \mu_p\, \hat{\boldsymbol{u}}_p\, \boldsymbol{k}_p;\; \mu_{p'}\, \hat{\boldsymbol{u}}_{p'}\, \boldsymbol{k}_{p'} | \rho^{(m,\,\text{1-DPI})} | n_{2^+}\; N_{2^+}\; \Lambda_{2^+};\; \mu'_p\, \hat{\boldsymbol{u}}_p\, \boldsymbol{k}_p;$$
$$\times\, \mu'_{p'}\, \hat{\boldsymbol{u}}_{p'}\, \boldsymbol{k}_{p'}\rangle \;=\; \frac{\mathrm{d}^3\sigma^{(m,\,\text{1-DPI})}(m_r)}{\mathrm{d}\epsilon_1\, \mathrm{d}\hat{\boldsymbol{k}}_p\, \mathrm{d}\hat{\boldsymbol{k}}_{p'}}\, \sigma^{(m,\,\text{1-DPI})}(S_0;\; S_{2^+};\; \hat{\boldsymbol{u}}_p;\; \hat{\boldsymbol{u}}_{p'})_{\mu_p,\mu_{p'};\mu'_p,\mu'_{p'}}. \tag{9.5}$$

The first term, present in (9.5), is [213]

$$\frac{\mathrm{d}^3\sigma^{(m,\,\text{1-DPI})}(m_r)}{\mathrm{d}\epsilon_1\, \mathrm{d}\hat{\boldsymbol{k}}_p\, \mathrm{d}\hat{\boldsymbol{k}}_{p'}} = (-1)^{m_r+N_0+N_{2^+}}\, \frac{\mathcal{K}_p\,(2N_{2^+}+1)}{4\pi\,(2-\delta_{0\,\Lambda_0})}$$
$$\times \sum_{\substack{\mathfrak{p}_0\, \ell_p\, \ell'_p\, L_p\, \ell_{p_t}\, \ell'_{p_t}\, N\, N' \\ \mathfrak{p}_{2^+}\, \ell_{p'}\, \ell'_{p'}\, L_{p'}\, L_r\, M}} (-1)^{\ell'_p+\ell'_{p'}+\ell_{p_t}+L_r}(2L_r+1)$$
$$\times \sqrt{(2L_p+1)(2L_{p'}+1)}\, \left[Y^{M}_{L_p}(\hat{\boldsymbol{k}}_p)\right]^* \left[Y^{-M}_{L_{p'}}(\hat{\boldsymbol{k}}_{p'})\right]^*$$
$$\times \begin{pmatrix} \ell_p & \ell'_p & L_p \\ 0 & 0 & 0 \end{pmatrix} \begin{pmatrix} \ell_{p'} & \ell'_{p'} & L_{p'} \\ 0 & 0 & 0 \end{pmatrix} \begin{pmatrix} 1 & 1 & L_r \\ m_r & -m_r & 0 \end{pmatrix} \begin{pmatrix} L_p & L_{p'} & L_r \\ M & -M & 0 \end{pmatrix} \begin{Bmatrix} 1 & 1 & L_r \\ N' & N & N_0 \end{Bmatrix}$$
$$\times \begin{Bmatrix} \ell_{p_t} & \ell'_{p_t} & L_r \\ N' & N & N_{2^+} \end{Bmatrix} \begin{Bmatrix} \ell_p & \ell'_p & L_p \\ \ell_{p'} & \ell'_{p'} & L_{p'} \\ \ell_{p_t} & \ell'_{p_t} & L_r \end{Bmatrix} \tag{9.6a}$$
$$\times \langle n_{2^+} \Lambda_{2^+} N_{2^+} \mathfrak{p}_{2^+};\; ((\ell_p \ell_{p'})\ell_{p_t} N_{2^+})N |\, F\, | n_0 \Lambda_0 N_0 \mathfrak{p}_0;\; (N_0 1)N\rangle$$
$$\times \langle n_{2^+} \Lambda_{2^+} N_{2^+} \mathfrak{p}_{2^+};\; ((\ell'_p \ell'_{p'})\ell'_{p_t} N_{2^+})N' |\, F\, | n_0 \Lambda_0 N_0 \mathfrak{p}_0;\; (N_0 1)N'\rangle^*,$$

where

$$\langle n_{2^+} \Lambda_{2^+} N_{2^+} \mathfrak{p}_{2^+};\; ((\ell_p \ell_{p'})\ell_{p_t} N_{2^+})N |\, F\, | n_0 \Lambda_0 N_0 \mathfrak{p}_0;\; (N_0 1)N\rangle$$
$$= \frac{1}{2}(-1)^N \left[1 - (-1)^{\mathfrak{p}_0+\mathfrak{p}_{2^+}+\ell_p+\ell_{p'}+N_0+N_{2^+}}\right]$$
$$\times (2\ell_{p_t}+1)(2N+1)\sqrt{(2\ell_p+1)(2\ell_{p'}+1)}(-i)^{\ell_p+\ell_{p'}}\, e^{i(\sigma_{\ell_p}+\sigma_{\ell_{p'}})}$$
$$\times \sum_{\substack{\lambda_p\, \lambda_{p'}\, \lambda_{p_t} \\ \lambda_r\, \Lambda_N}} (-1)^{\lambda_{p_t}} \begin{pmatrix} \ell_p & \ell_{p'} & \ell_{p_t} \\ \lambda_p & \lambda_{p'} & \lambda_{p_t} \end{pmatrix} \begin{pmatrix} \ell_{p_t} & N_{2^+} & N \\ -\lambda_{p_t} & \Lambda_{2^+} & \Lambda_N \end{pmatrix}$$
$$\times \Bigg[\begin{pmatrix} N_0 & 1 & N \\ \Lambda_0 & \lambda_r & \Lambda_N \end{pmatrix} \langle n_{2^+} \Lambda_{2^+};\; \ell_p \lambda_p;\; \ell_{p'} \lambda_{p'} |\, F_p\, | n_0 \Lambda_0;\; 1 \lambda_r\rangle$$

$$\times (-1)^{p_0} \begin{pmatrix} N_0 & 1 & N \\ -\Lambda_0 & \lambda_r & \Lambda_N \end{pmatrix} \langle n_{2+} \Lambda_{2+};\ \ell_p \lambda_p;\ \ell_{p'} \lambda_{p'} |\ F_p\ | n_0 - \Lambda_0;\ 1\lambda_r \rangle. \Bigg] \tag{9.6b}$$

The second term on the right-hand side of the density matrix (9.5) is given by [213]

$$\sigma^{(m,\,\text{1-DPI})}(S_0;\ S_{2+};\ \hat{\boldsymbol{u}}_p;\ \hat{\boldsymbol{u}}_{p'})_{\mu_p,\mu_{p'};\mu'_p,\mu'_{p'}} = (-1)^{\mu'_p+\mu'_{p'}}$$

$$\times \sum_{\substack{s\ s_{p_t} \\ \mu_1\ \mu_2\ \mu}} (-1)^{s_{pt}-\mu}\,(2s+1) \begin{pmatrix} 1/2 & 1/2 & s \\ \mu_p & -\mu'_p & \mu_1 \end{pmatrix} \begin{pmatrix} 1/2 & 1/2 & s \\ \mu_{p'} & -\mu'_{p'} & \mu_2 \end{pmatrix} \begin{Bmatrix} 1/2 & 1/2 & s \\ 1/2 & 1/2 & s_{p_t} \end{Bmatrix}$$

$$\times \left[\mathcal{D}^s_{\mu_1\,\mu}(\omega_p)\right]^* \left[\mathcal{D}^s_{\mu_2,-\mu}(\omega_{p'})\right]^*. \tag{9.7}$$

9.1.1.2 Properties

The density matrix (9.5), needed to study non-local correlation between the spins of photoelectrons $(e_p,\ e_{p'})$ simultaneously emitted from a RLM in the 1-DPI process (1.2) in Hund's coupling scheme (b) in the absence of SDIs, is a product of two independent terms given in (9.6) and (9.7). The first [i.e., (9.6)] of these two terms depends upon both the kinematics—on account of the presence of the momenta $(\boldsymbol{k}_p,\ \boldsymbol{k}_{p'})$ of the outgoing electrons $(e_p,\ e_{p'})$—and on dynamics—due to the ionization amplitudes (9.6b)—of 1-DPI. All the quantum numbers present in (9.6) are those which are not related to the spins of either of the states $|0\rangle$ and $|2^+\rangle$ of the molecular species $\mathfrak{T}$ and $\mathfrak{T}^{2+}$, respectively, participating in the 1-DPI process (1.2). Moreover, any variables pertaining to the spins of the outgoing electrons $(e_p,\ e_{p'})$ are also not present in (9.6). Hence, the first term $d^3\sigma^{(m,\,\text{1-DPI})}(m_r)/d\epsilon_1\, d\hat{\boldsymbol{k}}_p\, d\hat{\boldsymbol{k}}_{p'}$ in the density matrix (9.5) describes purely angular correlation between two photoelectrons simultaneously emitted from a RLM in the 1-DPI process (1.2). It, thus, always has a single positive value for $(e_p,\ e_{p'})$ with given $(\boldsymbol{k}_p,\ \boldsymbol{k}_{p'})$, but for all directions $(\hat{\boldsymbol{u}}_p,\ \hat{\boldsymbol{u}}_{p'})$ of quantization of the spins of two photoelectrons in the SF *OXYZ* shown in Fig. 9.1.

The second term (9.7) [i.e., $\sigma^{(m,\,\text{1-DPI})}(S_0;\ S_{2+};\ \hat{\boldsymbol{u}}_p;\ \hat{\boldsymbol{u}}_{p'})_{\mu_p,\mu_{p'};\mu'_p,\mu'_{p'}}$] in the density matrix (9.5) is not only totally independent of all the kinematics as well dynamics of the 1-DPI process (1.2) which control the angular correlation (9.6), but also completely determined by the projections $(\mu_p,\ \mu'_p)$ and $(\mu_{p'},\ \mu'_{p'})$ of the spins of e_p and $e_{p'}$ along the quantization directions $\hat{\boldsymbol{u}}_p$ and $\hat{\boldsymbol{u}}_{p'}$, respectively. Hence, $\sigma^{(m,\,\text{1-DPI})}(S_0;\ S_{2+};\ \hat{\boldsymbol{u}}_p;\ \hat{\boldsymbol{u}}_{p'})_{\mu_p,\mu_{p'};\mu'_p,\mu'_{p'}}$ in (9.5) represents correlation between only the spins of the members of an outgoing pair of $(e_p,\ e_{p'})$.

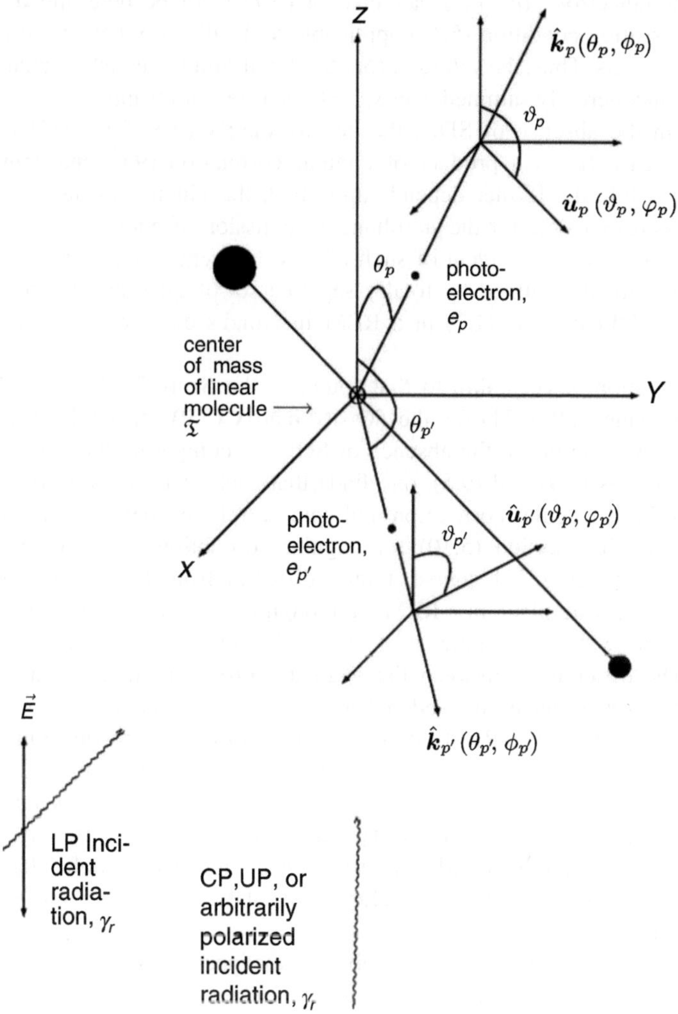

Fig. 9.1 Coordinate system showing the space-fixed frame of reference $OXYZ$ and the propagation directions ($\hat{\boldsymbol{k}}_p$, $\hat{\boldsymbol{k}}_{p'}$) as well as the spin quantization directions ($\hat{\boldsymbol{u}}_p$, $\hat{\boldsymbol{u}}_{p'}$) of the two photoelectrons (e_p, $e_{p'}$) simultaneously ejected from the rotating linear molecule $\mathfrak{T}$ in the 1-DPI process (1.2) following the absorption of a single photon γ_r in $E1$ approximation. Each electron enters its own Mott detector [285], oriented along $\hat{\boldsymbol{u}}_p(\vartheta_p, \varphi_p)$ for e_p and $\hat{\boldsymbol{u}}_{p'}(\vartheta_{p'}, \varphi_{p'})$ for $e_{p'}$. These detectors record whether a photoelectron's spin is up or down with respect to its own quantization direction. (First published in [213].)

The reasons for the explicit presence of (S_0, S_{2+}) of ($\mathfrak{T}$, $\mathfrak{T}^{2+}$) among the arguments on the left-hand side of the spin correlation matrix $\sigma^{(m,\,\text{1-DPI})}$, although neither of these two spin quantum numbers is seen on the right-hand side of (9.7), is that how many and which of the two allowed values (0 and 1) the total spin

s_{p_t} of photoelectrons $(e_p, e_{p'})$ can take in (9.7) is to be determined from the spin conservation condition (5.1c) applicable to 1-DPI in either of the atomic or molecular targets. Thus, the values of (S_0, S_{2+}) will simultaneously decide that (9.7) should be incoherently summed over $s_{p_t} = 0$, or 1, or both 0 and 1.

Thus, in the absence of SDIs, the density matrix (9.5) for 1-DPI in a RLM in Hund's case (b) is a product of angular correlation (9.6) and spin correlation (9.7); while the former depends upon both the kinematics and dynamics of the process responsible for the simultaneous emission of photoelectrons $(e_p, e_{p'})$, the later is totally independent of such effects. Consequently, these two different correlations are determined by totally separate set of quantum numbers characterizing a 1-DPI process (1.2) in a RLM in Hund's case (b) in the absence of SDIs.

This situation is very akin to that found in the Sects. 5.1.1 (pages 115–119) and 5.1.2 (pages 119–121) for the density matrix (5.9) for 1-DPI in an atomic target in L-S coupling in the absence of SOI. On comparing the two Coulombic density matrices (5.9) and (9.5), one finds that each of these is a product of the parts describing angular correlation and spin correlation between photoelectrons $(e_p, e_{p'})$. But, expression (5.10) for angular correlation in the case of 1-DPI in an atomic target is, of course, totally different from (9.6) applicable to the same process taking place in a RLM—although both of these expressions involve kinematics as well as dynamics of 1-DPI taking place in two different kinds of targets. The differences between the angular correlations (5.10) and (9.6) arise because structure of an atom—and, hence, the motion of its constituents—is totally different from that of a RLM. It is for this reason that the quantum numbers needed for representing an electronic state in an atom are completely different from those used in the case of a RLM. Hence, the differences between the respective expressions (5.10) and (9.6), representing angular correlation between $(e_p, e_{p'})$ coming simultaneously out from an atomic and a molecular target in the 1-DPI process (1.2) on the exclusion of SDIs, are neither unforeseen nor unexpected.

However, what is startling and, probably, unexpected about the density matrices (5.9) and (9.5), needed for studying spin-entanglement between $(e_p, e_{p'})$ in 1-DPI in two totally different types of targets, is the fact that the expressions (5.11) and (9.7) describing spin correlation in the two respective cases are completely identical to each other. This, in other words, means that a change from the central to the cylindrical nuclear force experienced by electrons moving in a RLM, in addition to the rotation of its nuclei, has no effect whatsoever on the spin correlation between two photoelectrons simultaneously departing in a 1-DPI process in the absence of SDIs. This is a very interesting result showing that as far as the non-separability of the spin-state of $(e_p, e_{p'})$ in the absence of SDIs is concerned, both atomic and molecular targets generate pairs of photoelectrons, which have identical entanglement properties in a 1-DPI experiment for given values of (S_0, S_{2+}) of $(\mathfrak{T}, \mathfrak{T}^{2+})$.

9.1.2 Discussions

From the discussion given on pages 206–208 herein, one immediately concludes that the analysis of the Sect. 5.2 (pages 121–133) for a study of the spin-entanglement properties of photoelectrons $(e_p, e_{p'})$ produced in 1-DPI of an atomic target in the absence of SOI is exactly applicable in the present case of a RLM as well. Following (5.13), we, therefore, write

$$\begin{aligned}\langle n_{2+}\, N_{2+}\, \Lambda_{2+};\ \mu_p\, \hat{\boldsymbol{u}}_p\, \boldsymbol{k}_p;\ \mu_{p'}\, \hat{\boldsymbol{u}}_{p'}\, \boldsymbol{k}_{p'}|\rho^{(m,\,\text{1-DPI})}|n_{2+}\, N_{2+}\, \Lambda_{2+};\ \mu'_p\, \hat{\boldsymbol{u}}_p\, \boldsymbol{k}_p;\\ \mu'_{p'}\, \hat{\boldsymbol{u}}_{p'}\, \boldsymbol{k}_{p'}\rangle \equiv \sigma^{(m,\,\text{1-DPI})}(S_0;\ S_{2+};\ \hat{\boldsymbol{u}}_p;\ \hat{\boldsymbol{u}}_{p'})_{\mu_p,\mu_{p'};\mu'_p,\mu'_{p'}}\\ = \sigma^{(\text{1-DPI})}(S_0;\ S_{2+};\ \hat{\boldsymbol{u}}_p;\ \hat{\boldsymbol{u}}_{p'})_{\mu_p,\mu_{p'};\mu'_p,\mu'_{p'}}\end{aligned} \quad (9.8)$$

in view of expressions [(9.5), (9.7)], and [(5.9), (5.11)]. The remaining part of the present section contains a brief discussion of the non-local correlation between the spins of photoelectrons $(e_p, e_{p'})$ with appropriate examples of experiments on 1-DPI of a RLM which can readily be performed in a laboratory in the absence of SDIs in Hund's case (b).

(i) Photoelectrons $(e_p, e_{p'})$ in the Bell State (2.17b)

According to Sect. 5.2.1 (pages 122–129), for $(e_p, e_{p'})$ to be in a singlet spin state, each of the molecule $\mathfrak{T}$ and its residual photo-dication $\mathfrak{T}^{2+}$ should be in its singlet electronic state (i.e., $S_0 = S_{2+} = 0$). Then $\sigma^{(m,\,\text{1-DPI})}(S_0;\ S_{2+};\ \hat{\boldsymbol{u}}_p;\ \hat{\boldsymbol{u}}_{p'}) \longrightarrow \sigma_1(\hat{\boldsymbol{u}}_p;\ \hat{\boldsymbol{u}}_{p'})$. The spin state $\sigma_1(\hat{\boldsymbol{u}}_p;\ \hat{\boldsymbol{u}}_{p'})$ of photoelectrons $(e_p, e_{p'})$ is given by (5.14) which is equivalent to the pure and maximally entangled Bell state (2.17b) [or (A.30b)].

The simplest possible example here is 1-DPI of H_2 in its ground electronic state $^1\Sigma_g^+$. The simultaneously coming out photoelectrons in the process (1.2) are not only correlated, but also form an EPR pair. Some of the other examples for this case of 1-DPI are [213]:

$$\text{Li}_2\, (1\sigma_g^2 1\sigma_u^2 2\sigma_g^2\, {}^1\Sigma_g^+) \xrightarrow{\gamma_r} \text{Li}_2^{2+}\, (1\sigma_g^2 1\sigma_u^2 2\sigma_g^0\, {}^1\Sigma_g^+),$$

$$\text{CO}\, (1\sigma^2\, 2\sigma^2\, 3\sigma^2\, 4\sigma^2\, 1\pi^4\, 5\sigma^2\, {}^1\Sigma^+) \xrightarrow{\gamma_r} \text{CO}^{2+}\, (1\sigma^2\, 2\sigma^2\, 3\sigma^2\, 4\sigma^2\, 1\pi^4\, 5\sigma^0\, {}^1\Sigma^+), etc$$

in the ground states of Li_2, CO, *etc* molecules. On account of the fact that each of these two diatomics and their residual photo-dications Li_2^{2+} and CO^{2+}, respectively, are in their singlet electronic states, the SDIs make no contributions at all in either of these cases.

(*ii*) Photoelectrons (e_p, $e_{p'}$) in a mixed, separable state

One learns from Sect. 5.2.2 (pages 129–131) that whenever spins (S_0, S_{2+}) of molecular species ($\mathfrak{T}$, $\mathfrak{T}^{2+}$), participating in the process (1.2) differ by $\pm$ 1, only s_{p_t} = 1 need to be considered in (9.7) [i.e., (9.8)]. The spin state of two photoelectrons (e_p, $e_{p'}$) is now given y [213]

$$\sigma^{(m,\,\text{1-DPI})}(S_0;\, S_{2+} = \left|S_0 \pm 1\right|;\, \hat{\boldsymbol{u}}_p;\, \hat{\boldsymbol{u}}_{p'}) = -\frac{1}{3}\sigma_1(\hat{\boldsymbol{u}}_p;\, \hat{\boldsymbol{u}}_{p'}) + \frac{4}{3}\sigma_0, \tag{9.9}$$

which is nothing but (5.29). Applying the discussion given on pages 129–131 for the atomic targets to the present case of RLMs, one can immediately say that (9.9) represents a mixed, separable Coulombic state of electronic qubits (e_p, $e_{p'}$) emitted in the 1-DPI process (1.2) in a molecule $\mathfrak{T}$ in the Hund's coupling scheme (b) with no SDIs taken in to account. Equation (9.9) does not, however, qualify to be a Werner state [121] because, according to the discussion given on pages 33–35, the mixing parameter p in (9.9) is less than zero. Thus, 1-DPI in the present case is unable to entangle (e_p, $e_{p'}$), although these two electrons were certainly correlated when in the bound state $|0\rangle$ of $\mathfrak{T}$ otherwise they would not have come out together during the molecular transition $\mathfrak{T} \longrightarrow \mathfrak{T}^{2+}$ caused by the one-body $E1$ operator .

Some of the lightest possible RLM which can generate spin-state (9.9) in a 1-DPI experiment in a Laboratory with negligible SDIs are B_2 ($1\sigma_g^2\, 1\sigma_u^2\, 2\sigma_g^2\, 2\sigma_u^2\, 1\pi_u^2\, {}^3\Sigma_g^-$) and CO ($1\sigma^2\, 2\sigma^2\, 3\sigma^2\, 4\sigma^2\, 1\pi^4\, 5\sigma^2\, {}^1\Sigma^+$). If, during 1-DPI of these molecules, the photo-dications formed are B_2^{2+} ($1\sigma_g^2\, 1\sigma_u^2\, 2\sigma_g^2\, 2\sigma_u^2\, 1\pi_u^0\, {}^1\Sigma_g^+$) and CO^{2+} ($1\sigma^2\, 2\sigma^2\, 3\sigma^2\, 4\sigma^2\, 1\pi^2\, 5\sigma^2\, {}^3\Sigma^-$), respectively, the photoelectrons (e_p, $e_{p'}$) in each of these cases are in the state (9.9). These two electrons were certainly correlated while bound to either of these two molecules, however, after leaving their parents they are not entangled. Thus, electron–electron correlation found in a many-body interacting system—which may either be an atom or a RLM—is not always capable of generating in the 1-DPI process (1.2) bipartite, non-separable sates of electronic qubits under the influence of purely Coulomb forces.

(*iii*) Photoelectrons (e_p, $e_{p'}$) in a mixed, separable, Werner state

The last possibility which may exist in a 1-DPI experiment corresponds to a situation when the electronic states [(3.48a), (3.57a)] of the molecular species [$\mathfrak{T}$, $\mathfrak{T}^{2+}$] have equal multiplicities, but neither of these is singlet. In view of the spin conservation condition (5.1c) applicable to Coulombic 1-DPI, now both values (i.e., 0 and 1) in the incoherent sum over s_{p_t} in (9.7) need to be included. According to Sect. 5.2.3 (pages 131–133), the Coulombic density matrix representing the spin-state of (e_p, $e_{p'}$) is now given by [213]

$$\sigma^{(m,\,\text{1-DPI})}\Big(S_0 \geq \frac{1}{2};\ S_{2+} = S_0;\ \hat{\boldsymbol{u}}_p;\ \hat{\boldsymbol{u}}_{p'}\Big) = \frac{1}{3}\sigma_1(\hat{\boldsymbol{u}}_p;\ \hat{\boldsymbol{u}}_{p'}) + \frac{2}{3}\sigma_0, \tag{9.10}$$

which is identical to (5.32a). Its partial transpose with respect to the photoelectron e_p is given by (5.32b); whereas (5.33) contain their respective eigenvalues.

It has been discussed in detail on pages 131–133 that (9.10) represents a mixed, separable, Werner state of two qubits. Similar to the paragraphs (i) and (ii) on pages 209 & 210, (9.10) too is a general result applicable to photoelectrons $(e_p, e_{p'})$ emitted in the 1-DPI of a RLM $\mathfrak{T}$ in the absence of SDIs in Hund's coupling scheme (b). Two of the possible experiments on 1-DPI of RLMs, which can readily be performed in a laboratory for producing bipartite spin-states (9.10) are

$$\text{CN}\,(1\sigma^2\,2\sigma^2\,3\sigma^2\,4\sigma^2\,1\pi^4\,5\sigma^1\;{}^2\Sigma^+) \xrightarrow{\gamma_r} \text{CN}^{2+}\,(1\sigma^2\,2\sigma^2\,3\sigma^2\,4\sigma^1\,1\pi^4\,5\sigma^0\;{}^2\Sigma^+)$$

and

$$\text{NO}\,(1\sigma^2\,2\sigma^2\,3\sigma^2\,4\sigma^2\,5\sigma^2\,1\pi^4\,2\pi^1\;{}^2\Pi)$$
$$\xrightarrow{\gamma_r} \text{NO}\,(1\sigma^2\,2\sigma^2\,3\sigma^2\,4\sigma^2\,5\sigma^1\,1\pi^4\,2\pi^0\;{}^2\Sigma^+).$$

Results presented in (i)–(iii), starting from page 209 to the present point, are completely general and independent of all dynamical models except that these refer to a RLM without SDIs in Hund's case (b). In each of these three cases, photoelectrons $(e_p, e_{p'})$—before their emission—were an integral part of the same system (in the present case a RLM). While bound in the molecular target $\mathfrak{T}$, $(e_p, e_{p'})$ were necessarily correlated with each other as well as with other sister electrons in the system; otherwise, these two photoelectrons would not have come out together following the absorption of a single photon γ_r in the 1-DPI process (1.2). In addition, $(e_p, e_{p'})$ are prepared neither independently nor in total isolation of each other. But, spins of $(e_p, e_{p'})$ are entangled only in (i); in the remaining two cases (ii) and (iii), these photoelectrons always form separable spin-state.

Thus, the studies performed on 1-DPI of atoms in Chap. 5, and of RLMs in the present section, show that electron–electron correlation and spin-entanglement are two independent properties. Although two electrons simultaneously emitted in 1-DPI (1.2) are necessarily correlated, however, they may or may not have their spin angular momenta entangled. In other words, electron–electron correlation, present in a many-body interacting system which may either be an atom or a RLM, is not always sufficient to provide the spin-entanglement to the emitted electrons.

Secondly, as far as 1-DPI in the absence of SDIs is concerned, the spin angular momenta of two photoelectrons, simultaneously emitted from an atom or a RLM, have identical entanglement properties. Features (e.g., cylindrical nature of the Coulomb field generated by the nuclei, rotation of nuclei, *etc*), always present in a linear molecule but never possible in an atom, have no effects at all on the Coulombic entanglement of the spins of two photoelectrons $(e_p, e_{p'})$, which have simultaneously departed from a RLM following the absorption of a single photon γ_r.

9.2 Fine-Structure Entanglement between the Spins of Molecular Photoelectrons $(e_p, e_{p'})$

Let us now investigate the effects of the SDIs on the Coulombic spin-entanglement of photoelectrons $(e_p,\ e_{p'})$ generated in 1-DPI process (1.2) of a RLM. On discarding the effects of nuclear spin, SRI and SOI are the two SDIs affecting the motions of electrons inside a RLM. It is, however, not necessary that both SRI and SOI are either simultaneously present or equally important in any given linear molecule. One should, therefore, study, in principle, the roles played by these two SDIs separately as well as jointly in entangling the spins of $(e_p,\ e_{p'})$ simultaneously departing from a RLM $\mathfrak{T}$ after the absorption of a single photon. Chandra and Ghosh [213] have carried out such investigations by taking only the C+SRIs as well as C+SDIs into account. For the reasons that the theoretical treatments for studying the entanglement between the spin angular momenta of $(e_p,\ e_{p'})$ in 1-DPI of a RLM in the presence of either C+SRIs or of C+SDIs are almost similar, in this monograph we discuss only the later situation which is more important of the two investigated in [213].

9.2.1 Density Matrix

Conditions applicable to the conservation of energy and total angular momentum when both C+SDIs are taken into account in the 1-DPI of a RLM are also those given, respectively, by (5.1a) and (7.8). It is already explained in Sect. 3.3.2 (pages 84–90) that the inclusion of C+SDIs in a RLM can properly be accounted for in the Hund's coupling scheme (a) [68, 69, 185]. Equations (3.62) is the case (a) representation of the state $|0\rangle$ of the RLM $\mathfrak{T}$ participating in the 1-DPI process (1.2); whereas state $|2^+\rangle$ for the residual dication $\mathfrak{T}^{2+}$ is obtained using the prescription explained in footnote (17) on page 79. The density operator (3.5b), which is applicable to any 1-step process caused by the absorption of a single photon γ_r in the $E1$ approximation, in the present case becomes

$$\rho^{(m,\,\text{1-DPI})} = \frac{\mathcal{K}_p}{(2-\delta_{0\Lambda_0})(2J_0+1)} \sum_{\mathfrak{p}_0\, M_0} F_p \left(|0;\ 1m_r\rangle\langle 0;\ 1m_r|\right) F_p^{\dagger}. \qquad (9.11)$$

The quantum numbers present in (9.11) have their meanings already explained in (3.62). Density matrix for the transition $|0\rangle \longrightarrow |2^+\rangle$ from the unpolarized RLM $\mathfrak{T}$ to its photo-dication $\mathfrak{T}^{2+}$ can be written from (9.11) in the following form

$$\langle J_{2^+};\ \mu_p\,\hat{\boldsymbol{u}}_p\,\boldsymbol{k}_p;\ \mu_{p'}\,\hat{\boldsymbol{u}}_{p'}\,\boldsymbol{k}_{p'}|\rho^{(m,\,\text{1-DPI})}|J_{2^+};\ \mu'_p\,\hat{\boldsymbol{u}}_p\,\boldsymbol{k}_p;\ \mu'_{p'}\,\hat{\boldsymbol{u}}_{p'}\,\boldsymbol{k}_{p'}\rangle$$

$$= \frac{\mathcal{K}_p}{(2-\delta_{0\Lambda_0})(2J_0+1)}$$

$$\times \sum_{\mathfrak{p}_0\, \mathfrak{p}_{2^+}\, M_0\, M_{2^+}} \langle 2^+;\ \mu_p\, \hat{\boldsymbol{u}}_p\, \boldsymbol{k}_p;\ \mu_{p'}\, \hat{\boldsymbol{u}}_{p'}\, \boldsymbol{k}_{p'} | F_p | 0;\ 1 m_r \rangle$$
$$\times \langle 2^+;\ \mu'_p\, \hat{\boldsymbol{u}}_p\, \boldsymbol{k}_p;\ \mu'_{p'}\, \hat{\boldsymbol{u}}_{p'}\, \boldsymbol{k}_{p'} | F_p | 0;\ 1 m_r \rangle^* \tag{9.12}$$

Here, $|0;\ 1m_r\rangle = |0\rangle\, |1m_r\rangle$ for the non-interacting $(\gamma_r + \mathfrak{T})$-system is readily obtained by combining (3.50) and (3.62a). Product of the spin orbitals (3.64) and $|\mu_{p'}\hat{\boldsymbol{u}}_{p'}\boldsymbol{k}_{p'}\rangle$ [obtained on making the replacement $p \longrightarrow p'$ in (3.64)] for the respective photoelectrons e_p and $e_{p'}$ with the bound state $|2^+\rangle$ of the photo-dication $\mathfrak{T}^{2+}$—each in Hund's coupling scheme (a)—will give us, for a $(\mathfrak{T}^{2+} + e_p + e_{p'})$-system, the state $|2^+;\ \mu_p\, \hat{\boldsymbol{u}}_p\, \boldsymbol{k}_p;\ \mu_{p'}\, \hat{\boldsymbol{u}}_{p'}\, \boldsymbol{k}_{p'}\rangle$ to be used in evaluating the transition amplitudes needed in the density matrix (9.12).

The calculation of $\langle 2^+;\ \mu_p\, \hat{\boldsymbol{u}}_p\, \boldsymbol{k}_p;\ \mu_{p'}\, \hat{\boldsymbol{u}}_{p'}\, \boldsymbol{k}_{p'} | F_p | 0;\ 1m_r\rangle$ is a lengthy process requiring a heavy use of Racah algebra.[2] The final form of this long expression can be gotten by following the procedures similar to those given, for example, in [213, 215, 310]. For calculating the required density matrix, one now needs to substitute in (9.12), the simplified form [213, 215, 310] of the Dirac's bracket $\langle 2^+;\ \mu_p\, \hat{\boldsymbol{u}}_p\, \boldsymbol{k}_p;\ \mu_{p'}\, \hat{\boldsymbol{u}}_{p'}\, \boldsymbol{k}_{p'} | F_p | 0;\ 1m_r\rangle$ and its Hermitian conjugate. After a large number of simplifications in the angular momentum algebra,[3] one finally arrives at [213]

$$\langle J_{2^+};\ \mu_p\, \hat{\boldsymbol{u}}_p\, \boldsymbol{k}_p;\ \mu_{p'}\, \hat{\boldsymbol{u}}_{p'}\, \boldsymbol{k}_{p'} | \rho^{(m,\,\text{1-DPI})} | J_{2^+};\ \mu'_p\, \hat{\boldsymbol{u}}_p\, \boldsymbol{k}_p;\ \mu'_{p'}\, \hat{\boldsymbol{u}}_{p'}\, \boldsymbol{k}_{p'}\rangle = (-1)^{\mu'_p + \mu'_{p'}}$$
$$\times \sum_{\substack{S_p\, M_{S_p}\, N_{S_p} \\ S_{p'}\, M_{S_{p'}}\, N_{S_{p'}}}} (-1)^{M_{S_p} + M_{S_{p'}}} \begin{pmatrix} \frac{1}{2} & \frac{1}{2} & S_p \\ \mu_p & -\mu'_p & M_{S_p} \end{pmatrix} \begin{pmatrix} \frac{1}{2} & \frac{1}{2} & S_{p'} \\ \mu_{p'} & -\mu'_{p'} & M_{S_{p'}} \end{pmatrix}$$
$$\times\, \mathcal{D}^{S_p}_{-M_{S_p}\, N_{S_p}}(\omega_p)\, \mathcal{D}^{S_{p'}}_{-M_{S_{p'}}\, N_{S_{p'}}}(\omega_{p'})\, A^{S_p S_{p'}}_{N_{S_p} N_{S_{p'}}}(m_r;\ \boldsymbol{k}_p, \boldsymbol{k}_{p'}). \tag{9.13a}$$

Here, we have defined

$$A^{S_p S_{p'}}_{N_{S_p} N_{S_{p'}}}(m_r;\ \boldsymbol{k}_p, \boldsymbol{k}_{p'}) = (-1)^{1+m_r+S_p+S_{p'}} \frac{(2S_p+1)(2S_{p'}+1)\mathcal{K}_p}{4\pi(2-\delta_{0\Lambda_0})}$$
$$\times \sum_{\substack{\mathfrak{p}_0\, \ell_p\, \ell'_p\, L_p\, M_{L_p}\, j_p\, j'_p\, J_p\, M_J\, L_r\, j_t \\ \mathfrak{p}_{2^+}\, \ell_{p'}\, \ell'_{p'}\, L_{p'}\, M_{L_{p'}}\, j_{p'}\, j'_{p'}\, J_{p'}\, j_{p_t}\, j'_{p_t}}}$$

---

[2] For an interested reader, these and other relevant calculational details are available on pages 291–298 in [215].

[3] For details of this simplification procedure, see footnote (3) in the present Chap. 9.

$$\times (-1)^{\ell'_p+\ell'_{p'}+j_{pt}+j_t+L_r}(2J_p+1)(2J_{p'}+1)$$

$$\times (2L_r+1)(2j_t+1)\sqrt{(2L_p+1)(2L_{p'}+1)}\begin{pmatrix} \ell_p & \ell'_p & L_p \\ 0 & 0 & 0 \end{pmatrix}\begin{pmatrix} \ell_{p'} & \ell'_{p'} & L_{p'} \\ 0 & 0 & 0 \end{pmatrix}$$

$$\times \begin{pmatrix} L_p & S_p & J_p \\ M_{L_p} & N_{S_p} & M_J \end{pmatrix}\begin{pmatrix} L_{p'} & S_{p'} & J_{p'} \\ M_{L_{p'}} & N_{S_{p'}} & -M_J \end{pmatrix}\begin{pmatrix} 1 & 1 & L_r \\ m_r & -m_r & 0 \end{pmatrix}\begin{pmatrix} J_p & J_{p'} & L_r \\ M_J & -M_J & 0 \end{pmatrix}$$

$$\times \begin{Bmatrix} 1 & 1 & L_r \\ j_{pt} & j'_{pt} & j_t \end{Bmatrix}\begin{Bmatrix} \ell_p & 1/2 & j_p \\ \ell'_p & 1/2 & j'_p \\ L_p & S_p & J_p \end{Bmatrix}\begin{Bmatrix} \ell_{p'} & 1/2 & j_{p'} \\ \ell'_{p'} & 1/2 & j'_{p'} \\ L_{p'} & S_{p'} & J_{p'} \end{Bmatrix}\begin{Bmatrix} j_p & j_{p'} & j_{pt} \\ j'_p & j'_{p'} & j'_{pt} \\ J_p & J_{p'} & L_r \end{Bmatrix}$$

$$\times Y_{L_p}^{M_{L_p}}(\hat{\boldsymbol{k}}_p)\, Y_{L_{p'}}^{M_{L_{p'}}}(\hat{\boldsymbol{k}}_{p'})$$

$$\times \langle J_{2+}\mathfrak{p}_{2+};\left(\left(\ell_p\tfrac{1}{2}\right)j_p\left(\ell_{p'}\tfrac{1}{2}\right)j_{p'}\right)j_{pt}|F(j_t)|J_0\mathfrak{p}_0;\,1\rangle$$

$$\times \langle J_{2+}\mathfrak{p}_{2+};\left(\left(\ell'_p\tfrac{1}{2}\right)j'_p\left(\ell'_{p'}\tfrac{1}{2}\right)j'_{p'}\right)j'_{pt}|F(j_t)|J_0\mathfrak{p}_0;\,1\rangle^*. \tag{9.13b}$$

The dynamical amplitudes, present in the last expression (9.13b), are [213]

$$\langle J_{2+}\mathfrak{p}_{2+};\left(\left(\ell_p\tfrac{1}{2}\right)j_p\left(\ell_{p'}\tfrac{1}{2}\right)j_{p'}\right)j_{p_t}|F(j_t)|J_0\mathfrak{p}_0;\,1\rangle$$

$$= \frac{1}{2}\left[1+(-1)^{\mathfrak{p}_0+\mathfrak{p}_{2+}+\ell_p+\ell_{p'}-J_0+J_{2+}}\right](-i)^{\ell_p+\ell_{p'}}$$

$$e^{i(\sigma_{\ell_p j_p}+\sigma_{\ell_{p'}j_{p'}})}(2j_p+1)(2j_{p'}+1)(2j_{p_t}+1)$$

$$\sqrt{(2\ell_p+1)(2\ell_{p'}+1)}(2J_{2+}+1)$$

$$\Sigma_{J_T}(2J_T+1)\begin{Bmatrix} 1 & J_0 & J_T \\ J_{2+} & j_{p_t} & j_t \end{Bmatrix} \tag{9.14a}$$

$$\times \sum_{\substack{\lambda_p\sigma_p m_{j_p} m_{j_{pt}} \\ \lambda_{p'}\sigma_{p'} m_{j_{p'}}}} (-1)^{m_{j_{pt}}}\begin{pmatrix} \ell_p & 1/2 & j_p \\ -\lambda_p & -\sigma_p & m_{j_p} \end{pmatrix}\begin{pmatrix} \ell_{p'} & 1/2 & j_{p'} \\ -\lambda_{p'} & -\sigma_{p'} & m_{j_{p'}} \end{pmatrix}$$

$$\times \begin{pmatrix} j_p & j_{p'} & j_{p_t} \\ -m_{j_p} & -m_{j_{p'}} & m_{j_{pt}} \end{pmatrix}\begin{pmatrix} j_{p_t} & J_{2+} & J_T \\ m_{j_{pt}} & \Omega_{2+} & M_T \end{pmatrix}\sum_{\lambda_r}(-1)^{\lambda_r}\left[\begin{pmatrix} J_0 & 1 & J_T \\ \Omega_0 & \lambda_r & M_T \end{pmatrix}\right.$$

$$\times \langle n_{2+}\Lambda_{2+}\Sigma_{2+}\Omega_{2+}\mathfrak{p}_{2+};\ \ell_p\lambda_p\sigma_p;\ \ell_{p'}\lambda_{p'}\sigma_{p'}|\ F_p\ |n_0\Lambda_0\Sigma_0\Omega_0\mathfrak{p}_0;\ 1\lambda_r\rangle$$
$$+ (-1)^{\mathfrak{p}_0+2\Omega_0}\begin{pmatrix} J_0 & 1 & J_T \\ -\Omega_0 & \lambda_r & M_T \end{pmatrix}$$
$$\times \langle n_{2+}\Lambda_{2+}\Sigma_{2+}\Omega_{2+}\mathfrak{p}_{2+};\ell_p\lambda_p\sigma_p;\ell_{p'}\lambda_{p'}\sigma_{p'}|\ F_p\ |n_0 - \Lambda_0 - \Sigma_0 - \Omega_0\mathfrak{p}_0;1\lambda_r\rangle \tag{9.14b}$$

with the definitions[4]

$$\Big(|n_0\ \Lambda_0\rangle|S_0\ \Sigma_0\rangle|J_0\ \Omega_0\ M_0\rangle\Big)|1\lambda_r\rangle \equiv |n_0\ \Lambda_0\ \Sigma_0\ \Omega_0\ \mathfrak{p}_0;\ 1\lambda_r\rangle\ |J_0\ \Omega_0\ M_0\rangle \tag{9.14c}$$

and[5]

$$\Big(|n_{2+}\ \Lambda_{2+}\rangle|S_{2+}\ \Sigma_{2+}\rangle|J_{2+}\ \Omega_{2+}\ M_{2+}\rangle\Big)\ |F^-_{\ell_p\lambda_p\sigma_p}(k_p)\rangle\ |F^-_{\ell_{p'}\lambda_{p'}\sigma_{p'}}(k_{p'})\rangle$$
$$\equiv |n_{2+}\ \Lambda_{2+}\ \Sigma_{2+}\ \Omega_{2+}\ \mathfrak{p}_{2+};\ \ell_p\lambda_p\sigma_p;\ \ell_{p'}\lambda_{p'}\sigma_{p'}\rangle\ |J_{2+}\ \Omega_{2+}\ M_{2+}\rangle \tag{9.14d}$$

for the $(\gamma_r + \mathfrak{T})$- and $(\mathfrak{T}^{2+} + e_p + e_{p'})$-system, respectively.

9.2.2 Discussions

The density matrix (9.13) is completely general and without any approximations, whatsoever, for studying entanglement between the spin angular momenta of photoelectrons $(e_p,\ e_{p'})$ simultaneously emitted in the 1-DPI process (1.2) taking place in a RLM $\mathfrak{T}$ of point symmetry group $C_{\infty v}$ or $D_{\infty h}$ in the presence of C+SDIs in Hund's coupling scheme (a). Chandra and Ghosh [213] have shown that if, on the other hand, one decides to take in this 1-DPI process only the C+SRIs, one then needs to work with Hund's scheme (b). Even in such studies, the form of the density matrix (9.13) remains unchanged, except that some of the phase factors and, of course, the dynamical amplitudes $\langle n_{2+}\Lambda_{2+}\Sigma_{2+}\Omega_{2+}\mathfrak{p}_{2+};\ \ell_p\lambda_p\sigma_p;\ \ell_{p'}\lambda_{p'}\sigma_{p'}|\ F_p|n_0\Lambda_0\Sigma_0\Omega_0\mathfrak{p}_0;\ 1\lambda_r\rangle$ [presently referring to the inclusion of C+SDIs in case (a) in (9.13b) and defined by (9.14)], need to be replaced by those[6] which will correspond to the presence of C+SRIs in the Hund's coupling (b).[7] Furthermore, the diagonal elements (*ie*, $\mu_p = \mu'_p$

[4]For equation (9.14c), see (3.50) and (3.62a).

[5]For (9.14d), see (3.64), footnote (23) on page 89, and the spin–orbital for the photoelectron $e_{p'}$ obtained on replacing p by p' in (3.64).

[6]That is, (B5a) and (B5b) in [213].

[7]For a detailed derivation of the density matrix for the 1-DPI process (1.2) in a RLM in the presence of C+SRIs in Hund's case (b) and a discussion of its similarities and/or differences with that [i.e.,

and $\mu_{p'} = \mu'_{p'}$) of the DM (9.13), describing spin-resolved angular distribution of photoelectrons $(e_p, e_{p'})$ in 1-DPI of a RLM in the Hund's coupling scheme (a), are identical to those given in Eq. (37) in Reference [310]; whereas, the non-diagonal terms (*ie*, $\mu_p = \mu'_p$ and/or $\mu_{p'} \neq \mu'_{p'}$ represent coherence effect in the DM (9.14).

Another interesting property of (9.13), in addition to its being non-separable—unlike the Coulombic expression (9.5)—in angular and spin parts, is that it involves the total angular momenta (J_0, J_{2+}) of the states $(|0\rangle, |2^+\rangle)$ of the molecular species $(\mathfrak{T}, \mathfrak{T}^{2+})$ participating in 1-DPI (1.2). The quantum numbers (J_0, J_{2+}) determine [68, 69, 185], even in a RLM, the fine-structure splitting of the energy levels in $(\mathfrak{T}, \mathfrak{T}^{2+})$. Hence, (9.13) give us expression for a density matrix needed for studying fine-structure entanglement between the spins of photoelectrons $(e_p, e_{p'})$ simultaneously emitted in the 1-DPI process (1.2) taking place in a RLM in the presence of C+SDIs, or C+SRIs [see footnote (8) on page 215], in the appropriate Hund's coupling scheme.

Thus, similar to the fine-structure entanglement in 1-DPI of atomic targets discussed in Sect. 7.2 on pages 160–169, the quantal non-separability (i.e., entanglement) of the spin-state (9.13) of photoelectrons $(e_p, e_{p'})$ generated in 1-DPI of a RLM in the presence of C+SDIs (or C+SRIs, for that matter) is very much governed by both the kinematics and dynamics of this process. Experimental geometry used to simultaneously observe two photoelectrons in a laboratory, including the frequency ν_r and polarization (specified by the parameter m_r) of the incident photon γ_r—ionizing $\mathfrak{T}$ in (1.2)—do play important roles as well.

Let us now compare the density matrix (7.13), one will use for studying fine-structure entanglementt in atomic 1-DPI, with (9.13) needed for investigating spin-entanglement between $(e_p, e_{p'})$ from a RLM in the presence of C+SDIs [or, C+SRIs]. The two density matrices (7.13) and (9.13) are remarkably identical in their structures with, of course, different dynamical amplitudes (7.13c) or (9.14) [or, the ones referred to in the footnote (7) on page 215] present in the two respective expressions. It, in other words, means that the discussion given in Sect. 7.2 on pages 164–169 for the properties of fine-structure entanglement between the spins of photoelectrons $(e_p, e_{p'})$ simultaneously emitted in 1-DPI of a generic atom becomes exactly applicable also when the target $\mathfrak{T}$ in (1.2) is a RLM after replacing in (7.13b) the dynamical amplitude (7.13c) for atomic 1-DPI with that given in (9.14) for the same process with C+SDIs in Hund's case (a) $\{$or that referred to in the footnote (7) on page 215 for 1-DPI in the presence of C+SRIs in the Hund's coupling scheme (b) [68, 69, 185]$\}$. Thus, the density matrices which determine the properties of the fine-structure entanglement of a bipartite spin-state of photoelectrons $(e_p, e_{p'})$ produced in 1-DPI of an atom or a RLM are almost of identical forms with, of course, different dynamical amplitudes corresponding to the respective targets.

However, identical nature of the density matrices (7.13) and (9.13) does not in any way imply that the quantum informatic properties of a bipartite state of electronic qubits $(e_p, e_{p'})$ generated in 1-DPI of an atom will necessarily be similar

---

the density matrix (9.13)] obtained herein, see, for example, Sect. IV and the Appendices B & D in [213], or Appendix E to Chap. 5 in [215].

to those of a state of two photoelectrons simultaneously ejected from a RLM following the absorption of a single photon in an experimental set-up identical to that used in the former case. The properties of the fine-structure entanglement in the two cases will, in general, be naturally dissimilar for the simple reason that the dynamical amplitudes to be used for 1-DPI of an atom will be completely different from those appropriate for a RLM, in spite of the fact that the other physical quantities (e.g., ν_r, m_r, E_0, E_{2^+}, $\boldsymbol{k}_p$, $\boldsymbol{k}_{p'}$, $\hat{\boldsymbol{u}}_p$, $\hat{\boldsymbol{u}}_{p'}$, J_0, J_{2^+}) may have identical values.

Chapter 10
Two-Step Double Photoionization of Molecules

Chapter 6 and Sect. 7.3 (on pages 169–178) contain discussions of the Coulombic and fine-structure entanglement, respectively, of a bipartite state of electronic qubits emitted sequentially from an atom in the two-step process (1.3). Such processes were therein called as 2-DPI. Sequential emission of two electrons can equally well take place also from a molecule—both linear and non-linear. Due to the presence of highly repulsive nuclear forces, molecular ions, in general, those formed after the loss of two or more bound electrons, in particular, are very unstable and, hence, prone to fragmentation. It is, therefore, always probable, unlike in an atom, that a molecule with an inner-shell vacancy may fragment (i.e., loses one or more of its neutral and/or ionic atoms) before the emission of a secondary (i.e., Auger) electron in the former's non-radiative spontaneous decay takes place. In this monograph—including the present Chap. 10—we will consider, unless stated otherwise, only those 2-DPI processes (1.3) taking place in a RLM $\mathfrak{T}$ wherein the life-time for the spontaneous non-radiative decay [in the second-step (1.3b)] of the state $|1^{+*}\rangle$ of the excited molecular photoion $\mathfrak{T}^{1+*}$ [formed in the first-step (1.3a)] is much less than that for its fragmentation.

Studies on Auger spectroscopy of molecules began, naturally, with experimental measurements and theoretical calculations of energies, intensities, and rates for a given non-radiative transition (see, for example, [183] and references therein). Probably, the very first study of angular distribution of Auger electrons from a randomly oriented linear molecule was performed by Dill et al. [323]. This theoretical investigation was inspired by two observations on atomic Auger spectroscopy: First, angular distributions provide information both about the magnitudes and phases of transition amplitudes. Second, Flugge et al. [300] had found a non-isotropic angular distribution of Auger electrons emitted in the decay of a vacancy caused by photoionization in an other than inner K-shell of an atom. It was shown in [323] that due to the highly anisotropic nature of photon–molecule interactions, vacancies generated in a molecule due to photoabsorption behave

N. Chandra and R. Ghosh, *Quantum Entanglement in Electron Optics*, Springer Series on Atomic, Optical, and Plasma Physics 67, DOI 10.1007/978-3-642-24070-6_10,

differently from their atomic counter parts. This general theoretical prediction has since then been experimentally verified by Hemmers et al. [324]. Subsequently, Zähringer et al. [325] had calculated Auger electron angular distributions for LiF and HF molecules.

However, in order to perform a complete quantum mechanical study of Auger spectra, it is necessary to observe also the spin polarization of electrons emitted in non-radiative decay of an inner-shell vacancy in these targets. Such angle- and spin-resolved studies in Auger spectroscopy were first performed by Chandra and Chakraborty [210, 211]. The formulations developed in [210, 211] were fully based on powerful methods of group theory and, hence, were equally applicable to any linear or non-linear molecule belonging to one of the 32 point groups [70]. Later on, Bonhoff et al. [326] and Lehmann and Blum [327] developed theoretical frameworks for angular distribution of spin-unresolved Auger electrons from linear or non-linear molecules, respectively, wherein excited molecular ion $\mathfrak{T}^{1+^*}$ is created by electron impact ionization of the target $\mathfrak{T}$. The discussions in [326, 327] were based upon *order parameters*, in place of the group theoretical treatment given in [210, 211], and the coherence properties of the primary ionization, producing inner-shell vacancy due to electron impact, were taken into account.

Meanwhile, development of high-resolution electron spectrometers had enabled (see, for example, [322] and references therein) to resolve photoelectrons arising from production of diatomic ions in different rotational states of H_2^{1+}, N_2^{1+}, HO^{1+}, O_2^{1+}, *etc.*, which have relatively large rotational constants. These developments naturally lead one to think that such high-resolution spectrometers can, in principle, be equally used in not too distant a future to study angle- and spin-resolved Auger electron spectroscopy of such light linear molecules with their rotational motion properly taken into account. Motivated with these thoughts, Chandra and Sen [214, 218, 328, 329] performed theoretical studies of angle-resolved Auger electron spectroscopy of RLM with [214, 218, 328] and without [329] observing spins of outgoing particles.

Although a 2-DPI process (1.3) in a RLM, similar to that in an atom, provides a natural source for generating bipartite states of electronic qubits, no investigations until [216] on the entanglement properties of a spin state of a molecular (e_p, e_a)-pair were, however, reported in the literature. Chandra and Ghosh [216] and Ghosh et al. [217] have performed, probably, the very first studies of the presence, or otherwise, of the non-local correlation between the spins of electrons (e_p, e_a) emitted sequentially from a RLM $\mathfrak{T}$ without or with SDIs. The Coulombic and fine-structure entanglement between the spins of a photoelectron e_p and an Auger electron e_a, departed sequentially from a RLM $\mathfrak{T}$ and its excited photoion $\mathfrak{T}^{1+^*}$, are discussed in the following two Sects. 10.1 and 10.2, respectively.

10.1 Coulombic Entanglement Between the Spins of Molecular (Photo, Auger)-Electrons (e_p, e_a)

If a 2-DPI process (1.3) takes place in a RLM $\mathfrak{T}$ while its electrons and nuclei interact with each other only through the Coulomb forces, the conservation conditions applicable in Hund's coupling scheme (b) are [(3.46a), (3.46b)] for the photoionization (1.3a) of $\mathfrak{T}$ and [(3.47a), (3.47b)] for the spontaneous decay (1.3b) of the excited molecular photoion $\mathfrak{T}^{1+^*}$; whereas (6.1) express energy conservation for the two respective steps of the process (1.3). In such kinds of physical situations wherein Coulomb interactions are only taken into account and no SDIs contribute, a RLM belonging to either of the $C_{\infty v}$ and $D_{\infty h}$ point groups, is adequately described [68, 69, 185] by Hund's coupling scheme (b) of angular momenta.

10.1.1 Density Matrix

This section briefly describes derivation and properties of the Coulombic density matrix needed for studying entanglement between the spins of the photoelectron e_p and Auger electron e_a, sequentially emitted in the 2-DPI process (1.3) in a RLM $\mathfrak{T}$ in the absence of SDIs in Hund's coupling scheme (b).

10.1.1.1 Derivation

Equation (3.15) is the general expression for the density matrix for the 2-step process (1.3) taking place in an atom or a molecule. In the present case of target $\mathfrak{T}$ being a RLM in Hund's case (b), (3.15) takes the following form

$$\begin{aligned}
&\langle n_{2^+}\, N_{2^+}\, \Lambda_{2^+};\, \mu_p\, \hat{\boldsymbol{u}}_p\, \boldsymbol{k}_p;\, \mu_a\, \hat{\boldsymbol{u}}_a\, \boldsymbol{k}_a|\rho^{[(b),\,\text{2-DPI}]}|n_{2^+}\, N_{2^+}\, \Lambda_{2^+};\, \mu'_p\, \hat{\boldsymbol{u}}_p\, \boldsymbol{k}_p;\, \mu'_a\, \hat{\boldsymbol{u}}_a\, \boldsymbol{k}_a\rangle \\
&= \frac{\mathcal{K}_{ps}}{(2-\delta_{0\Lambda_0})(2N_0+1)(2S_0+1)} \sum_{\substack{\mathfrak{p}_0\, M_{N_0}\, M_{S_0}\, \mathfrak{p}_{2^+}\, M_{N_{2^+}}\, M_{S_{2^+}} \\ \mathfrak{p}_{1^{+*}}\, M_{N_{1^{+*}}}\, M_{S_{1^{+*}}}\, \mathfrak{p}'_{1^{+*}}\, M'_{N_{1^{+*}}}\, M'_{S_{1^{+*}}}}} \\
&\quad\times \langle 2^+;\, \mu_a\, \hat{\boldsymbol{u}}_a\, \boldsymbol{k}_a|\, F_s\, |1^{+^*}\rangle\langle 1^{+^*};\, \mu_p\, \hat{\boldsymbol{u}}_p\, \boldsymbol{k}_p|\, F_p\, |0;\, 1m_r\rangle \\
&\quad\times \langle 0;\, 1m_r|\, F_p^\dagger\, |1'^{+^*};\, \mu'_p\, \hat{\boldsymbol{u}}_p\, \boldsymbol{k}_p\rangle\langle 1'^{+^*}|\, F_s^\dagger\, |2^+;\, \mu'_a\, \hat{\boldsymbol{u}}_a\, \boldsymbol{k}_a\rangle.
\end{aligned} \tag{10.1}$$

Here, $\mathcal{K}_{ps}$ is defined immediately after (6.2); all other symbols have their meanings well explained in Sect. 3.3.1 on pages 75–84. Next, one substitutes (3.56) and (3.61) in (10.1). The consequent expression for the density matrix can be simplified[1] to the following final form [216]

[1] Details of this simplification procedure can be found in [215]

$$\langle n_{2^+}\, N_{2^+}\, \Lambda_{2^+};\, \mu_p\, \hat{\boldsymbol{u}}_p\, \boldsymbol{k}_p;\, \mu_a\, \hat{\boldsymbol{u}}_a\, \boldsymbol{k}_a | \rho^{[(b),\,\text{2-DPI}]} | n_{2^+}\, N_{2^+}\, \Lambda_{2^+};\, \mu_p'\, \hat{\boldsymbol{u}}_p\, \boldsymbol{k}_p;\, \mu_a'\, \hat{\boldsymbol{u}}_a\, \boldsymbol{k}_a \rangle$$
$$= \frac{d^3\sigma^{(m,\,\text{2-DPI})}(m_r)}{d\epsilon_r\, d\hat{\boldsymbol{k}}_p\, d\hat{\boldsymbol{k}}_a}\, \sigma^{(m,\,\text{2-DPI})}(S_0;\, S_{1^{+*}};\, S_{2^+};\, \hat{\boldsymbol{u}}_p;\, \hat{\boldsymbol{u}}_a)_{\mu_p,\mu_a;\mu_p',\mu_a'} \tag{10.2}$$

Here, we have defined [216]

$$\frac{d^3\sigma^{(m,\,\text{2-DPI})}(m_r)}{d\epsilon_r\, d\hat{\boldsymbol{k}}_p\, d\hat{\boldsymbol{k}}_a} = (-1)^{N_0+N_{1^{+*}}-N_{2^+}+m_r}\, \frac{(2N_{1^{+*}}+1)^2(2N_{2^+}+1)}{4\pi(2-\delta_{0\Lambda_0})}\, \mathcal{K}_{ps}$$
$$\times \sum_{\substack{\mathfrak{p}_0\, \mathfrak{p}_{2^+}\, \ell_p\, \ell_p'\, L_p\, L_a N\, N' \\ \mathfrak{p}_{1^{+*}}\, \mathfrak{p}_{1'^{+*}}\, \ell_a\, \ell_a'\, L_r\, M}} (-1)^{\ell_p'+L_p+L_r+N}(2L_r+1)\sqrt{(2L_p+1)(2L_a+1)}$$
$$\times \begin{pmatrix} \ell_p & \ell_p' & L_p \\ 0 & 0 & 0 \end{pmatrix} \begin{pmatrix} \ell_a & \ell_a' & L_a \\ 0 & 0 & 0 \end{pmatrix} \begin{pmatrix} 1 & 1 & L_r \\ m_r & -m_r & 0 \end{pmatrix} \begin{pmatrix} L_p & L_a & L_r \\ M & -M & 0 \end{pmatrix} \begin{Bmatrix} 1 & 1 & L_r \\ N & N' & N_0 \end{Bmatrix}$$
$$\times \begin{Bmatrix} \ell_a & \ell_a' & L_a \\ N_{1^{+*}} & N_{1^{+*}} & N_{2^+} \end{Bmatrix} \begin{Bmatrix} \ell_p & \ell_p' & L_p \\ N & N' & L_r \\ N_{1^{+*}} & N_{1^{+*}} & L_a \end{Bmatrix} \left[Y_{L_p}^{M}(\hat{\boldsymbol{k}}_p)\right]^* \left[Y_{L_a}^{-M}(\hat{\boldsymbol{k}}_a)\right]^*$$
$$\times \langle n_{1^{+*}}\, \Lambda_{1^{+*}}\, \mathfrak{p}_{1^{+*}};\, (N_{1^{+*}}\ell_p)N |\, \mathcal{F}_p^{(b)}\, | n_0 \Lambda_0 \mathfrak{p}_0;\, (N_0 1)N \rangle$$
$$\times \langle n_{1^{+*}}\, \Lambda_{1^{+*}}\, \mathfrak{p}_{1^{+*}}';\, (N_{1^{+*}}\ell_p')N' |\, \mathcal{F}_p^{(b)}\, | n_0 \Lambda_0 \mathfrak{p}_0;\, (N_0 1)N' \rangle^*$$
$$\times \langle n_{2^+}\, \Lambda_{2^+}\, \mathfrak{p}_{2^+};\, (N_{2^+}\ell_a)N_{1^{+*}} |\, \mathcal{F}_a^{(b)}\, | n_{1^{+*}}\, \Lambda_{1^{+*}}\, \mathfrak{p}_{1^{+*}};\, N_{1^{+*}} \rangle$$
$$\times \langle n_{2^+}\, \Lambda_{2^+}\, \mathfrak{p}_{2^+};\, (N_{2^+}\ell_a')N_{1^{+*}} |\, \mathcal{F}_a^{(b)}\, | n_{1^{+*}}\, \Lambda_{1^{+*}}\, \mathfrak{p}_{1'^{+*}};\, N_{1^{+*}} \rangle^*, \tag{10.3}$$

where photoionization and Auger amplitudes
$\langle n_{1^{+*}}\, \Lambda_{1^{+*}}\, \mathfrak{p}_{1^{+*}};\, (N_{1^{+*}}\ell_p)N |\, \mathcal{F}_p^{(b)}\, | n_0 \Lambda_0 \mathfrak{p}_0;\, (N_0 1)N \rangle$ and
$\langle n_{2^+}\, \Lambda_{2^+}\, \mathfrak{p}_{2^+};\, (N_{2^+}\ell_a)N_{1^{+*}} | \mathcal{F}_a^{(b)} | n_{1^{+*}}\, \Lambda_{1^{+*}}\, \mathfrak{p}_{1^{+*}};\, N_{1^{+*}} \rangle$
are given by (3.54) and (3.60), respectively.

The second term on the right-hand side of the density matrix (10.2) is [216]

$$\sigma^{(m,\,\text{2-DPI})}(S_0;\, S_{1^{+*}};\, S_{2^+};\, \hat{\boldsymbol{u}}_p;\, \hat{\boldsymbol{u}}_a)_{\mu_p,\mu_a;\mu_p',\mu_a'}$$
$$= (-1)^{S_0+2S_{1^{+*}}+S_{2^+}+\mu_p'+\mu_a'}(2S_{1^{+*}}+1)$$
$$\times \sum_{s\,\mu\,\mu_1\,\mu_2} (-1)^{s+\mu}\,(2s+1) \begin{pmatrix} 1/2 & 1/2 & s \\ \mu_p & -\mu_p' & \mu_1 \end{pmatrix} \begin{pmatrix} 1/2 & 1/2 & s \\ \mu_a & -\mu_a' & \mu_2 \end{pmatrix}$$
$$\times \begin{Bmatrix} 1/2 & 1/2 & s \\ S_{1^{+*}} & S_{1^{+*}} & S_0 \end{Bmatrix} \begin{Bmatrix} 1/2 & 1/2 & s \\ S_{1^{+*}} & S_{1^{+*}} & S_{2^+} \end{Bmatrix} \left[\mathcal{D}^s_{\mu_1\,\mu}(\omega_p)\right]^* \left[\mathcal{D}^s_{\mu_2,\,-\mu}(\omega_a)\right]^* \tag{10.4}$$

10.1.1.2 Properties

Similar to the atomic case discussed in Sect. 6.1 (pages 136–140), one finds that in the absence of SDIs, the density matrix (10.2), which represents a spin-state of a (e_p, e_a) pair generated in the 2-DPI process (1.3) taking place in a RLM, is again a product of two independent terms representing the angular correlation (10.3) and the spin correlation (10.4). The angular part (10.3) is different from that obtained in (6.4) for the same process of 2-DPI taking place in an atom with the exclusion of SOI. The differences between the two expressions (6.4) and (10.3) for angular correlation are, of course, natural as each of them contains both the kinematics and the dynamics of photoionization as well as of Auger decay taking place in the respective atomic and molecular targets, which are governed by the Coulomb forces among their respective constituents and possess entirely different geometrical structures of nuclei and electrons, including their degrees of freedom.

Expression (10.4), on the other hand, represents correlation between the spin angular momenta of a photoelectron e_p and an Auger electron e_a in the density matrix (10.2). It is a 4×4 matrix whose each of the sixteen elements is totally independent of the quantities (including the dynamics) present in the angular correlation (10.3). The former is completely determined by the spins (S_0, $S_{1^{+*}}$, S_{2^+}) of the electronic parts of the states ($|0\rangle$, $|1^{+*}\rangle$, $|2^+\rangle$) of the molecular species ($\mathfrak{T}$, $\mathfrak{T}^{1+^*}$, $\mathfrak{T}^{2+}$) participating in the 2-DPI process (1.3), in addition to the projections (μ_p, μ'_p) and (μ_a, μ'_a) of the spins of photoelectron e_p and Auger electron e_a along their respective directions of quantization $\hat{\boldsymbol{u}}_p$ and $\hat{\boldsymbol{u}}_a$.

Moreover, the present expression (10.4) for spin-correlation is identical to its atomic counter part (6.5). Although it is, to some extent, surprising; it, however, seems to be physically very logical. In the absence of SDIs, each of the two expressions (6.5) and (10.4) for spin-correlation is always completely independent of the kinematics as well as dynamics of the process of 2-DPI responsible for ejecting two electrons from an atom and a RLM, respectively. Moreover, conditions (3.20) for spin conservation are also identical and equally applicable to a 2-DPI process (1.3) taking place in either of these two completely different kinds of targets. The identical nature of the atomic and molecular expressions (6.5) and (10.4) for Coulombic correlation between the spins of the members of a (e_p, e_a) pair is, therefore, a natural consequence of the simultaneous presence of these two circumstances in a 2-DPI process (1.3) in these two entirely different kinds of targets in the absence of SDIs.

10.1.2 Discussions

The properties of the Coulombic entanglement between the spins of photoelectron e_p and Auger electron e_a, sequentially emitted in 2-DPI of a RLM, will be determined by the spin-correlation matrix (10.4). In view of the fact that the present spin-correlation matrix (10.4) is exactly identical to its atomic counter part (6.5), analysis given in Sect. 6.2 on pages 140–148 for 2-DPI in an atomic target becomes exactly applicable even in the present case of a RLM as well. The following contains some examples of experiments on 2-DPI in a RLM, which can be performed in a laboratory for the possible cases of Coulombic entanglement between the spins of the electronic qubits (e_p, e_a), earlier discussed in Sects. 6.2.1 (pages 141–148) and 6.2.2 (page 148) for atomic targets.

10.1.2.1 $S_0 = S_{2^+} = \left|S_{1^{+*}} \pm \frac{1}{2}\right|$

With these restrictions, arising from the conservation condition (3.20), on the spins (S_0, $S_{1^{+*}}$, S_{2^+}) of the states ($|0\rangle$, $|1^{+*}\rangle$, $|2^+\rangle$) of the molecular species ($\mathfrak{T}$, $\mathfrak{T}^{1+^*}$, $\mathfrak{T}^{2+}$) participating in a 2-DPI process (1.3), one needs to consider the four following possibilities for the Coulombic entanglement between the spins of photoelectron e_p and Auger electron e_a coming out of a RLM.

$S_0 = S_{2^+} = S_{1^{+*}} + \frac{1}{2}$ with $S_{1^{+*}} \geq 0$.

It corresponds to the situation discussed on pages 141–146 in Sect. 6.2.1. In these cases, values of the mixing parameter p, needed in the Werner state (6.7), are always determined from (6.9a) and are shown by Curve A in Fig. 6.2 on page 142. These are all mixed and separable Coulombic states of the spins of (e_p, e_a) formed in the 2-DPI process (1.3) taking place in a RLM in the absence of SDIs. Herein, properties of the bipartite states of electronic qubits are always independent of the kinematics, dynamics, as well as geometry of a 2-DPI experiment. Let us consider two interesting situations separately:

$S_0 = S_{2^+} = \frac{1}{2}$, $S_{1^{+*}} = 0$.

This corresponds to the case wherein mixing parameter p ($S_0 = 1/2$, $S_{1^{+*}} = 0$, $S_{2^+} = 1/2$) $= 0$ for the Werner state (6.7) is obtained from (6.9a). Now, spins of (e_p, e_a) form a maximally chaotic [22, 122], separable state σ_0 defined in (5.22). It is a mixed state with no entanglement at all.

One of the simplest possible examples of this result can be the 2-DPI $\mathrm{BeH}(1\sigma^2\, 2\sigma^2\, 3\sigma^1\ {}^2\Sigma^+) \longrightarrow \mathrm{BeH}^{1+^*}(1\sigma^1\, 2\sigma^2\, 3\sigma^1\ {}^1\Sigma^+) \longrightarrow \mathrm{BeH}^{2+}(1\sigma^2\, 2\sigma^1\, 3\sigma^0\ {}^2\Sigma^+)$ in the ground electronic configuration of a BeH molecule. Another example for this interesting case is $\mathrm{MgH}(1\sigma^2\, 2\sigma^2\, 3\sigma^2 1\pi^4\, 4\sigma^2\, 5\sigma^1\, {}^2\Sigma^+) \longrightarrow \mathrm{MgH}^{1+^*}(1\sigma^1\, 2\sigma^2\, 3\sigma^2 1\pi^4\, 4\sigma^2\, 5\sigma^1\, {}^1\Sigma^+) \longrightarrow \mathrm{MgH}^{2+}(1\sigma^2\, 2\sigma^2\, 3\sigma^2 1\pi^4\, 4\sigma^1\, 5\sigma^0\ {}^2\Sigma^+)$. MgH is obviously not as light as BeH. However, neither BeH^{1+^*} nor

MgH^{1+^*} will have any SRI in their singlet spin states; the SRI will be absent [68, 69, 185] in the ground rotational states of each of BeH, MgH, BeH^{2+}, and MgH^{2+} as well. Also, there is no SOI in the Σ state of a RLM.

$S_0 = S_{2+} = S_{1+^*} + \frac{1}{2}$ with $S_{1+^*} > 0$.

With these values of the spins (S_0, S_{1+^*} S_{2+}) of the electronic states ($|0\rangle$, $|1^{+^*}\rangle$, $|2^+\rangle$) of the molecular species ($\mathfrak{T}$, $\mathfrak{T}^{1+^*}$, $\mathfrak{T}^{2+}$) participating in the 2-DPI process (1.3) on the exclusion of SDIs, the mixing parameter $1/3 > \ p \ > 0$ is given by (6.9a). According to the eigenvalues of the density matrix (10.4) and of its partial transpose—obtained, respectively, from (5.25) and (5.26) [or, from (6.11a) and (6.13a)]—such values of p will always generate mixed and separable Werner states (6.7) of (e_p, e_a). However, none of these will ever be a maximally chaotic state σ_0, discussed in the preceding paragraph.

$S_0 = S_{2+} = S_{1+^*} - \frac{1}{2}$ with $S_{1+^*} \geq \frac{1}{2}$.

Equation (6.9b) now gives the values of the mixing parameter p needed for representing a spin state of (e_p, e_a) in the form of the Werner state (6.7). These values are $1/3 \leq \ p \ \leq 1$, always, and shown by the Curve B in Fig. 6.2 on page 142. With such values of p, spins of (e_p, e_a) are always entangled by the Coulomb forces inside a RLM with its concurrence and entanglement of formation also given in Fig. 6.2. Here, two following interesting cases arise:

$S_0 = S_{2+} = 0$, $S_{1+^*} = \frac{1}{2}$.

Equation (6.9b) now immediately gives $p \ = \ 1$ and, hence, according to (6.7), (e_p, e_a) are in a pure as well as maximally entangled state σ_1 which, in the present case, is the singlet state (5.14) of two spin-$\frac{1}{2}$ particles. This property is completely independent of the kinematics as well as dynamics of the 2-DPI process (1.3) taking place in a RLM in the absence of SDIs.

Among the several possible simple examples for this particular case are $Li_2(1\sigma_g^2\,1\sigma_u^2\ \ 2\sigma_g^2\,{}^1\Sigma_g^+) \longrightarrow Li_2^{1+^*}(1\sigma_g^1\,1\sigma_u^2\ \ 2\sigma_g^2\,{}^2\Sigma_g^+) \longrightarrow Li_2^{2+}(1\sigma_g^2\,1\sigma_u^2\ 2\sigma_g^0\,{}^1\Sigma_g^+)$; $NaH(1\sigma^2\,2\sigma^2\,3\sigma^2\,1\pi^4\,4\sigma^2\ \ {}^1\Sigma^+) \longrightarrow NaH^{1+^*}(1\sigma^1\,2\sigma^2\,3\sigma^2\,1\pi^4\,4\sigma^2\ {}^2\Sigma^+) \longrightarrow NaH^{2+}(1\sigma^2\,2\sigma^2\,3\sigma^2 1\pi^4\,4\sigma^0\,{}^1\Sigma^+)$; $N_2(1\sigma_g^2\,1\sigma_u^2 2\sigma_g^2\,2\sigma_u^2 1\pi_u^4\,3\sigma_g^2\,{}^1\Sigma_g^+) \longrightarrow N_2^{1+^*}(1\sigma_g^1 1\sigma_u^2\,2\sigma_g^2\,2\sigma_u^2\,1\pi_u^4\,3\sigma_g^2\ {}^2\Sigma_g^+) \longrightarrow N_2^{2+}(1\sigma_g^2\,1\sigma_u^2\,2\sigma_g^2\,2\sigma_u^2\,1\pi_u^4\,3\sigma_g^0\ {}^1\Sigma_g^+)$; *etc*. Although NaH and N_2 molecules are certainly heavier than Li_2, SDIs are, however, not present in any of the examples given herein as each of the electronic states of these molecules (or of their derivatives) participating in the 2-DPI process (1.3) is Σ. Consequently, the photoelectron e_p and Auger electron

e_a emitted in these transitions do form maximally entangled, pure and singlet spin state.

$S_0 = S_{2+} = S_{1+^*} - \frac{1}{2}$ with $S_{1+^*} > \frac{1}{2}$.

For these values of the spins of $(\mathfrak{T}, \mathfrak{T}^{1+^*}, \mathfrak{T}^{2+})$ participating in 2-DPI (1.3), mixing parameter obtained from (6.9b) is always $1/3 < p < 1$. With p in this range, eigenvalues (5.25) and (5.26) correspond to Werner states (6.7), which are always mixed and have Coulombic entanglement. [One can calculate these eigenvalues also from the respective (6.11b) and (6.13b) as well.] Negativity/Concurrence of such states is calculated from (6.14).

One of the examples of such a 2-DPI process is $CN(1\sigma^2\, 2\sigma^2\, 3\sigma^2\; 4\sigma^2\, 1\pi^4\, 5\sigma^1\; {}^2\Sigma^+) \longrightarrow CN^{1+^*}(1\sigma^1\, 2\sigma^2\, 3\sigma^2\; 4\sigma^2\, 1\pi^4\, 5\sigma^1\; {}^3\Sigma^+) \longrightarrow CN^{2+}(1\sigma^2\, 2\sigma^2\, 3\sigma^2\; 4\sigma^1\; 1\pi^4\, 5\sigma^0\; {}^2\Sigma^+)$ with $p(S_0 = 1/2, S_{1+^*} = 1, S_{2+} = 1/2) = 2/3$ obtained from (6.9b). This gives, in view of the eigenvalues (5.25) and (5.26) [or, equivalently, (6.11b) and (6.13b)], a Werner state (6.7) of (e_p, e_a), which is mixed and has Coulombic entanglement with negativity/concurrence $= \frac{1}{2}$ obtained from (6.14).

10.1.2.2 $|S_0\, (= |S_{1+^*} \pm \frac{1}{2}|) - S_{2+}\, (= |S_{1+^*} \mp \frac{1}{2}|)| = 1$

Here, mixing parameter $p = -1/3$ always, irrespective of the value of the spin S_{1+^*} of the excited photoion $\mathfrak{T}^{1+^*}$ participating in the 2-DPI process (1.3). The spin state of (e_p, e_a) is now represented by (6.15). It is a mixed and separable state of the electronic qubits. Moreover, as the value of p is now negative, the state of (e_p, e_a) does not qualify to be a Werner state. A relevant example for the present case can be $CO(1\sigma^2\, 2\sigma^2\, 3\sigma^2\; 4\sigma^2\, 1\pi^4\, 5\sigma^2\; {}^1\Sigma^+) \longrightarrow CO^{1+^*}(1\sigma^1\, 2\sigma^2\, 3\sigma^2\; 4\sigma^2\, 1\pi^4\, 5\sigma^2\; {}^2\Sigma^+) \longrightarrow CO^{2+}(1\sigma^2\, 2\sigma^2\, 3\sigma^2\; 4\sigma^1\, 1\pi^4\, 5\sigma^1\; {}^3\Sigma^+)$

10.2 Fine-Structure Entanglement Between the Spins of Molecular (Photo, Auger)-Electrons (e_p, e_a)

This section investigates effects on the Coulombic entanglement in a (e_p, e_a) pair, discussed in the previous Sect. 10.1, of all those interactions, which may arise due to spin of electrons present in a RLM $\mathfrak{T}$—excluding, of course, the hyperfine interaction [10, 58, 60]. Such SDIs include mainly SRI and SOI. Hund's coupling scheme (a) [68, 69, 185] provides a proper description of a RLM on the inclusion of both C+SDIs. The following, therefore, contains a discussion [217] of the properties of entanglement between the spins of a photoelectron e_p and an Auger electron e_a sequentially emitted in the 2-DPI process (1.3) taking place in a RLM $\mathfrak{T}$ in Hund's scheme (a) on the inclusion of C+SDIs. A detailed discussion of entanglement in a (e_p, e_a) pair in the presence of C+SRIs in Hund's case (b) is also available in [215].

10.2.1 Density Matrix

10.2.1.1 Derivation

On adapting the density matrix (3.15) to the present problem, one writes

$$\begin{aligned}
&\langle n_{2^+}\, J_{2^+};\, \mu_p\, \hat{\boldsymbol{u}}_p\, \boldsymbol{k}_p;\, \mu_a\, \hat{\boldsymbol{u}}_a\, \boldsymbol{k}_a|\rho^{[(a),\,\text{2-DPI}]}|n_{2^+}\, J_{2^+};\, \mu_p'\, \hat{\boldsymbol{u}}_p\, \boldsymbol{k}_p;\, \mu_a'\, \hat{\boldsymbol{u}}_a\, \boldsymbol{k}_a\rangle \\
&= \frac{\mathcal{K}_{ps}}{(2-\delta_{0\Lambda_0})(2J_0+1)} \sum_{\substack{\mathfrak{p}_0\, M_0\, \mathfrak{p}_{2^+}\, M_{2^+} \\ \mathfrak{p}_{1^{+*}}\, M_{1^{+*}}\, \mathfrak{p}'_{1^{+*}}\, M'_{1^{+*}}}} \\
&\quad \times \langle 2^+;\, \mu_a\, \hat{\boldsymbol{u}}_a\, \boldsymbol{k}_a|\, F_s\, |1^{+^*}\rangle\langle 1^{+^*};\, \mu_p\, \hat{\boldsymbol{u}}_p\, \boldsymbol{k}_p|\, F_p\, |0;\, 1m_r\rangle \\
&\quad \times \langle 0;\, 1m_r|\, F_p^\dagger\, |1'^{+^*};\, \mu_p'\, \hat{\boldsymbol{u}}_p\, \boldsymbol{k}_p\rangle\langle 1'^{+^*}|\, F_s^\dagger\, |2^+;\, \mu_a'\, \hat{\boldsymbol{u}}_a\, \boldsymbol{k}_a\rangle.
\end{aligned} \tag{10.5}$$

All the symbols present in (10.5) pertain to the Hund's case (a) parity adapted states [68,69,75] $(|0\rangle, |1^{+^*}\rangle, |2^+\rangle)$ of the RLM species $(\mathfrak{T}, \mathfrak{T}^{1^{+^*}}, \mathfrak{T}^{2+})$ participating in the 2-DPI process (1.3) in the presence of C+SDIs and have their meanings already explained in Sect. 3.3.2 on pages 84–90. In order to proceed further, we now substitute in (10.5), expressions (3.66a) and (3.68a) for the products of each of the two photoionization and two Auger decay amplitudes, respectively. These substitutions yield

$$\begin{aligned}
&\langle n_{2^+}\, J_{2^+};\, \mu_p\, \hat{\boldsymbol{u}}_p\, \boldsymbol{k}_p;\, \mu_a\, \hat{\boldsymbol{u}}_a\, \boldsymbol{k}_a|\rho^{[(a),\,\text{2-DPI}]}|n_{2^+}\, J_{2^+};\, \mu_p'\, \hat{\boldsymbol{u}}_p\, \boldsymbol{k}_p;\, \mu_u'\, \hat{\boldsymbol{u}}_a\, \boldsymbol{k}_a\rangle \\
&= \frac{\mathcal{K}_{ps}}{(2-\delta_{0\Lambda_0})} \\
&\quad \times \sum_{\substack{\mathfrak{p}_{1^{+*}}\, M_{1^{+*}} \\ \mathfrak{p}'_{1^{+*}}\, M'_{1^{+*}}}} \left[\frac{1}{2J_0+1} \sum_{\mathfrak{p}_0\, M_0} \mathcal{P}^{(m,a)}_{0\longrightarrow 1^{+*}} \left(\mathcal{P}'^{(m,a)}_{0\longrightarrow 1^{+*}}\right)^\dagger\right] \\
&\quad \times \left[\sum_{\mathfrak{p}_{2^+}\, M_{2^+}} \mathcal{A}^{(m,a)}_{1^{+*}\longrightarrow 2^+} \left(\mathcal{A}'^{(m,a)}_{1^{+*}\longrightarrow 2^+}\right)^\dagger\right].
\end{aligned} \tag{10.6}$$

The final form of the desired density matrix is obtained by substituting (3.66b) and (3.68b) in (10.6) and simplifying the consequent expression. One can then write [216, 217]

$$
\begin{aligned}
&\langle n_{2^+}\, J_{2^+};\ \mu_p\, \hat{\boldsymbol{u}}_p\, \boldsymbol{k}_p;\ \mu_a\, \hat{\boldsymbol{u}}_a\, \boldsymbol{k}_a | \rho^{[(a),\,\text{2-DPI}]} | n_{2^+}\, J_{2^+};\ \mu_p'\, \hat{\boldsymbol{u}}_p\, \boldsymbol{k}_p;\ \mu_a'\, \hat{\boldsymbol{u}}_a\, \boldsymbol{k}_a \rangle \\
&= (-1)^{\mu_p'+\mu_a'} \\
&\quad \times \sum_{\substack{S_p\, M_{S_p}\, N_{S_p} \\ S_a\, M_{S_a}\, N_{S_a}}} \begin{pmatrix} \frac{1}{2} & \frac{1}{2} & S_p \\ \mu_p & -\mu_p' & M_{S_p} \end{pmatrix} \begin{pmatrix} \frac{1}{2} & \frac{1}{2} & S_a \\ \mu_a & -\mu_a' & M_{S_a} \end{pmatrix} \\
&\quad \times \left[\mathcal{D}^{S_p}_{M_{S_p}\, N_{S_p}}(\omega_p) \right]^* \left[\mathcal{D}^{S_a}_{M_{S_a}\, N_{S_a}}(\omega_a) \right]^* \\
&\quad \times A^{S_p S_a}_{N_{S_p} N_{S_a}}(m_r;\, \boldsymbol{k}_p, \boldsymbol{k}_a)
\end{aligned} \tag{10.7a}
$$

with

$$
\begin{aligned}
A^{S_p S_a}_{N_{S_p} N_{S_a}}(m_r;\, \boldsymbol{k}_p, \boldsymbol{k}_a) \;&=\; (-1)^{1+m_r+J_0+2J_{1^{+*}}+J_{2^+}}\, \frac{\mathcal{K}_{ps}}{4\pi(2-\delta_{0\Lambda_0})} \\
&\times (2S_p+1)(2S_a+1)(2J_{1^{+*}}+1)(2J_{2^+}+1) \\
&\times \sum_{\substack{\mathfrak{p}_0\, \mathfrak{p}_{2^+}\, \ell_p\, \ell_p'\, L_p\, M_{L_p}\, j_p\, j_p'\, J_p\, j_t\, j_t'\, L_r \\ \mathfrak{p}_{1^{+*}}\, \mathfrak{p}_{1^{+*}}'\, \ell_a\, \ell_a'\, L_a\, M_{L_a}\, j_a\, j_a'\, J_a\, Q}} \\
&\times (-1)^{\ell_p'+\ell_a-j_a'+j_t}\, (2J_p+1)(2J_a+1)(2L_r+1)\sqrt{(2L_p+1)(2L_a+1)} \\
&\times \begin{pmatrix} \ell_p & \ell_p' & L_p \\ 0 & 0 & 0 \end{pmatrix} \begin{pmatrix} \ell_a & \ell_a' & L_a \\ 0 & 0 & 0 \end{pmatrix} \begin{pmatrix} L_p & S_p & J_p \\ M_{L_p} & N_{S_p} & Q \end{pmatrix} \\
&\times \begin{pmatrix} L_a & S_a & J_a \\ M_{L_a} & N_{S_a} & -Q \end{pmatrix} \begin{pmatrix} 1 & 1 & L_r \\ m_r & -m_r & 0 \end{pmatrix} \begin{pmatrix} J_p & J_a & L_r \\ Q & -Q & 0 \end{pmatrix} \\
&\times \begin{Bmatrix} J_{1^{+*}} & J_{1^{+*}} & J_a \\ j_t & j_t' & J_0 \end{Bmatrix} \begin{Bmatrix} J_{1^{+*}} & J_{1^{+*}} & J_a \\ j_a & j_a' & J_{2^+} \end{Bmatrix} \begin{Bmatrix} \ell_p & \ell_p' & L_p \\ 1/2 & 1/2 & S_p \\ j_p & j_p' & J_p \end{Bmatrix} \\
&\times \begin{Bmatrix} \ell_a & \ell_a' & L_a \\ 1/2 & 1/2 & S_a \\ j_a & j_a' & J_a \end{Bmatrix} \begin{Bmatrix} 1 & 1 & L_r \\ j_p & j_p' & J_p \\ j_t & j_t' & J_a \end{Bmatrix} \left[Y^{M_{L_p}}_{L_p}(\hat{\boldsymbol{k}}_p) \right]^* \left[Y^{M_{L_a}}_{L_a}(\hat{\boldsymbol{k}}_a) \right]^* \\
&\times \langle \mathfrak{p}_{1^{+*}}\, J_{1^{+*}};\ (\ell_p \tfrac{1}{2})\, j_p | F(j_t) | \mathfrak{p}_0 J_0;\ 1 \rangle \langle \mathfrak{p}_{1'^{+*}}\, J_{1^{+*}};\ (\ell_p' \tfrac{1}{2})\, j_p' | F(j_t) | \mathfrak{p}_0 J_0;\ 1 \rangle^* \\
&\times \langle \mathfrak{p}_{2^+}\, J_{2^+};\ (\ell_a \tfrac{1}{2})\, j_a | \mathcal{F}^{(a)}_a | \mathfrak{p}_{1^{+*}}\, J_{1^{+*}} \rangle \langle \mathfrak{p}_{2^+}\, J_{2^+};\ (\ell_a' \tfrac{1}{2})\, j_a' | \mathcal{F}^{(a)}_a | \mathfrak{p}_{1'^{+*}}\, J_{1^{+*}} \rangle^*.
\end{aligned} \tag{10.7b}
$$

Here, the photoionization amplitude $\langle \mathfrak{p}_{1^{+*}} J_{1^{+*}};\ (\ell_p \frac{1}{2}) j_p | F(j_t) | \mathfrak{p}_0 J_0;\ 1\rangle$ and the Auger amplitude $\langle \mathfrak{p}_{2^+} J_{2^+};\ (\ell_a \frac{1}{2}) j_a | \mathcal{F}_a^{(a)} | \mathfrak{p}_{1^{+*}} J_{1^{+*}}\rangle$ are those defined in (3.65b) and (3.67b), respectively.

10.2.1.2 Properties

Unlike the density matrix (10.2) applicable to study entanglement in 2-DPI of a RLM on the exclusion of SDIs, the present expression (10.7) is not a product of two terms, which represent, in the former, angular-correlation and spin-correlation separately between the photoelectron e_p and Auger electron e_a emitted sequentially in the 2-step process (1.3). Expression (10.7) for the density matrix, on the other hand, contains every thing which may affect a 2-DPI process (1.3) taking place in a RLM. Thus, the presence of SDIs has coupled the non-local correlation between the spins of a $(e_p,\ e_a)$ pair with the kinematics as well dynamics, in addition to other things [e.g. frequency and polarization of the photon γ_r absorbed for ionization in (1.3a)], of a DPI process taking place in a RLM. Moreover, expression (10.7) very much depends upon the total angular momenta $(J_0,\ J_{1^{+*}},\ J_{2^+})$ which determine [68, 69, 75, 185] fine-structure splitting of an energy-level in the states $(|0\rangle,\ |1^{+^*}\rangle,\ |2^+\rangle)$ of the rotating linear molecular species $(\mathfrak{T},\ \mathfrak{T}^{1+^*},\ \mathfrak{T}^{2+})$ participating in the 2-DPI process (1.3). Thus, the density matrix (10.7) describes fine-structure entanglement between the spins of $(e_p,\ e_a)$. Unlike the Coulombic entanglement, the non-local correlation between $(e_p,\ e_a)$ cannot be predetermined without a knowledge of the properties of the absorbed photon, geometry of the experiment prepared to observe angle- and spin-resolved $(e_p,\ e_a)$, amplitudes (3.65b) for photoionization (1.3a) and (3.67b) for Auger decay (1.3b).

10.2.2 A Generic Example

It is obvious from the discussions given in the last paragraph of the preceding Sect. 10.2.1.2 that, on account of its dependence on dynamics, among other things, the fine-structure entanglement in a $(e_p,\ e_a)$ pair will be different for different molecules participating in the 2-DPI process (1.3), although each of the other physical variables needed to determine the density matrix (10.7) may have identical values for two different molecular targets. Similarly, the non-local correlation between the spins of $(e_p,\ e_a)$ should change with a change only in one or more of $(J_0,\ J_{1^{+*}},\ J_{2^+})$, although the remaining physical variables, including the RLM target $\mathfrak{T}$, present in (10.7) may remain same.

In this section, we briefly describe results obtained by Ghosh et al. [215, 217] for the spin-entanglement in a $(e_p,\ e_a)$ pair emitted in a generic 2-DPI process involving the following states

$$\left.\begin{array}{l} |0\rangle \equiv |\mathfrak{p}_0,\, J_0 = 0,\, M_0 = 0\rangle, \\ |1^{+*}\rangle \equiv |\mathfrak{p}_{1^{+*}},\, J_{1^{+*}} = \frac{1}{2},\, M_{1^{+*}} = \pm\frac{1}{2}\rangle, \\ |2^{+}\rangle \equiv |\mathfrak{p}_{2^{+}},\, J_{2^{+}} = 0,\, M_{2^{+}} = 0\rangle \end{array}\right\} \tag{10.8}$$

of the molecular species ($\mathfrak{T}$, $\mathfrak{T}^{1+^*}$, $\mathfrak{T}^{2+}$) belonging to a $C_{\infty v}$ or $D_{\infty h}$ point group and participating in (1.3) with nuclear rotation as well as C+SDIs taken into account.

10.2.2.1 Selection Rules

In order to have the density matrix (10.7a), one needs to calculate from (10.7b) all the sixteen coefficients $A^{S_p S_a}_{N_{S_p} N_{S_a}}(m_r; \boldsymbol{k}_p, \boldsymbol{k}_a)$. The requirement [65] that the second 6-j symbol, present in (10.7b), does not vanish identically gives (with $J_{1^{+*}} = \frac{1}{2}$, $J_{2^+} = 0$) $j_a = \frac{1}{2}$ for the total angular momentum of the Auger electron e_a in the conservation condition (3.63b) applicable to the non-radiative spontaneous decay (1.3b) of the excited photoion $\mathfrak{T}^{1+^*}$. For the existence [65] of the second 9-j symbol, present in the same expression (10.7b), the orbital angular momentum of e_a (for $j_a = \frac{1}{2}$) must consequently be $\ell_a = 0$ and 1, i.e., the Auger electron e_a should be represented only by the s- and/or p-partial waves in the example of 2-DPI specified by the states (10.8).

Again, first of the two 6-j symbols in (10.7b) suggests [65] that the total angular momentum transferred from the absorbed photon γ_r to the observed photoelectron e_p is $j_t = \frac{1}{2}$ (for $J_0 = 0$, $J_{1^{+*}} = \frac{1}{2}$). This, in turn, gives—in view of the third 9-j symbol in (10.7b)—$j_p = \frac{1}{2}, \frac{3}{2}$ for the total angular momentum of e_p. With these restrictions on j_p, one finds from the first 9-j symbol in (10.7b) that $\ell_p = 0$, 1, and 2 are the only values allowed for the orbital angular momentum of the photoelectron e_p emitted in the ionization step (1.3b) in the present application (10.8) of the 2-DPI process.

Next we consider the selection rules which will allow, according to (10.8), photoionization of $\mathfrak{T}$ in the $E1$ approximation and non-radiative decay of $\mathfrak{T}^{1+^*}$ to successively take place. For the photoionization in the Hund's case (a), the $E1$ transition amplitude (3.65b) must exist. This requires, from (3.65c), that $\frac{1}{2}+\mathfrak{p}_0+\mathfrak{p}_{1^{+*}}+\ell_p+J_0 - J_{1^{+*}}$ must be even. Consequently, the selection rule, applicable to the photoionizing transition in (10.8), becomes

$$\mathfrak{p}_0 + \mathfrak{p}_{1^{+*}} + \ell_p = \text{even}. \tag{10.9a}$$

According to (3.67b), the Auger transition in Hund's scheme (a) is possible provided $\frac{1}{2}$ + $\mathfrak{p}_{1^{+*}}$ + $\mathfrak{p}_{2^+}$ + ℓ_a + $J_{1^{+*}}$ - J_{2^+} is odd so that the corresponding amplitude (3.67a) does not identically vanish. For specializing this condition to our present example (10.8), one requires that

$$\mathfrak{p}_{1^{+*}} + \mathfrak{p}_{2^+} + \ell_a = \text{even} \tag{10.9b}$$

for the excited photoion $\mathfrak{T}^{1+^*}$ to decay non-radiatively.

Various experimental techniques have recently been developed [322, 330–332] for preparing a RLM in a specific parity state according to the chosen values of $(\mathfrak{p}_0, \mathfrak{p}_{1^{+*}}, \mathfrak{p}_{2^+})$ as well as in a desired rotational state specified by the quantum numbers $(J_0, J_{1^{+*}}, J_{2^+})$ in the Hund's coupling scheme (a). At the same time, high-resolution spectrometers are now available [322, 330–332], which can distinguish electrons differing in energies corresponding to various rotational state of a linear molecule. In view of these developments, for selected $\mathfrak{p}_0 + \mathfrak{p}_{1^{+*}}$ and $\mathfrak{p}_{1^{+*}} + \mathfrak{p}_{2^+}$, values of ℓ_p and ℓ_a of definite parities will only contribute to the respective (10.9a) and (10.9b) in the Hund's coupling scheme (a).

Using the selection rules (10.9) as well the discussion presented in the preceding paragraph, one can readily conclude that in the current example (10.8) of 2-DPI, photoelectron e_p be represented by the s- and d-, or simply by a single p-, partial wave, depending upon whether the sum $\mathfrak{p}_0 + \mathfrak{p}_{1^{+*}}$, present in (10.9a), is even or odd, respectively. The Auger electron e_a, on the other hand, needs always be represented in the present example of (10.8) by a single partial wave only—namely, s- or p—determined by the parity of the sum $\mathfrak{p}_{1^{+*}} + \mathfrak{p}_{2^+}$ present in the selection rule (10.9b). In order to keep their discussion physically as transparent and simple as possible, in addition to avoiding unnecessary numerical complications, Ghosh et al. [215, 217] represented both photoelectron e_p and the Auger electron e_a by partial waves of odd parity which, in the case of (10.8), correspond to taking each of ℓ_p and $\ell_a = 1$. The other advantage of this physically acceptable choice is that it does not cause the SOI in the continua of (e_p, e_a) to vanish identically.

10.2.2.2 Transition Amplitudes

It is obvious from the first 9-j symbol present in (10.7b) that if the photoelectron e_p in the example (10.8) is represented by a p-wave, then one needs two following photoionization amplitudes

$$\left.\begin{array}{r} \langle \mathfrak{p}_{1^{+*}}, \tfrac{1}{2};\ (1\tfrac{1}{2})\tfrac{1}{2} |\ F(\tfrac{1}{2})\ |\mathfrak{p}_0,\ 0;\ 1\rangle \ \equiv\ d_1\, e^{i\Delta_1} \\ \text{and} \qquad\qquad\qquad\qquad\qquad\qquad \\ \langle \mathfrak{p}_{1^{+*}}, \tfrac{1}{2};\ (1\tfrac{1}{2})\tfrac{3}{2} |\ F(\tfrac{1}{2})\ |\mathfrak{p}_0,\ 0;\ 1\rangle \ \equiv\ d_2\, e^{i\Delta_2}, \end{array}\right\} \tag{10.10a}$$

with $\mathfrak{p}_0 + \mathfrak{p}_{1^{+*}} = \text{odd}$ from the selection rule (10.9a). However, on account of the second 6-j symbol in (10.7b)—and remembering, from (10.8), that $J_{1^{+*}} = \frac{1}{2}$ and $J_{2^+} = 0$—only following single Auger amplitude

$$\langle \mathfrak{p}_{2^+}, 0;\ (1\frac{1}{2})\frac{1}{2} |\ \mathcal{F}_a^{(a)})\ |\mathfrak{p}_{1^{+*}},\ \frac{1}{2}\rangle \tag{10.10b}$$

contributes to $A^{S_p S_a}_{N_{S_p} N_{S_a}}(m_r;\, \boldsymbol{k}_p, \boldsymbol{k}_a)$ in (10.7b). Here, one must satisfy the selection rule (10.9b), i.e., have $\mathfrak{p}_{1+^*} + \mathfrak{p}_{2+} =$ odd, otherwise the Auger amplitude (10.10b) will identically vanish.

Amplitudes (10.10) can be determined in a hierarchy of approximations starting from semi-empirical or phenomenological methods to sophisticated *ab initio* techniques well explained in the concerned literature (see, for example, [322] and [177, 183] for photoionization and for non-radiative decay, respectively). Determination of these amplitudes from first principles may itself be the goal [183, 322] of a calculational study of any 2-DPI process. With an aim to provide, among other things, a qualitative analysis, as well as to develop a framework for the calculations and analysis of measurements, of the spin-entanglement between a photoelectron and an Auger electron sequentially ejected in the 2-DPI process (1.3) taking place in a RLM in the presence of C+SDIs in Hund's coupling scheme (a), Ghosh et al. [215, 217] performed a model calculation taking, quite arbitrarily,

$$d_1 = d_2 = 1.0 \quad \text{and} \quad \Delta_1 - \Delta_2 = 1.53\,\text{rad} \tag{10.11}$$

in (10.10a). Unlike the photoionization amplitudes (10.10a), only a single Auger amplitude (10.10b) contributes in the present example (10.8) of 2-DPI. The contributions of the later amplitude cancels out during the trace-normalization[2] of the density matrix (10.7). No value of the Auger amplitude (10.10b) is, therefore, needed in the present model calculation specified by the states (10.8).

10.2.2.3 Experimental Configuration

Electrons $(e_p,\, e_a)$ can leave their molecular parents $(\mathfrak{T}, \mathfrak{T}^{1+^*})$ in an infinite number of possible directions $(\hat{\boldsymbol{k}}_p,\, \hat{\boldsymbol{k}}_a)$. Moreover, these directions of their emission are quite independent of each other. In addition, spins of each of the photoelectron e_p and the Auger electron e_a can also be quantized in all possible directions $(\hat{\boldsymbol{u}}_p,\, \hat{\boldsymbol{u}}_a)$ in space, quite independently of each other. But, entanglement is well known [17, 24, 34, 43, 99, 132] to be independent of the local choices of the bases and coordinate system. Ghosh et al. [215, 217] performed their generic study of spin-entanglement in a $(e_p,\, e_a)$-pair emitted in the 2-DPI of a RLM in the presence of C+SDIs in the experimental geometry specified by (7.22) and previously used in [188] for investigating the fine-structure non-local correlation between the spin angular momenta of photoelectron e_p and Auger electron e_a sequentially emitted from an atomic target. Geometry (7.22) means, in other words, that the line joining the longitudinally polarized electrons $(e_p,\, e_a)$ always lie in the vertical plane containing the polar OZ-axis of the SF shown in Fig. 10.1. This vertical plane has elsewhere been called as an emission plane (see, e.g., page 174).

[2]See, for example, (3.18) and footnote (4) on page 56.

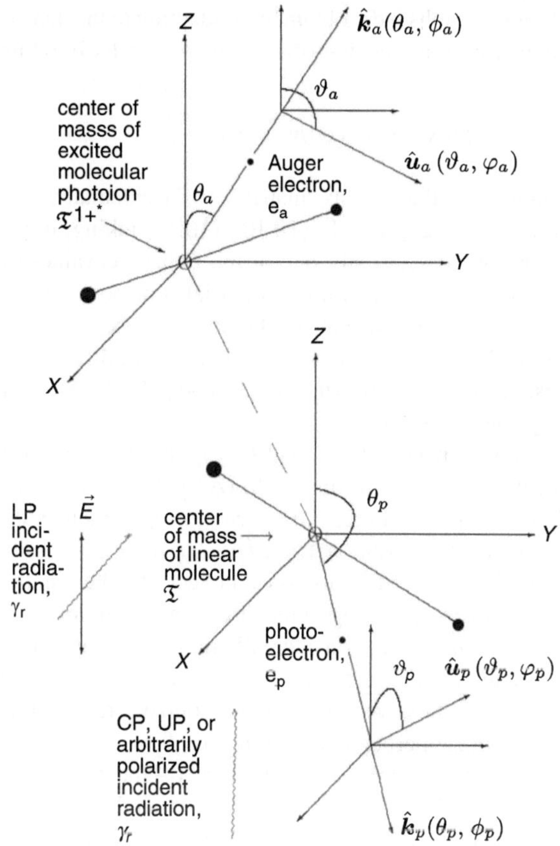

Fig. 10.1 Coordinate system showing the space-fixed frame of reference $OXYZ$, linear molecule $\mathfrak{T}$ as well as its excited photoion $\mathfrak{T}^{1+^*}$, both the propagation directions $(\hat{\boldsymbol{k}}_p\ \hat{\boldsymbol{k}}_a)$ and the spin quantization directions $(\hat{\boldsymbol{u}}_p, \hat{\boldsymbol{u}}_a)$ of the photo-Auger electron pair (e_p, e_a). The molecular axis $\hat{\boldsymbol{R}}$ (not shown in this figure) is along the line containing all the nuclei of the linear molecule $\mathfrak{T}$ (or, its excited photoion $\mathfrak{T}^{1+^*}$). The origin "O" of the space-frame $OXYZ$ is at the centre of mass of the molecule/ion. Each of the directions $\left[\text{i.e., } (\hat{\boldsymbol{k}}_p, \hat{\boldsymbol{k}}_a), (\hat{\boldsymbol{u}}_p, \hat{\boldsymbol{u}}_a), \hat{\boldsymbol{R}}\right]$ is defined with respect to the SF. Following the absorption, in $E1$ approximation, of a single photon γ_r in the 2-DPI process (1.3), the photoelectron e_p and the Auger electron e_a are sequentially ejected from $\mathfrak{T}$ and $\mathfrak{T}^{1+^*}$, respectively. [The molecular frame (MF), not shown in this figure, is concentric with the SF and has its polar axis along $\hat{\boldsymbol{R}}$.]

10.2.2.4 Results

The following is a brief description of the properties of fine-structure entanglement of a (e_p, e_a) pair produced in a 2-DPI experiment (3.3) on a RLM with specifications given in (10.8), (10.10), (10.11), and (7.22). Ghosh et al. [215,217] have considered four different kinds of polarization of the ionizing electromagnetic radiation γ_r

in (1.3). The following results were obtained by adapting to the density matrix (10.7) the generic computer programme described in the Appendix E to this monograph.

Ionizing Radiation Linearly Polarized ($m_r = 0$)

Let $\rho_0(\theta, \phi)$ stands for the density matrix (10.7) calculated according to the specifications given by (7.22), (10.8), (10.10), (10.11) taking, in addition, $m_r = 0$ for the LP ionizing photon γ_r. Only one of the four eigenvalues of $\rho_0(\theta, \phi)$ was found [215, 217] to be non-zero, while its partial transpose was negative. These characteristics of $\rho_0(\theta, \phi)$ mean that, in the present case, (e_p, e_a) are always in a pure state [61] and are entangled [124, 125] as well. All the three measures of entanglement discussed in the following were found [215,217] to be independent of the azimuthal angle ϕ defined in (7.22).

Figure 10.2 shows negativity $\mathcal{N}(\rho_0(\theta, \phi))$ (2.45c), concurrence $\mathcal{C}(\rho_0(\theta, \phi))$ (2.46), and entanglement of formation $\mathcal{E}_F(\rho_0(\theta, \phi))$ (2.47) as a function of the polar angle θ defined in (7.22). In the present case, in agreement with the theoretical prediction made in [142], $\mathcal{N}(\rho_0(\theta, \phi)) = \mathcal{C}(\rho_0(\theta, \phi))$ for all possible values of θ. Thus, unlike in the absence of SDIs discussed in Sect. 10.1 herein, the entanglement properties now very much depend upon the directions of motion of (e_p, e_a). All the three measures of entanglement, considered herein, have their maximum possible values of unity for the diametric emission of two electrons (e_p, e_a) either in the directions of the electric field vector of the LP ionizing radiation (i.e., $\theta = 0/\pi$ in Fig. 10.1), or in a plane perpendicular to it (i.e., $\theta = \pi/2$). For other values of the polar angle θ, these measures vary significantly—their minimum values being at $\theta \simeq \pi/4$ and $3\pi/4$, where (e_p, e_a) are barely entangled. Thus, SDIs make the entanglement between (e_p, e_a) highly anisotropic, and tunable, with respect to the directions of emission of two electrons. Further, the entanglement shown in Fig. 10.2 is symmetric about the direction $\theta = \pi/2$. This example, further shows that a pure bipartite state of qubits may have any amount of entanglement—from being a separable to a maximally entangled state.

Ionizing Radiation Circularly Polarized ($m_r = \pm 1$)

In order to study the spin-entanglement properties of (e_p, e_a) when the ionizing radiation in the example (10.8) is circularly polarized, one needs to calculate the density matrix (10.7), with specifications given in [(7.22), (10.8), (10.10), (10.11)], taking $m_r = \pm 1$ in (10.7b). This density matrix has herein been denoted by $\rho_{m_r}(\theta, \phi)$. Ghosh et al. [215, 217] found that, in the present example (10.8) of 2-DPI, the spin-entanglement between (e_p, e_a) has identical properties both for LCP (with $m_r = -1$) and RCP (with $m_r = +1$) ionizing radiation, i.e., $\rho_{+1}(\theta, \phi) = \rho_{-1}(\theta, \phi)$. This, in other words, means that in the example (10.8), there is no circular dichroism [see footnote (3) in Chapter 11] in the spin-entanglement of (e_p, e_a).

Further, similar to the ionization by LP light, each of the four eigenvalues of $\rho_{\pm 1}(\theta, \phi)$ is independent of the azimuthal angle ϕ and (e_p, e_a) are always in a pure

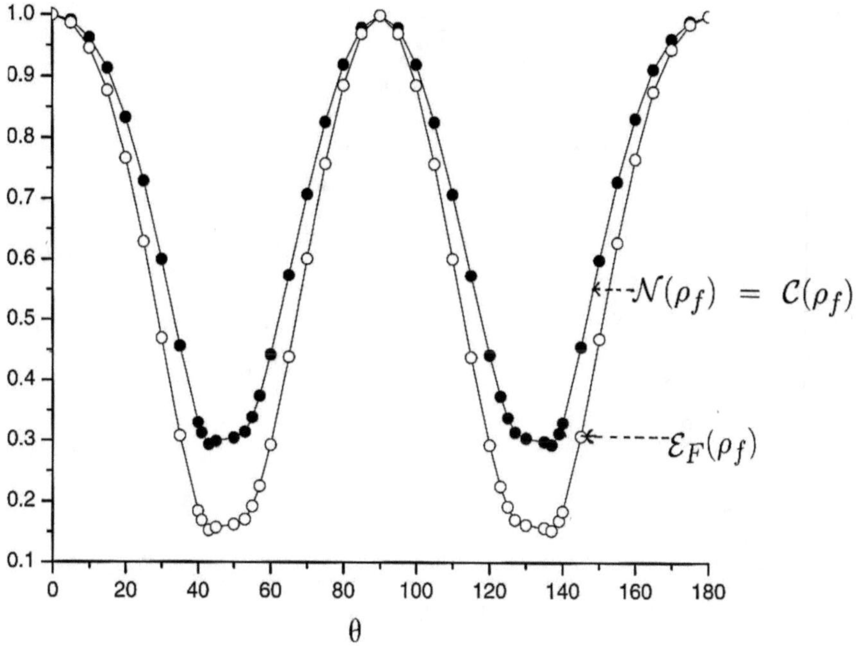

Fig. 10.2 Variations of each of the three measures of entanglement {namely: negativity $\mathcal{N}(\rho_0(\theta, \phi))$ (2.45c), concurrence $\mathcal{C}((\rho_0(\theta, \phi))$ (2.46), and entanglement of formation $\mathcal{E}_F(\rho_0(\theta, \phi))$ (2.47)} with the angle θ. These three measures are calculated for the spin-state $\rho_0(\theta, \phi)$ {obtained on using [(7.22), (10.8), (10.10), (10.11)] in the density matrix (10.7) of a photo-Auger electron pair sequentially emitted from a RLM in the 2-DPI process (1.3) in the presence of C+SDIs in Hund's coupling scheme (a). The angle θ, defined in (7.22a), is the direction of motion of (e_p, e_a) in the emission plane containing the OZ-axis of the SF shown in Fig. 10.1 and oppositely moving photoelectron e_p and Auger electron e_a. Spin angular momenta of each of the emitted electrons (e_p, e_a) are polarized longitudinally to their respective directions of motion [according to (7.22c)]. For the presently considered experimental geometry of diametric emission of (e_p, e_a), the results shown in this figure are independent of the azimuthal angle ϕ [see (7.22a)] which specifies orientation of the emission plane with respect to the OX-axis in Fig. 10.1. All the three measures in this figure are calculated when the ionizing radiation in (1.3a) is LP [i.e., $m_r = 0$ in the density matrix (10.7)]. On account of the pure nature of the state $\rho_0(\theta, \phi)$ of (e_p, e_a), negativity and concurrence in this case have identical values. (First published in [217].)

spin-state. However, unlike in the case of $\rho_0(\theta, \phi)$, none of the four eigenvalues of the partial transpose of $\rho_{\pm 1}(\theta, \phi)$ was less than zero for $\theta = 0$ or π; otherwise, partial transpose of $\rho_{\pm 1}(\theta, \phi)$ was always negative.

These properties of the present entanglement in a (e_p, e_a) pair are evident also from Fig. 10.3, which contains the negativity, concurrence, and entanglement of formation for the present example of 2-DPI (10.8) by LCP or RCP light. There is no entanglement at all for the diametric motion of photoelectron e_p and Auger electron e_a parallel to the direction of incidence of the CP ionizing radiation. These two flying particles, on the other hand, have their spins maximally entangled when they are moving in a plane perpendicular to incident γ_r. In other direction of motions,

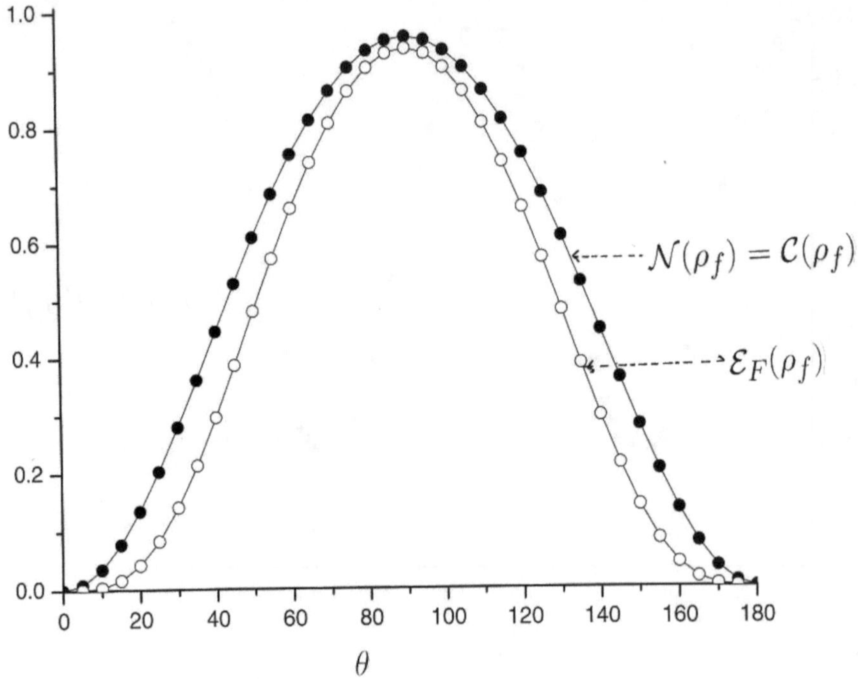

Fig. 10.3 Same as Fig. 10.2, but for ionization by an LCP (i.e., $m_r = -1$) photon γ_r. Here too, both negativity and concurrence have equal values since (e_p, e_a) are again in a pure state. Also, as discussed in the text on pages 234 and 235, this figure represents values of all the three measures of entanglement for ionization in (1.3) caused by the absorption of an RCP (i.e., $m_r = +1$) γ_r as well

spin-entanglement between oppositely moving (e_p, e_a) varies between these two extreme values.

Ionizing Radiation Unpolarized

An UP electromagnetic wave can be taken to be an even mixture of LCP and RCP radiations. The density matrix for the present case is, therefore, $\rho_{\text{UP}}(\theta, \phi) \equiv [\rho_{+1}(\theta, \phi) + \rho_{-1}(\theta, \phi)]/2$, where $\rho_{+1}(\theta, \phi)$ and $\rho_{-1}(\theta, \phi)$ are those calculated for Fig. 10.3. The diagonalization of $\rho_{\text{UP}}(\theta, \phi)$ showed that it never represents a pure spin-state of (e_p, e_a); although the partial transpose of the density matrix $\rho_{\text{UP}}(\theta, \phi)$ is negative for most of the values of θ shown in Fig. 10.4. Hence, in the present case, (e_p, e_a) are always in a mixed entangled state except when they are moving in opposite directions parallel to the incident unpolarized light, along $\theta \simeq \pi/3$, or $\theta \simeq 2\pi/3$. Thus, use of UP, in place of LP or CP, light in (10.8) has changed (e_p, e_a) from being in a pure to a mixed state.

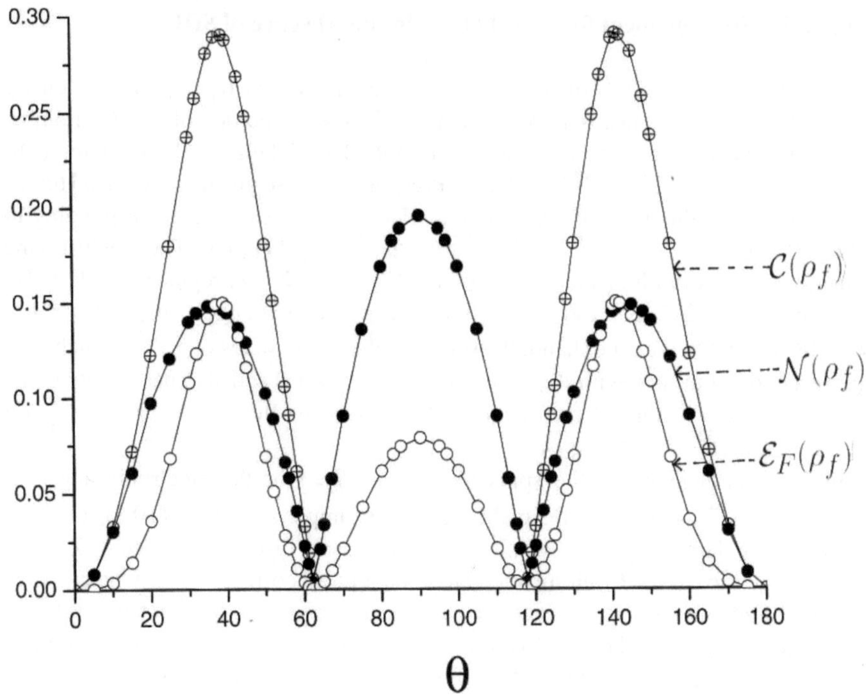

Fig. 10.4 Same as Fig. 10.2, but for ionization by UP photon γ_r. In this case, unlike that of ionization by LP, RCP, or RCP photon, $(e_p,\, e_a)$ are always in a mixed spin-state. Consequently, negativity and concurrence do not always have same values for the whole range of $0^0 \leq \theta \leq 180^0$

Next, Ghosh et al. [215,217] calculated $\mathcal{N}\big(\rho_{\text{UP}}(\theta,\phi)\big)$, $\mathcal{C}\big(\rho_{\text{UP}}(\theta,\phi)\big)$, and $\mathcal{E}_F\big(\rho_{\text{UP}}(\theta,\phi)\big)$. Figure 10.4 shows the variation of each of these three measures of entanglement with respect to the polar angle θ. On account of mixed nature of the density matrix in the present case, negativity and concurrence are not necessarily equal for the whole range of θ in Fig. 10.4. When $(e_p,\, e_a)$ move diametrically in directions $\theta \simeq= 0^0$, 60^0, 120^0, or 180^0, they are in product spin-states. Each of the three measures in Fig. 10.4 attains its maximum value simultaneously at $\theta \simeq 40^0$, 90^0, 140^0. But the largest values achieved at these three angles is much less than the maximum allowed values of unity of each of negativity, concurrence, and entanglement of formation. Thus, unlike in the case of ionization by LP or CP radiation, discussed on pages 234–236, $(e_p,\, e_a)$ in the present case of ionization by UP light will never be in a pure or a maximally entangled state. Furthermore, a comparison of the three measures in Figs. 10.2, 10.3, and 10.4 shows that the degree of spin-entanglement between $(e_p,\, e_a)$ is minimum when the incident light is UP. Finally, similar to Figs. 10.2 and 10.3, all the three measures of entanglement in Fig. 10.4 are also symmetric about $\theta = \frac{\pi}{2}$.

10.2.2.5 Entanglement for 2-DPI (10.8) in the Absence of SDIs

In order to see the effects of SDIs on spin-entanglement of (e_p, e_a) in the generic example (10.8), one needs to know [215, 217] the properties of the Coulombic entanglement of the same two electrons emitted in 2-DPI involving states $(|0\rangle, |1^{+^*}\rangle, |2^+\rangle)$ of $(\mathfrak{T}, \mathfrak{T}^{1+^*}, \mathfrak{T}^{2+})$, which correspond to those given in (10.8) in Hund's case (a), but without SDIs. The possible electronic states of $\mathfrak{T}$ corresponding to $J_0 = 0$ are: ${}^{(1,3,5,\ldots)}\Sigma_0$; ${}^{(3,5,7,\ldots)}\Pi_0$; ${}^{(5,7,9,\ldots)}\Delta_0$; ${}^{(7,9,11,\ldots)}\Phi_0$; *etc*. These will be the electronic states (with the subscript "0" on Σ, Π, Δ, Φ, *etc*, replaced by 2^+) also for the dication $\mathfrak{T}^{2+}$ as it too is shown to possess the total angular momentum $J_{2^+} = 0$ in (10.8). On the other hand, the possible electronic states of $\mathfrak{T}^{1+^*}$, which may correspond to the state with $J_{1^{+*}} = \frac{1}{2}$ in Hund's case (a) and also participate in the 2-DPI process (1.3) in a RLM without SDIs, are given by: ${}^{(2,4,6,\ldots)}(\Sigma_{1^{+*}}, \Pi_{1^{+*}})$; ${}^{(4,6,8,\ldots)}\Delta_{1^{+*}}$; ${}^{(6,8,8,\ldots)}\Phi_{1^{+*}}$; *etc*.

With a knowledge of the spins $(S_0, S_{1^{+*}}, S_{2^+})$ of the electronic states of $(\mathfrak{T}, \mathfrak{T}^{1+^*}, \mathfrak{T}^{2+})$ participating in the generic example of 2-DPI (10.8), one can now readily determine as to which of the two cases, discussed in Sect. 10.1.2 on pages 223–226, of the Coulombic entanglement these correspond to. One can then trivially find out the pure/mixed as well as the entangled/product character of the state of (e_p, e_a) when no SDIs are taken into account in the 2-DPI example corresponding to that given by (10.8). For example, the negativity and concurrence of these Coulombic states is also gotten from (6.14). The consequent properties of the fine-structure entanglement and of Coulombic entanglement of the spins in a (e_p, e_a)-pair, ejected in the example (10.8) of 2-DPI with or without SDIs are then easily compared. One thing which is very obvious from the discussions presented in Sect. 10.1 on pages 221–226 is that, unlike the fine-structure non-local correlation between the spins of (e_p, e_a), whenever these two electrons are sequentially ejected in a 2-DPI process taking place in a RLM in the absence of SDIs, they are in a state of definite entanglement which is completely independent of the polarization of the ionizing radiation γ_r, directions of ejection and of spin polarization of (e_p, e_a), in addition to photoionization as well as of the Auger decay dynamics.

Part III
Miscellaneous

Chapter 11
Conclusions and Prospectives

An essential ingredient to both the theoretical studies in, and technological applications of, quantum information science is easy availability of entangled states of two or more qubits. Hitherto, these qubits have primarily been photons and/or atoms. Photons are excellent carriers of information—nothing can travel faster than an electromagnetic wave. Also, it is relatively easy to generate states of two or more photonic qubits entangled with respect to either their (linear or circular) polarization, or amplitude and phase. In addition, properties of a beam of light can easily be measured and/or manipulated in a laboratory using various readily available, and usually inexpensive, optical devices. Although a photon is easy to detect, detection of a photon, nevertheless, always leads to its destruction unless some special kinds of extremely difficult experiments (e.g., quantum non-demolition) are performed. Further, hitherto developed quantum communication protocols usually require storing and saving qubits so that they can later be measured jointly with other qubits. But, a photon cannot be stored and saved for an arbitrarily long time without affecting its quantum state.

An atomic qubit, on the other hand, can be stored for as long a period as one wants without corrupting its quantum state. Many, relatively easy techniques in atom optics have been suggested for entangling two internal states of two or more neutral/ionized atoms. However, state of a neutral atom can not be directly detected. One needs to perform a fluorescence or an ionization type experiment for detecting a state of an atom, which leads to a change in its (i.e., atom's) entangled state. Moreover, it will be extremely difficult to make an atom to travel with a high speed required for long distance quantum communication.

The practical implementations of almost all of the hitherto developed quantum communication protocols require entangled states of qubits, which are not only separately addressable at distant locations, but can also be, saved, stored, and detected—without changing their quantum states and/or entanglement properties—for using them at later times. An electron is a stable particle and, due to its charge, directly detectable by an electron spectrometer. With a suitable combination of electric and magnetic fields, it can readily be made to fly with sufficiently high

N. Chandra and R. Ghosh, *Quantum Entanglement in Electron Optics*, Springer Series on Atomic, Optical, and Plasma Physics 67, DOI 10.1007/978-3-642-24070-6_11,

speeds to a desired location. Also, an electron can be stored for as long a time as one wants with a minimal change in its quantal properties. It is, probably, on account of their these and other similar properties, electrons have since long been used to carry and/or store information in the form of electric signals. In addition, it has already been suggested that because the spin angular momentum of an electron is one-half which, according to space quantization (see, for example, [58]) in quantum mechanics, can have only two distinct spatial orientations, an electron will be a very suitable qubit needed for quantum information science. Nevertheless, in spite of all these properties of an electron, few studies were ever made on the generation and characterization of the entangled states of two or more electronic qubits.

In this monograph, we have shown as to how bipartite states of flying electronic qubits, electronic plus photonic qubits, or electronic qubit plus ionic qudit can be generated in simple experiments regularly performed in a laboratory for atomic and molecular physics. The properties of these states have been investigated in detail using the tools hitherto developed in quantum information science. Chapter 1 lists those processes, among others, wherein absorption of a single photon in the $E1$ approximation by an atom or a RLM can liberate two electrons either simultaneously in a single step, or sequentially in two different steps. These processes have herein been called as 1-DPI (one-step double photoionization) and 2-DPI (two-step double photoionization), respectively. The pairs of outgoing electrons in these two respective processes have been designated as $(e_p, e_{p'})$ and (e_p, e_a). In the sequential two-step process, however, a photon γ_d, in place of an electron e_a, may also leave the target in the second step. The entanglement properties of such a hybrid (e_p, γ_d)-pair of two different kinds of qubits, one fermion and the other a boson, have also been investigated in this monograph. In addition, entanglement between the spins of a photoelectron and a photoion, generated in the simplest possible process—namely, one-step single photoionization (1-SPI)—in atomic physics has been discussed as well. We show that each of these processes generates pairs of particles with widely different entanglement properties, which depend upon the states of the participating target and of its ionic species.

In both, one- and two-step processes, entanglement between the emitted particles is generated primarily by the forces present inside an atomic or a molecular target. These forces are mainly of two types: Coulomb forces due to the electric charges of nuclei as well as of electrons and the forces arising from the interaction of the electronic spin magnetic momenta with the magnetic fields generated by the motion of charged particles constituting an atom or a molecule. The non-local correlation between the two free particles in either of the one- and two-step processes taking place in an atom or a molecule in the presence of purely the Coulomb forces is named as Coulombic entanglement; whereas the one found on the inclusion of both the C+SDIs has been called as fine-structure entanglement. Although purely Coulombic forces in an atom are incapable of entangling spin of a photoelectron with the polarization of a photon emitted sequentially in a two-step process; nevertheless, fine-structure entanglement between this hybrid system of a fermion plus a boson does exist.

All the properties of the Coulombic entanglement of a bipartite state of particles—at least one of which is an electronic qubit—can exactly be determined always merely by a knowledge of the spins of the electronic states of an atomic or molecular target and of its ionic species taking part in a 1-SPI, 1-DPI, or 2-DPI processes; otherwise, it is completely independent of every thing (e.g., properties of the ionizing radiation, dynamics of photoionization and/or of Auger decay, kinematics of participating particles, experimental geometry, etc.) in this world. Consequently, spins of two electrons outgoing in 1-DPI or 2-DPI, or of a photoelectron and one photoion in 1-SPI, are always in a definite state of Coulombic entanglement which can readily be pre-determined and quantified theoretically.

The Coulombic $(e_p, \mathfrak{T}^{1+})$ entanglement, discussed in the Chap. 4, is probably the simplest possible method for generating, in a photoelectron spectroscopy experiment in a laboratory, the qubit-qubit, qubit-qutrit, or qubit-qudit spin-states possessing free or bound entanglement. It shows as to how one can realistically produce non-separable states of $(e_p, \mathfrak{T}^{1+})$ possessing negative or positive partial transpose and completely analyse their properties merely by knowing the spin multiplicities of the target atom $\mathfrak{T}$ and of its residual photoion $\mathfrak{T}^{1+}$ participating in a 1-SPI process (1.1).

In the case of two electronic qubits, on the other hand, the properties of the Coulombic entanglement generated in 1-DPI are widely different from that produced in 2-DPI; these, however, are found not to depend, in any way whatsoever, whether the two electrons come out in the same process taking place in an atom or a RLM. In other words, neither the presence of nuclear rotation nor changes in the nuclear field, which is spherical in an atom to the one possessing cylindrical symmetry in a RLM, has any effects at all on the Coulombic entanglement generated in either of the 1-DPI and 2-DPI processes.

The following Table 11.1 shows a comparison, in brief, of the properties of the Coulombic entanglement of $(e_p, e_{p'})$ and of (e_p, e_a) generated in the respective 1-DPI and 2-DPI of an atom or a RLM. It is obvious from this table that although the spin of the electronic states of the target atom/RLM and of its residual dication in 1-DPI may be the same as in the case of 2-DPI; the Coulombic entanglement generated in these two processes may, however, have very different properties. In general, 2-DPI is capable of generating tunable Coulombic entanglement in (e_p, e_a), which is not possible in the case of $(e_p, e_{p'})$ in 1-DPI. The participation of the excited photoion $\mathfrak{T}^{1+^*}$ in each of the two steps of a 2-DPI process, which is never formed in 1-DPI, makes the presence of spin-entanglement in a bipartite state of electronic qubits to be more frequent with various degrees of measures, which is not possible in the later process, although both of them may take place under identical physical conditions with $(\mathfrak{T}, \mathfrak{T}^{2+})$ being same in 1-DPI and 2-DPI.

The presence of SDIs, in addition to the Coulomb interaction, has a significant effect on the properties of the spin-entanglement generated in 1-SPI, 1-DPI, or 2-DPI. For example, unlike the Coulombic entanglement, the non-separability of a bipartite state is now determined by the total angular momentum $\boldsymbol{J}$, which defines a fine-structure level of an atom/RLM. It is, for this reason, that the non-

Table 11.1 Comparison of the properties of Coulombic entanglement generated in the respective 1-DPI and 2-DPI processes (1.2) and (1.3), taking place in an atom or a rotating linear molecule

	(S_0, S_{2+}): spins of the (target $\mathfrak{T}$, dication $\mathfrak{T}^{2+}$)	Bipartite spin-state (5.11) of $(e_p, e_{p'})$ formed in the 1-DPI process (1.2)	Bipartite spin-state (6.5) of (e_p, e_a) formed in the 2-DPI process (1.3) (S_{1+^*}: spin of the excited photoion $\mathfrak{T}^{1+^*}$)
1.	$S_0 = S_{2+} = 0$	$(e_p, e_{p'})$ in a singlet spin state (5.14) which is pure and maximally entangled [i.e., Bell state (2.17b)]	$S_{1+^*} = \frac{1}{2}$: (e_p, e_a) in a singlet spin state (5.14) which is pure and maximally entangled [i.e., Bell state (2.17b)]
2.	$S_0 = S_{2+} = \frac{1}{2}$	$(e_p, e_{p'})$ in mixed separable state (5.32a)	(a) $(S_{1+^*} = 0)$: (e_p, e_a) in maximally chaotic state σ_0 [(6.7) & (6.8a)/(6.9a)] (b) $(S_{1+^*} = 1)$: (e_p, e_a) in mixed, entangled state (6.7) with (6.8b)/(6.9b)
3.	$\left\| S_0 - S_{2+} \right\| = 1$	$(e_p, e_{p'})$ in mixed separable state (5.29)	($S_{1+^*} \geq \frac{1}{2}$ with $S_0 = \left\| S_{1+^*} \pm \frac{1}{2} \right\|$ and $S_{2+^*} = \left\| S_{1+^*} \mp \frac{1}{2} \right\|$): (e_p, e_a) in mixed, separable state (6.15)
4.	$S_0 = S_{2+} = S_{1+^*} + \frac{1}{2}$	Although, S_{1+^*} is not present in 1-DPI; this case, nevertheless, corresponds to $S_0 = S_{2+} \geq \frac{1}{2}$ for $S_{1+^*} \geq 0$. Then, according to Eq. (5.32a), (e_p, e_p') are always in a mixed separable state.	$(S_{1+^*} > 0)$: (e_p, e_a) in mixed, separable state (6.7) with (6.8b)/(6.9a)
5.	$S_0 = S_{2+} = S_{1+^*} - \frac{1}{2}$	Although, S_{1+^*} is not present in 1-DPI; this case, nevertheless, corresponds to $S_0 = S_{2+} = 0$ for $S_{1+^*} = \frac{1}{2}$ which is the same as this table's 1st entry. For $S_{1+^*} > \frac{1}{2}$, on the other hand, $S_0 = S_{2+} \geq \frac{1}{2}$, corresponding to the present table's 4th entry.	$(S_{1+^*} > \frac{1}{2})$: (e_p, e_a) in mixed, entangled state (6.7) with (6.8b)/(6.9b)

local correlation generated in the presence of both Coulomb and spin-dependent interactions in a $(e_p, \mathfrak{T}^{1+})$, $(e_p, e_{p'})$, or (e_p, e_a) pair in 1-SPI, 1-DPI, or 2-DPI, respectively, has been called as fine-structure entanglement. The inclusion of SDIs, couples the entanglement with the kinematics as well as with the dynamics of a one-step or two-step process taking place in an atom/RLM; moreover, the $(e_p, \mathfrak{T}^{1+})$, $(e_p, e_{p'})$, or (e_p, e_a) entanglement is now affected also by a change in the polarization and/or frequency of the photon γ_r absorbed for primary ionization during which photoelectron e_p is emitted. This coupling of the non-local correlation in a $(e_p, \mathfrak{T}^{1+})$, $(e_p, e_{p'})$, or a (e_p, e_a) pair not only with the geometrical and dynamical parameters of a 1-SPI, 1-DPI, or a 2-DPI process, but also with the properties of the absorbed photon, makes the entanglement anisotropic, i.e., varying with the directions of outgoing electrons in different ways for different polarization of γ_r. Consequently, while $(e_p, \mathfrak{T}^{1+})$, $(e_p, e_{p'})$, or (e_p, e_a) may be in a completely separable state in certain directions of their motions, but maximally entangled in other directions.

The other important consequence of the inclusion of SOI in a two-step process taking place in an atomic target is that it may generate fermion-boson entanglement. Namely, the spin of a photoelectron e_p may now be entangled with the polarization of a photon γ_d—the two particles which were otherwise in a separable state in the absence of SOI. Similar to the fine-structure entanglement in 1-SPI, 1-DPI, and 2-DPI, the (e_p, γ_d) entanglement too is determined by the total angular momenta $(\boldsymbol{J}_0, \boldsymbol{J}_{1^{+*}}, \boldsymbol{J}_{2+})$ of the species $(\mathfrak{T}, \mathfrak{T}^{1+*}, \mathfrak{T}^{1+})$ participating in a two-step process—in addition to depending upon the geometry of such experiments and frequency as well as polarization of the ionizing radiation. One of the important differences between the fine-structure (e_p, γ_d) entanglement and the one generated in a 1-DPI or 2-DPI process is that the former involves dynamics of emission of only a single particle (namely, the photoelectron e_p), unlike the other two which require dynamics for the emission of both of the particles [i.e., of $(e_p, e_{p'})$ in 1-DPI and of (e_p, e_a) in 2-DPI]. Thus, dynamics of spontaneous emission of γ_d plays no role whatsoever in entangling the spin of the photoelectron e_p with the polarization of the former in the presence of C+SOI—although entanglement of the spin of photoelectron e_p with a LP γ_d is very different from that when γ_d is CP. Moreover, for the experimental configurations discussed in this monograph, there is no (e_p, γ_d) fine-structure entanglement when the ionizing radiation γ_r in (1.4) has linear polarization.

Thus, in order to study fine-structure entanglement between the angular momenta of two particles generated in any of the four (1.1)–(1.4) processes investigated in this monograph, one must know the dynamical amplitudes for that process. Properties of a Coulombic entanglement, on the other hand, can readily be predicted merely with a knowledge of the spins $[(S_0, S_{1+})$ of $(\mathfrak{T}, \mathfrak{T}^{1+})]$, $[(S_0, S_{2+})$ of $(\mathfrak{T}, \mathfrak{T}^{2+})]$, or $[(S_0, S_{1^{+*}}, S_{2+})$ of $(\mathfrak{T}, \mathfrak{T}^{1+^*}, \mathfrak{T}^{2+})]$ obtained by measuring simply energies of e_p, $(e_p, e_{p'})$, or of (e_p, e_a) in a respective 1-SPI, 1-DPI, or 2-DPI experiment in $L-S$ coupling. But, a verification of the properties of either or both of the Coulombic and fine-structure entanglement, obtained using the theoretical procedures described herein, will necessarily require experimental measurements. It involves [24] full quantum state tomography, or use of an entanglement witness or any other such protocols hitherto developed in quantum information.

Separation of a Coulombic density matrix into its two parts describing purely angular and purely spin correlations, obtained in the absence of SDIs in each of the Chaps. 4–6 on atomic processes and in Sects. 9.1 and 10.1 on molecular processes, is completely rigorous and independent of all dynamical models. In the absence of SDIs, the state of an atom or a molecule is well known to be a product of its spatial and spin parts (see, for example, [10, 58–60, 68–70, 72, 184–186], etc) in Russell–Saunders coupling for atoms, or in Hund's scheme (b) (e.g., [68, 69, 75, 185]) for linear molecules. A density matrix also represents state of a system (see, for example, [60–62] etc). Hence, separation of angular and spin parts of the Coulombic density matrices for the 1-SPI, 1-DPI, 2-DPI processes in the Chaps. 4–6 and in Sects. 9.1 and 10.1 in this monograph, as well as for the respective 2-TPI and 3-TPI processes (1.6) and (1.7) in [56, 57], is a consequence of the exclusion of SDIs, i.e., consideration of purely the Coulomb forces present in an atom or a molecule.

On the other hand, a simple product of spatial and angular parts is no longer an eigenstate of the Hamiltonian of an atom or a molecule which contains both the C+SDIs (see, for example, [10, 58–60, 68–70, 72, 75, 184–186], etc). The only eigenstates of such a Hamiltonian are those in which spatial and spin parts are coupled [e.g., the $j - j$ coupling for atoms or Hund's coupling scheme (a) for linear molecules]. Consequently, the fine-structure density matrices obtained in Chap. 7 and 8, and in Sects. 9.2 and 10.2, are not separable in angular and spin parts, or in any other way whatsoever. Moreover, the presence of SDIs, in addition to the Coulomb interaction, has coupled in a fine-structure density matrix in a very complicated way the kinematics and dynamics of a process with each other as well as with the characteristics (i.e., energy and polarization) of the radiation used to initiate these processes by appropriately ionizing an atomic or a molecular target.

The studies described in this monograph can be further developed in several important directions. In the remaining part of this concluding chapter, we discuss some of those possible investigations which can be performed in near future. These investigations will, hopefully, further show as to how the studies in photon-induced electron optics in atomic and molecular physics, using the tools developed in quantum information, are beneficial to both of these important fields of research in pure as well as applied sciences.

Suppose photon γ_d, detected in the process (1.4b) and entangled with the photoelectron e_p, is absorbed by another atom, say $\mathfrak{T}'$. Let us assume that γ_d has sufficient energy for any of the processes (1.1)–(1.4) to take place in $\mathfrak{T}'$ and, hence, generating any of the entangled bipartite states $(e'_p, \mathfrak{T}'^{1+})$, $(e'_p, e'_{p'})$, (e'_p, e'_a) and (e'_p, γ'_d), respectively. If e_p of the original, non-separable (e_p, γ_d)-pair is entangled with any of these bipartite systems, it then provides a simple and powerful sequential procedure for generating multipartite entangled states involving photonic qubits, flying electronic qubits, and/or ionic qudits.

In the process (1.1)–(1.8), the target $\mathfrak{T}$ is assumed to absorb the ionizing photon γ_r in its unpolarized state $|0\rangle$.[1] It is, nevertheless, possible by optical pumping or laser excitation, for example, to prepare an atom in a pure state corresponding to a specific value of $|0\rangle = |J_0 M_{J_0}\rangle$ in j-j coupling. The density matrix for such an atom is simply given by $|0\rangle\langle 0| \equiv |J_0 M_{J_0}\rangle\langle J_0 M_{J_0}|$, without involving any sum over M_{J_0}.[2] Several, both theoretical (see, for example, [173, 333–336], *etc*) and experimental (e.g., [334, 337–339]) studies have already been performed on photoionization of such state-selected, polarized atoms. The advantages of performing experiments with such atoms are that these provide an opportunity to observe in the photoelectron angular distribution various kinds of dichroism[3] which, in turn, helps in performing a "complete" experiment making it possible to extract from experimental measurements full information[4] about the amplitudes contributing to a given photoionization process.

Can the studies reported in this monograph be performed if the atom $\mathfrak{T}$ in the processes (1.1)–(1.4) is in a particular state? Can such studies lead to the observation of any new physical phenomena? It is obvious that Coulombic entanglement is, probably, not possible in a state-selected atom because it now becomes necessary (see, for example, [173,333–339], *etc*) to take both C+SOI into account for preparing the atom $\mathfrak{T}$ in one of its fine-structure state $|0\rangle = |J_0 M_{J_0}\rangle$. On using a state-selected atom in the processes (1.1)–(1.4), there will primarily be two things to observe: First, how is the fine-structure $(e_p, \mathfrak{T}^{1+})$, $(e_p, e_{p'})$, (e_p, e_a), (e_p, γ_d) entanglement, studied herein for an atom $\mathfrak{T}$ with a random orientation of $\boldsymbol{J}_0$ in space, is affected? Second, to what extent and in what way, the optical and/or magnetic dichroism manifest themselves in entanglement; will both or any one of the two dichroisms be present in the entanglement generated in all or some of the processes (1.1)–(1.4); what, if any, may be the differences in the dichroism observed in a separable and in an entangled state, *etc*?

The processes discussed in this monograph are initiated by the absorption of a photon in the $E1$ approximation. On account of its dipole nature, only a

---

[1] In this monograph, these states for atoms, for example, have been specified by $(L_0, S_0, M_{L_0}, M_{S_0})$ in L-S coupling, or by (J_0, M_{J_0}) in j-j coupling. It is for this reason that, in the mathematical expressions used herein, the atomic density matrix $|0\rangle\langle 0|$ has been averaged over (M_{L_0}, M_{S_0}), or M_{J_0}, as the case may be (See Chap. 3 for a detailed discussion of this point.)

[2] See, for example, discussion given on pages 48–49.

[3] it is the dependence of the light–matter interaction on the polarization of the former and/or anisotropy of the later. The differences observed due to a change in the helicity (direction of the electric field vector) of a circularly (linearly) polarized light—but, without any change in the nature of the anisotropy of the matter which in the case of a state-selected atom means, for example, that the direction of quantization of $\boldsymbol{J}_0$ should remain unaltered—have come to be known as circular [196, 340] (linear [341]) "optical" dichroism. The "magnetic" dichroism, on the other hand, arises (see, for example, [342–344]) due to a change in the anisotropy of the matter while polarization of the interacting light remains the same. The magnetic dichroism is called [334] circular or linear depending upon whether the light interacting with the matter has circular or linear polarization.

[4] It consists of magnitudes of the transition amplitudes and differences (with signs) of their phases.

restricted number of partial waves representing the photoelectron e_p contribute to the ionization of the target which absorbs the incident photon. Electron impact inner-shell excitation or ionization of atoms too has very extensively been studied both theoretically and experimentally (see, for example, [183]). Let us represent this incident electron by e_{inc}. Then one set of possible processes relevant to the studies in quantum information science and resulting from inner-shell excitation of the target atom/RLM $\mathfrak{T}$, is

$$e_{\text{inc}} + \mathfrak{T}' \longrightarrow \mathfrak{T}'^{*} + e'_s \tag{11.1a}$$

followed by

$$\mathfrak{T}'^{*} \longrightarrow \mathfrak{T}' + \gamma'_d \tag{11.1b}$$

or

$$\mathfrak{T}'^{*} \longrightarrow \mathfrak{T}'^{1+} + e'_a. \tag{11.1c}$$

Here, unless stated otherwise, e'_s stands for the scattered electron which is nothing but e_{inc} after it has lost some of its energy during its inelastic collision with $\mathfrak{T}'$ forming this target's specie $\mathfrak{T}'^{*}$ with an inner-shell excitation. A primed (unprimed) symbol in above equations and hereafter means that the particle represented by it is emitted due to the collision (absorption) of the electron (photon) e_{inc} (γ_r) with (by) the target $\mathfrak{T}'$ ($\mathfrak{T}$). For example, e_a (e'_a) represents an Auger electron emitted in the non-radiative decay of an inner-shell vacancy created on supplying energy to the target $\mathfrak{T}$ ($\mathfrak{T}'$) by the absorption of γ_r (inelastic collision of e_{inc}), *etc*.

Suppose, e_{inc} in (11.1) is the photoelectron e_p of one of the pairs (e_p, $\mathfrak{T}^{1+}$), (e_p, $e_{p'}$), (e_p, e_a), (e_p, γ_d) generated in any of the respective processes (1.1)–(1.4). Assuming e_p has sufficient energy for (11.1a) to take place, then what is the relationship among the particles [$\mathfrak{T}^{1+}$, e'_s, γ'_d] or [$\mathfrak{T}^{1+}$, e'_s, e'_a] generated in the consecutive processes [(1.1), (11.1a), (11.1b)] or [(1.1), (11.1a), (11.1c)], respectively. Similarly, one can study entanglement properties of the tripartite states consisting of [$e_{p'}$, e'_s, γ'_d], [$e_{p'}$, e'_s, e'_a], [e_a, e'_s, γ'_d], [e_a, e'_s, e'_a], [γ_d, e'_s, γ'_d], or [γ_d, e'_s, e'_a] produced in the sequential processes [(1.2), (11.1a), (11.1b)], [(1.2), (11.1a), (11.1c)], [(1.3), (11.1a), (11.1b)], [(1.3), (11.1a), (11.1c)], [(1.4), (11.1a), (11.1b)], or [(1.4), (11.1a), (11.1c)], respectively.

Are these, and those mentioned earlier in this chapter, three particle states entangled? Do they form a $|GHZ\rangle$ [345, 346] or $|W\rangle$ [347] class of tripartite states? If any of these tripartite states is entangled, then this process can easily be continued to generate multipartite non-separable states of flying electronic qubits or hybrid states of electronic plus photonic (ionic) qubits (qudits). Such multipartite states are very much needed in quantum communication. In order to investigate such multipartite entanglement, it is, however, necessary to study the entanglement generated in the electron impact processes [(11.1a), (11.1b)] and [(11.1a), (11.1c)].

It is obvious from the studies reported in this monograph that environment external to $\mathfrak{T}$ does not play any role, whatsoever, in the generation of the Coulombic or fine-structure entanglement in any of the processes (1.1)–(1.8); however, the interaction of the flying particles (i.e., photoelectrons e_p and $e_{p'}$, Auger electron e_a, photoion $\mathfrak{T}^{1+}$) with the background random fields, during their journey to their distant locations, is unavoidable. This qubit/qudit-environment interaction usually leads to the decay of the coherence not only of the individual particles (e.g., $\mathfrak{T}^{1+}$, e_p, $e_{p'}$, e_a, γ_d) of a composite system [e.g., (e_p, $\mathfrak{T}^{1+}$), (e_p, $e_{p'}$), (e_p, e_a), (e_p, γ_d)], but also of their joint state and, hence, resulting in the dissipation of this state's entanglement. Consequently, dynamics of quantum entanglement wherein one studies time evolution of an entangled state in a practical environment is very important in quantum information science. There have, recently, been several [160] theoretical (see, for example, [53,348–352], *etc*) as well as experimental [54,55,353] studies of environment-induced effects on the coherence and entanglement of a quantum state of a composite system consisting of two or three qubits. Probably, the most important results of these studies have so far been the sudden disappearance [53, 350] of entanglement and—in a non-Markavian noisy environment—its subsequent revival [352]. These studies have been performed for bipartite states of qubits of the same kind (e.g., two photonic or spin-$\frac{1}{2}$ qubits). Will the results obtained in [53,350,352] be applicable also to hybrid entanglement say, for example, between (e_p, $\mathfrak{T}^{1+}$) or (e_p, γ_d) generated in the 1-step and 2-step processes (1.1) and (1.4), respectively? These and other properties of entanglement dynamics of the bipartite and tripartite states of particles generated in the processes (1.1)–(1.4) and (1.5)–(1.8), respectively, need to be studied.

Appendix A
Density Operator and Density Matrix

In this appendix, we introduce density operator as well as density matrix and explain, in brief, some of their important properties relevant to quantum information in general, and to the studies reported in this monograph in particular.

A.1 Completely Random, Pure, and Mixed Systems

Let us consider a beam of silver atoms directly coming out of a hot oven in a Stern–Gerlach type experiment [58]. (This beam can also be called a system/collection/ensemble consisting of subsystems which, in the present case, are silver atoms.) Such a beam can be characterized by

$$p_{\uparrow} = \frac{N_{\uparrow}}{N_{\uparrow} + N_{\downarrow}} \qquad \text{and} \qquad p_{\downarrow} = \frac{N_{\downarrow}}{N_{\uparrow} + N_{\downarrow 1}}. \tag{A.1a}$$

Here, $N_{\uparrow}$ and $N_{\downarrow}$ are the number of silver atoms in the beam which have their spin up and down (i.e., are in the spin state $|0\rangle$ and $|1\rangle$), respectively. Then, $p_{\uparrow}$ and $p_{\downarrow}$ in (A.1a) represent atoms' fractional populations in the two spin states. An unpolarized beam is looked upon to be an even mixture of atoms in the states $|0\rangle \equiv |\uparrow\rangle$ and $|1\rangle \equiv |\downarrow\rangle$. This, in other words, means

$$N_{\uparrow} = N_{\downarrow} \qquad \Longrightarrow \qquad p_{\uparrow}, \quad p_{\downarrow} = 0.5 \tag{A.1b}$$

for a beam in which atoms are expected to have their spins oriented randomly in space.

In (A.1), $p_{\uparrow}$ and $p_{\downarrow}$ are simply two real numbers. In this equation, there is no information on the relative phase between the spin-up and spin-down kets $|0\rangle$ and $|1\rangle$, respectively. Such a collection of particles is usually referred to as an incoherent,

N. Chandra and R. Ghosh, *Quantum Entanglement in Electron Optics*, Springer Series on Atomic, Optical, and Plasma Physics 67, DOI 10.1007/978-3-642-24070-6,

even mixture of the $|\uparrow\rangle$ and $|\downarrow\rangle$ states. This is called unpolarized because there is no preferred direction for the orientation of the spin of a silver atom and, hence, is characterized by (A.1).

Such a beam, containing particles coming directly out of an oven, is an example of a completely random system. In contrast, a beam that has gone through a selective Stern–Gerlach type measurement [58] is an example of a pure system. This beam is said to be polarized because all members of the collection are characterized by a single common ket ($|0\rangle$ or $|1\rangle$) that describes a state with spin ($\uparrow$ or $\downarrow$) of each atom pointing in some definite direction.

Unpolarized and pure collections are two opposite extremes of a so-called mixed collection. In a mixed beam of silver atoms, for example, a certain fraction—say, 70%—of the members are characterized by the state $|0\rangle$, the remaining 30% by $|1\rangle$. Such a beam is said to be partially polarized. Although states $|0\rangle$ and $|1\rangle$ need not be orthogonal, they are, however, normalized. One can, for example, have 70% of the atoms with their spins in the positive OX direction and 30% of the spin in the negative OZ direction.

Thus, a pure system is defined to be a collection of physical subsystems such that each member is characterized by the same state $|\psi\rangle$. In a mixed system, on the other hand, a fraction of its members with relative population p_1 are in the normalized state $|\psi_1\rangle$, some other fraction with relative population p_2 in the normalized state $|\psi_2\rangle$, and so on. Alternatively, a mixed system can be viewed as a mixture of pure collections. The fractional populations of any mixed system must satisfy the condition

$$\sum_i p_i = 1. \tag{A.2}$$

The number of terms in (A.2) needs not coincide with the dimensionality "d" of the Hilbert space of the system being considered. For example, for a system consisting of spin-$\frac{1}{2}$ particles with $d = 2$, one may have 40% with spins in the positive OZ direction, 30% in the positive OX direction, and the remaining 30% with their spins in the negative OY direction.

State in (2.1), [or (2.2)], on the other hand, is a coherent, linear superposition wherein the phase relation (in the form of the azimuthal angle ϕ) between kets $|0\rangle$ and $|1\rangle$ contains vital information on the spin orientation of a spin-$\frac{1}{2}$ particle. Neither of $p_\uparrow$ and $p_\downarrow$ in (A.1) should, in general, be confused with the probabilities $|C_0|^2$ and $|C_1|^2$ in (2.1b), or r_0^2 and r_1^2 in (2.2a).

A.2 Averaged Value and Representations

According to one of the fundamental postulates [58] of quantum mechanics, the complete information about a system is contained in its wavefunction. But a quantum mechanical wavefunction can describe only a pure system. The density operator formalism, introduced by J. von Neumann [155, 354], is capable of quantitatively describing physical situations with mixed as well as pure collections.

The methods of density operator for mixed and/or pure states of a system were first applied to atomic physics, probably, by Fano [62]. These methods have since then been discussed in varying details in books on quantum mechanics (see, for example, [58]) as well as in the literature (e.g., [60, 61, 68], *etc*) pertaining to the atomic physics.

Suppose, one makes a measurement of some observable, say, Ω on the mixed collection defined in the following Sect. A.1 of this appendix. Then the averaged, measured value (i.e., system/collection average) of Ω for a large number of measurements is given by

$$\begin{aligned}[][\Omega] &= \sum_{i} p_i \langle \psi_i | \Omega | \psi_i \rangle \\ &= \sum_{i\,n\,m} p_i \langle \psi_i | \phi_n \rangle \langle \phi_n | \Omega | \phi_m \rangle \langle \phi_m | \psi_i \rangle \\ &= \sum_{i\,n\,m} p_i \, [\langle \phi_m | \psi_i \rangle \langle \psi_i | \phi_n \rangle] \, \langle \phi_n | \Omega | \phi_m \rangle . \end{aligned} \tag{A.3}$$

Here, the number of terms to be included in the sums over n (or, m) is just equal to those present in the complete set whose members are the orthonormal basis $\{|\phi_n\rangle, n = 1, 2, \ldots\}$ (or, $\{|\phi_m\rangle, m = 1, 2, \ldots\}$). The number of terms in the sum over i in (A.3) depends, on the other hand, on the decomposition of the mixed collection in pure subsystems. The quantity

$$\rho \equiv \sum_{i} p_i \, |\psi_i\rangle \langle \psi_i | \tag{A.4}$$

is called the density operator for our mixed system/collection. The (m, n)-th element of this operator's matrix representation in the orthonormal basis $\{|\phi_n\rangle\}$ is

$$\langle \phi_m | \rho | \phi_n \rangle = \sum_{i} p_i \, \langle \phi_m | \psi_i \rangle \, \langle \psi_i | \phi_n \rangle . \tag{A.5}$$

The density operator/matrix contains all the physically significant information about the system under consideration. Now the system-average (A.3) can be written as

$$\begin{aligned}[][\Omega] &= \sum_{m\,n} \langle \phi_m | \rho | \phi_n \rangle \, \langle \phi_n | \Omega | \phi_m \rangle \\ &= \sum_{m} \langle \phi_m | \rho \, \Omega | \phi_m \rangle \\ &= \mathrm{Tr}(\rho \, \Omega). \end{aligned} \tag{A.6}$$

As the trace of a matrix is known to be independent of its representation, any complete set $\{|\phi\rangle\}$ can, therefore, be used for calculating $\text{Tr}(\rho\,\Omega)$ in (A.6).

The Hermitian conjugate of the density operator (A.4) is

$$\begin{aligned}\rho^\dagger &= \left(\sum_i p_i\,|\psi_i\rangle\langle\psi_i|\right)^\dagger\\ &= \sum_i p_i\,|\psi_i\rangle\langle\psi_i| = \rho,\end{aligned} \tag{A.7}$$

remembering that each of the probabilities p_i in(A.2) is real. Thus, a physically acceptable density operator/matrix should always be Hermitian.

We further have

$$\begin{aligned}\text{Tr}(\rho) &= \sum_n \langle\phi_n|\rho|\phi_n\rangle\\ &= \sum_n \langle\phi_n|\left(\sum_i p_i\,|\psi_i\rangle\langle\psi_i|\right)|\phi_n\rangle\\ &= \sum_{i\,n} p_i\,\langle\phi_n|\,\psi_i\rangle\,\langle\psi_i|\,\phi_n\rangle\\ &= \sum_i p_i\left(\sum_n \langle\psi_i|\,\phi_n\rangle\,\langle\phi_n|\,\psi_i\rangle\right)\\ &= \sum_i p_i\,\langle\psi_i|\,\psi_i\rangle = \sum_i p_i\end{aligned} \tag{A.8}$$

assuming that the states $|\psi_i\rangle$ are normalized. Then in view of (A.2),

$$\text{Tr}(\rho) = 1. \tag{A.9}$$

Such a density operator is said to be normalized to unit trace. In situations wherein normalization (A.9) does not hold, the system-average of an operator is given by

$$[\Omega] = \frac{\sum_i p_i\,\langle\psi_i|\Omega|\psi_i\rangle}{\sum_i p_i}. \tag{A.10a}$$

Using relations (A.6) and (A.8), one can write

$$[\Omega] = \frac{\text{Tr}\,(\rho\Omega)}{\text{Tr}(\rho)}. \tag{A.10b}$$

Let us now calculate the trace of the square of a density operator, i.e.,

$$\begin{aligned}\mathrm{Tr}(\rho^2) &= \sum_m \langle\phi_m|\rho^2|\phi_m\rangle \\ &= \sum_{m\,n} \langle\phi_m|\rho|\phi_n\rangle\,\langle\phi_n|\rho|\phi_m\rangle.\end{aligned}$$

On substituting (A.4)

$$\begin{aligned}\mathrm{Tr}(\rho^2) &= \sum_{m\,n}\left[\langle\phi_m|\left(\sum_i p_i\,|\psi_i\rangle\langle\psi_i|\right)|\phi_n\rangle\right]\left[\langle\phi_n|\left(\sum_j p_j\,|\psi_j\rangle\langle\psi_j|\right)|\phi_m\rangle\right] \\ &= \sum_{i\,j\,m\,n} p_i\,p_j\,\langle\phi_m|\,\psi_i\rangle\,\langle\psi_i|\,\phi_n\rangle\,\langle\phi_n|\,\psi_j\rangle\,\langle\psi_j|\,\phi_m\rangle \\ &= \sum_{i\,j\,m} p_i\,p_j\,\langle\phi_m|\,\psi_i\rangle\left(\sum_n \langle\psi_i|\,\phi_n\rangle\,\langle\phi_n|\,\psi_j\rangle\right)\langle\psi_j|\,\phi_m\rangle.\end{aligned}$$

Assuming that the pure states $|\psi_i\rangle$ in (A.4) are orthonormal, one obtains

$$\begin{aligned}\mathrm{Tr}(\rho^2) &= \sum_{i\,j\,m} p_i\,p_j\,\langle\phi_m|\,\psi_i\rangle\,\delta_{i\,j}\,\langle\psi_j|\,\phi_m\rangle \\ &= \sum_i p_i^2\left(\sum_m \langle\psi_i|\,\phi_m\rangle\,\langle\phi_m|\,\psi_i\rangle\right) \\ &= \sum_i p_i^2\,\langle\psi_i|\,\psi_i\rangle \\ &= \sum_i p_i^2 \le \left(\sum_i p_i\right)^2 \end{aligned}\tag{A.11}$$

in view of the fact that $p_i \ge 0$, always. Thus, relations (A.8) and (A.11) suggest

$$\mathrm{Tr}(\rho^2) \le [\mathrm{Tr}(\rho)]^2 \tag{A.12}$$

provided the states $|\psi_i\rangle$ representing the i-th subsystem in the density operator (A.4) of the mixed system are orthonormal.

A pure ensemble is specified by $p_i = \delta_{i\,i_0}$ for some specific $|\psi_{i_0}\rangle$. The corresponding density operator

$$\rho = |\psi_{i_0}\rangle\langle\psi_{i_0}| \tag{A.13}$$

is readily obtained from (A.4). Clearly, the density operator for a pure ensemble is idempotent, i.e.,

$$\rho^2 = \rho \tag{A.14a}$$

or, equivalently

$$\rho(\rho - 1) = 0. \tag{A.14b}$$

Hence, for a pure ensemble

$$\mathrm{Tr}(\rho^2) = \mathrm{Tr}(\rho) = 1 \tag{A.15}$$

in view of (A.9). Equations (A.12) and (A.15) suggest that $\mathrm{Tr}(\rho^2)$ has its maximum value of one when the ensemble is pure and normalized; for a mixed ensemble, however, it [*i.e.*, $\mathrm{Tr}(\rho^2)$] is a positive number whose value is less than one.

Let us assume that the complete set $\{|\chi\rangle\}$ diagonalizes the density operator (A.13) for a pure ensemble, that is,

$$\langle \chi_m | \rho | \chi_n \rangle = \rho_m \, \delta_{mn} \tag{A.16a}$$

with ρ_m being one of its eigenvalues. Then relation (A.14) gives

$$\langle \chi_m | \rho^2 | \chi_n \rangle = \langle \chi_m | \rho | \chi_n \rangle$$

$$\text{or,} \quad \sum_k \langle \chi_m | \rho | \chi_k \rangle \, \langle \chi_k | \rho | \chi_n \rangle = \rho_m \, \delta_{mn}$$

$$\text{or,} \quad \sum_k \rho_m \, \delta_{mk} \, \rho_k \, \delta_{kn} = \rho_m \, \delta_{mn}$$

$$\text{i.e.,} \quad \rho_m^2 \, \delta_{mn} = \rho_m \, \delta_{mn}$$

$$\text{or,} \quad \rho_m (\rho_m - 1) \delta_{mn} = 0. \tag{A.16b}$$

This suggests that, for $m = n$ (i.e., for diagonal elements), $\rho_m = 0$ or 1. Remembering that $\mathrm{Tr}(\rho) = 1$ for a normalized density matrix and sum of the eigenvalues of a matrix is always equal to its trace, one concludes that, when diagonalized, the normalized density matrix for a pure system has only one non-zero eigenvalue which is always equal to one.[1]

It is obvious from (A.12) and (A.15) that

$$\mathrm{Tr}(\rho^2) = [\mathrm{Tr}(\rho)]^2 = 1 \tag{A.17a}$$

for a pure and normalized system. But for a mixed system

$$0 < \mathrm{Tr}(\rho^2) < [\mathrm{Tr}(\rho)]^2 \,, \quad \text{with} \quad \mathrm{Tr}(\rho^2) < 1. \tag{A.17b}$$

[1]The converse of this statement is proved at the end of (A.29).

Let us suppose that $|\varphi\rangle$ is any arbitrary state. Then the expectation value

$$\begin{aligned}\langle\varphi|\rho|\varphi\rangle &= \sum_i p_i \,\langle\varphi|\,\psi_i\rangle\langle\psi_i|\,\varphi\rangle \\ &= \sum_i p_i \,|\langle\varphi|\,\psi_i\rangle|^2 \geq 0\end{aligned} \tag{A.18}$$

of the density operator (A.4) in this state is a real, non-negative number. An operator whose expectation values are strictly greater than zero for all $|\varphi\rangle \neq 0$, is said to be a positive definite operator; whereas a positive operator can have zero eigenvalues as well. Hence, the density operator ρ—defined in (A.4)—is always, at least, a positive operator.

Different density operators in the form of (A.4), each describing a different mixture of subsystems, can be represented by the same density matrix. Consider, for example [123], a density matrix of the form $\rho = \frac{1}{2}\mathbb{I}_2$, where $\mathbb{I}_2$ is an unit matrix of size (2×2). Some of the possible systems in (A.4) which can be represented by this density matrix are [123]:

1. Mixture of two subsystems

$$|\psi_1\rangle = \begin{pmatrix}1\\0\end{pmatrix}, \; |\psi_2\rangle = \begin{pmatrix}0\\1\end{pmatrix}, \qquad \text{with probabilities:} \quad p_1,\; p_2 = \frac{1}{2}; \tag{A.19a}$$

2. Mixture of three subsystems

$$(a) \quad |\psi_1\rangle = \begin{pmatrix}1\\0\end{pmatrix}, \; |\psi_2\rangle = \frac{1}{2}\begin{pmatrix}1\\\sqrt{3}\end{pmatrix}, \; |\psi_3\rangle = \frac{1}{2}\begin{pmatrix}1\\-\sqrt{3}\end{pmatrix},$$

$$\text{with probabilities:} \quad p_1,\; p_2,\; p_3 = \frac{1}{3}; \tag{A.19b}$$

$$(b) \quad |\psi_1\rangle = \frac{1}{\sqrt{281}}\begin{pmatrix}9\\-i10\sqrt{2}\end{pmatrix}, \; |\psi_2\rangle = \frac{1}{\sqrt{194}}\begin{pmatrix}12\\-i5\sqrt{2}\end{pmatrix},$$

$$|\psi_3\rangle = \frac{1}{\sqrt{17}}\begin{pmatrix}-i3\\2\sqrt{2}\end{pmatrix},$$

$$\text{with probabilities:} \quad p_1 = \frac{281}{900},\; p_2 = \frac{97}{450},\; p_3 = \frac{17}{36}; \tag{A.19c}$$

etc. Here, each ket represents a column-matrix of size (2×1) (*i.e.*, a rectangular matrix with two rows and one column). One may construct an innumerable number of such examples. In the present example, however, all $|\psi_i\rangle$'s in (A.4), will always have only one column but as many rows as those occurring in the corresponding density matrix ρ.

A.3 Reduced Density Operator and Density Matrix

The concept of reduced density operator and of reduced density matrix is helpful in the analysis of composite quantum systems.

Let us consider two subsystems A and B described jointly by the density operator ρ^{AB}. A complete orthonormal basis set of size d_A for A is $\{|a_i\rangle, i = 1, 2, \ldots, d_A\}$ and that of size d_B for B is $\{|b_j\rangle, j = 1, 2, \ldots, d_B\}$. Then the basis set

$$|\psi_{ij}\rangle \equiv |a_i\rangle \otimes |b_j\rangle \equiv |a_i\, b_j\rangle \equiv |i\ j\rangle \tag{A.20}$$

for the composite system contains $d_A d_B$ members. A typical element of the $(d_A d_B \times d_A d_B)$ density matrix is now given by

$$\langle\psi_{ij}|\rho^{AB}|\psi_{i'j'}\rangle \equiv \langle ij|\rho^{AB}|i'j'\rangle. \tag{A.21}$$

Here, the first set of indices (i, i') refers to the state of the subsystem A and the second set (j, j') to that of the subsystem B. Suppose, an operator Ω_A acts only on the subsystem A. Then a typical matrix element of Ω_A in the basis (A.20) is given by

$$\begin{aligned}\langle\psi_{ij}|\Omega_A|\psi_{i'j'}\rangle &\equiv \langle ij|\Omega_A|i'j'\rangle\\ &\equiv \langle i|\Omega_A|i'\rangle\,\langle j|\,j'\rangle\\ &\equiv \langle i|\Omega_A|i'\rangle\,\delta_{jj'}.\end{aligned} \tag{A.22}$$

Let us calculate the trace of $\rho^{AB}\Omega_A$ in the basis set (A.20):

$$\begin{aligned}\mathrm{Tr}\left(\rho^{AB}\Omega_A\right) &= \sum_{i\,j} \langle\psi_{ij}|\rho^{AB}\Omega_A|\psi_{ij}\rangle\\ &= \sum_{i\,j\,i'\,j'} \langle\psi_{ij}|\rho^{AB}|\psi_{i'j'}\rangle\,\langle\psi_{i'j'}|\Omega_A|\psi_{ij}\rangle\\ &= \sum_{i\,j\,i'\,j'} \langle ij|\rho^{AB}|i'j'\rangle\,\langle i'|\Omega_A|i\rangle\,\delta_{j\,j'}\\ &= \sum_{i\,i'\,j} \langle ij|\rho^{AB}|i'j\rangle\,\langle i'|\Omega_A|i\rangle\end{aligned} \tag{A.23}$$

Here, we have used (A.22). Let us now define the reduced density matrix

$$\langle i|\rho^A|i'\rangle = \sum_j \langle ij|\rho^{AB}|i'j\rangle. \tag{A.24}$$

This definition facilitates rewriting of (A.23) in the following form

$$\begin{aligned}\mathrm{Tr}\left(\rho^{AB}\Omega_A\right) &= \sum_{i\,i'} \langle i|\rho^A|i'\rangle\,\langle i'|\Omega_A|i\rangle \\ &= \sum_{i} \langle i|\rho^A\Omega_A|i\rangle \\ &= \mathrm{Tr}\left(\rho^A\Omega_A\right) = [\Omega_A] \end{aligned} \tag{A.25}$$

Thus, according to the last of (A.25), the average value of the one-particle operator Ω_A may be obtained with the reduced density operator ρ^A [see, (A.26)] whose $(i,\ i')$-th matrix elements (A.24) are derivable from those of the density operator ρ^{AB} defined for the composite quantum system formed from the subsystems A and B. The important distinction between ρ^{AB} and ρ^A is that while the former refers to the composite system as a whole, the latter refers to a component (namely, subsystem A) of the composite quantum system. The operator Ω_A operates only on this part of the whole system. Hence, one can write

$$\rho^A \equiv \mathrm{Tr}_B\left(\rho^{AB}\right) \tag{A.26}$$

with its $(i,\ i')$-th matrix element given by (A.24). Also,

$$\rho^B \equiv \mathrm{Tr}_A\left(\rho^{AB}\right), \tag{A.27a}$$

with

$$\langle j|\rho^B|j'\rangle = \sum_i \langle i\,j|\rho^{AB}|i\,j'\rangle. \tag{A.27b}$$

Here, ρ^A in (A.26) and ρ^B in (A.27) are the reduced density operators for the respective subsystems A and B constituting the composite system $A \otimes B$ whose density operator is ρ^{AB}. The operators ρ^A and ρ^B are, of course, defined over the respective bases sets $\{|a_i\rangle\}$ and $\{|b_j\rangle\}$. Further, it is obvious from expressions [(A.24), (A.27b)] that [Tr_B, Tr_A] in [(A.26), (A.27a)] denote, respectively, the sum of the diagonal elements—associated with the respective subsystems [B, A]—of the density matrix of the composite system ρ^{AB}. This, in other words, means that if one is interested, for example, in the development of the subsystem A only, the pertinent density operator is ρ^A, which has no reference to the indices associated with its companion subsystem B. In this way, unwanted indices are eliminated. Further, dimensions of the matrices representing the reduced density operators [ρ^A, ρ^B] are [$d_A \times d_A$, $d_B \times d_B$].

A.4 Miscellaneous

(1) Matrix representations of the density matrix (A.4) will obviously be different in different orthonormal basis. If λ_i's are the eigenvalues of the density operator in (A.4) in the orthonormal basis $\{|i\rangle,\ i = 1, 2, \ldots\}$, then a spectral representation of ρ for the present basis is given by

$$\rho = \sum_i \lambda_i \ |i\rangle\langle i| \tag{A.28a}$$

and, hence,

$$\rho^2 = \sum_i \lambda_i^2 \ |i\rangle\langle i|. \tag{A.28b}$$

It is obvious from these relations that

$$\mathrm{Tr}\,(\rho) = \sum\nolimits_i \lambda_i = 1, \tag{A.29a}$$

where the last result holds for any normalized density matrix [or, from (A.9)]; and

$$\mathrm{Tr}\,(\rho^2) = \sum\nolimits_i \lambda_i^2. \tag{A.29b}$$

Equations (A.29) reinforces the statements of (A.12)–(A.15). Equation (A.29a) further shows that if only one, say m-th, of the eigenvalues is non-zero with $\lambda_m = 1$, then (A.28a) reduces to $\rho = |m\rangle\langle m|$, i.e., the system represented by the density operator ρ is in a pure state $|m\rangle$.

(2) A state of a system in quantum mechanics can be represented in more than one equivalent ways, e.g., by a state vector, wave function, density operator, or by a density matrix. For example, in view of the discussions presented in this appendix, the density operator and density matrix for first of the four Bell states (2.17) are

$$\rho_{\psi^+} \equiv \begin{cases} |\psi^+\rangle\langle\psi^+| = \frac{1}{2}\Big(|01\rangle\langle 01| + |01\rangle\langle 10| + |10\rangle\langle 01| + |10\rangle\langle 10|\Big) \quad \text{and} \\ \frac{1}{2}\begin{pmatrix} 0\,0\,0\,0 \\ 0\,1\,1\,0 \\ 0\,1\,1\,0 \\ 0\,0\,0\,0 \end{pmatrix}. \end{cases} \tag{A.30a}$$

respectively. Here, while $(|0\rangle, |1\rangle)$ are the bases for each of the two qubits in the state $|\psi^+\rangle$; but, the matrix representation of ρ_{ψ^+} is in the basis $(|00\rangle, |01\rangle, |10\rangle, |11\rangle)$. One can, however, calculate the matrix representation of ρ_{ψ^+} in the bases taken in $(|11\rangle, |10\rangle, |01\rangle, |00\rangle)$, or in some other order as well. Similarly, one obtains

$$\rho_{\psi^-} \equiv \begin{cases} |\psi^-\rangle\langle\psi^-| = \frac{1}{2}\Big(|01\rangle\langle 01| - |01\rangle\langle 10| - |10\rangle\langle 01| + |10\rangle\langle 10|\Big) \\ \frac{1}{2}\begin{pmatrix} 0 & 0 & 0 & 0 \\ 0 & 1 & -1 & 0 \\ 0 & -1 & 1 & 0 \\ 0 & 0 & 0 & 0 \end{pmatrix}, \end{cases} \tag{A.30b}$$

$$\rho_{\phi^+} \equiv \begin{cases} |\phi^+\rangle\langle\phi^+| = \frac{1}{2}\Big(|00\rangle\langle 00| + |00\rangle\langle 11| + |11\rangle\langle 00| + |11\rangle\langle 11|\Big) \\ \frac{1}{2}\begin{pmatrix} 1 & 0 & 0 & 1 \\ 0 & 0 & 0 & 0 \\ 0 & 0 & 0 & 0 \\ 1 & 0 & 0 & 1 \end{pmatrix}, \end{cases} \tag{A.30c}$$

and

$$\rho_{\phi^-} \equiv \begin{cases} |\phi^-\rangle\langle\phi^-| = \frac{1}{2}\Big(|00\rangle\langle 00| - |00\rangle\langle 11| - |11\rangle\langle 00| + |11\rangle\langle 11|,\Big) \\ \frac{1}{2}\begin{pmatrix} 1 & 0 & 0 & -1 \\ 0 & 0 & 0 & 0 \\ 0 & 0 & 0 & 0 \\ -1 & 0 & 0 & 1 \end{pmatrix}. \end{cases} \tag{A.30d}$$

for the remaining three Bell states (2.17b) and (2.17c).

(3) Suppose we have a composite system consisting of n parts such that $\Omega^{(u)}$ represents an operator for the u-th part of this system. If $\Omega^{(u)}$ is defined in the Hilbert space H_u of dimensions d_u, then operator

$$\Omega \equiv \sum_{u=1}^{n} \Omega^{(u)} \tag{A.31a}$$

for this composite system of n parts is given by

$$\begin{aligned}\Omega = \Omega^{(1)} \otimes I^{(2)} \otimes I^{(3)} \otimes \cdots \otimes I^{(n-1)} \otimes I^{(n)} +\\ I^{(1)} \otimes \Omega^{(2)} \otimes I^{(3)} \otimes \cdots \otimes I^{(n-1)} \otimes I^{(n)} +\\ \cdots\cdots\cdots \\ I^{(1)} \otimes I^{(2)} \otimes I^{(3)} \otimes \cdots \otimes \Omega^{(n-1)} \otimes I^{(n)} +\\ I^{(1)} \otimes I^{(2)} \otimes I^{(3)} \otimes \cdots \otimes I^{(n-1)} \otimes \Omega^{(n)}.\end{aligned} \tag{A.31b}$$

Here, $I^{(u)}$ is a unit operator defined in the Hilbert space of the u-th part of the composite system. Operator Ω in (A.31a) is defined in the Hilbert space

$$H \equiv H_1 \otimes H_2 \otimes \cdots \otimes H_n \tag{A.32a}$$

of size

$$d \equiv d_1 \, d_2 \, d_3 \cdots d_n. \tag{A.32b}$$

Thus, the u-th term $I^{(1)} \otimes I^{(2)} \otimes I^{(3)} \otimes \cdots \otimes I^{(u-1)} \otimes \Omega^{(u)} \otimes I^{(u+1)} \otimes \cdots \otimes I^{(n)}$ in (A.31b) represents $\Omega^{(u)}$ in the $(d \times d)$-dimensional Hilbert space (A.32) of the composite system Ω defined by (A.31a).

For example, (2.6) and (2.8) contain expressions for the three Cartesian components of the Pauli spin operator [58] in the matrix and Dirac notations, respectively. After multiplying the right-hand side of each of the two equations by $\frac{1}{2}$, these can be looked upon also as representations of the three Cartesian components of an operator for angular momentum $\frac{1}{2}$ in the $(|0\rangle, |1\rangle)$ bases. According to (A.31) and (A.32), the angular momentum operator for a composite system of two spin-$\frac{1}{2}$ particles is then

$$\boldsymbol{\sigma} \equiv \frac{1}{2}\left(\boldsymbol{\sigma}^{(1)} + \boldsymbol{\sigma}^{(2)}\right) = \frac{1}{2}\left(\boldsymbol{\sigma}^{(1)} \otimes I^{(2)} + I^{(1)} \otimes \boldsymbol{\sigma}^{(2)}\right), \tag{A.33}$$

where each of $I^{(1)}$ and $I^{(2)}$ is an unit operator defined in the (2×2)-dimensional Hilbert space of the first and the second spin-$\frac{1}{2}$ particles, respectively. Using (2.6), one now obtains

$$\sigma_x = \frac{1}{2}\begin{cases}\left(\sigma_x^{(1)} \otimes I^{(2)} + I^{(1)} \otimes \sigma_x^{(2)}\right), \\ \Big(|1\rangle\langle 0| + |0\rangle\langle 1|\Big)^{(1)} \otimes \Big(|0\rangle\langle 0| + |1\rangle\langle 1|\Big)^{(2)} + \\ \Big(|0\rangle\langle 0| + |1\rangle\langle 1|\Big)^{(1)} \otimes \Big(|1\rangle\langle 0| + |0\rangle\langle 1|\Big)^{(2)}, \\ \Big(|00\rangle + |11\rangle\Big)\Big(\langle 01| + \langle 10|\Big) + \Big(|10\rangle + |01\rangle\Big)\Big(\langle 00| + \langle 11|\Big), \\ \begin{pmatrix} 0&1&1&0 \\ 1&0&0&1 \\ 1&0&0&1 \\ 0&1&1&0 \end{pmatrix},\end{cases} \tag{A.34a}$$

$$\sigma_y = \frac{1}{2}\begin{cases} \left(\sigma_y^{(1)} \otimes I^{(2)} + I^{(1)} \otimes \sigma_y^{(2)}\right), \\ -i\left(|00\rangle - |11\rangle\right)\left(\langle 01| + \langle 10|\right) + i\left(|10\rangle + |01\rangle\right)\left(\langle 00| - \langle 11|\right), \\ \begin{pmatrix} 0 & -i & -i & 0 \\ i & 0 & 0 & -i \\ i & 0 & 0 & -i \\ 0 & i & i & 0 \end{pmatrix}, \end{cases} \tag{A.34b}$$

and

$$\sigma_z = \begin{cases} \frac{1}{2}\left(\sigma_z^{(1)} \otimes I^{(2)} + I^{(1)} \otimes \sigma_z^{(2)}\right), \\ |00\rangle\langle 00| - |11\rangle\langle 11|, \\ \begin{pmatrix} 1 & 0 & 0 & 0 \\ 0 & 0 & 0 & 0 \\ 0 & 0 & 0 & 0 \\ 0 & 0 & 0 & -1 \end{pmatrix} \end{cases} \tag{A.34c}$$

for the matrix representations [in the $(|00\rangle, |01\rangle, |10\rangle, |11\rangle)$ bases] of the three Cartesian components of the angular momentum operator $\boldsymbol{\sigma}$ in (A.33) for a composite system of two spin-$\frac{1}{2}$ particles. Furthermore, two times of each of the three operators in (A.34) are applicable to any bipartite system of two qubits as well.

Appendix B
Density Matrix and State Multipoles for a Photon Beam

In the theoretical considerations developed in the Chap. 3 and thereafter, the ionizing radiation γ_r has always been assumed to be in a definite state of its polarization (i.e., LP, RCP, LCP, or UP). However, in many experiments, the incident electromagnetic wave may be either partially or arbitrarily polarized. A light beam of any kind of polarization is properly described by three Stokes' parameters [61, 76]. The present appendix contains such description of γ_r used in (1.1)–(1.8) and shows as to how it can easily be introduced in the density matrices derived in this monograph to study entanglement properties of states of qubits generated in these processes when ionizing radiation γ_r has an arbitrary polarization. This appendix also gives the description of a photon beam in terms of its state multipoles (or statistical tensors) [61,68,173,355]. The following description is applicable when the incident radiation is being treated in the $E1$ approximation.

Let us represent by $\langle 1m_r|\, \rho_r\, |1m_r'\rangle$ an element of the density operator ρ_r, introduced in (3.3) for describing the ionizing radiation γ_r in the spherical bases states $|10\rangle$, $|1,+1\rangle$, and $|1\text{-}1\rangle$ defined on page 7 in Chap. 1. In order to describe an uncorrelated system of (arbitrarily polarized photon + polarized target), before the absorption of γ_r by $\mathfrak{T}$ takes place, expression (3.4b) for the density operator everywhere in this monograph be replaced by

$$\rho_i = \sum_{m_r\, m_r'} |0;\, 1m_r\rangle\langle 1m_r|\, \rho_r\, |1m_r'\rangle\langle 0;\, 1m_r'|. \tag{B.1a}$$

If, on the other hand, the target $\mathfrak{T}$ too is unpolarized before absorbing the photon, then the density operator

$$\rho_i = \frac{1}{n_{d_0}} \sum_{d_0\, m_r\, m_r'} |0;\, 1m_r\rangle\langle 1m_r|\, \rho_r\, |1m_r'\rangle\langle 0;\, 1m_r'|, \tag{B.1b}$$

N. Chandra and R. Ghosh, *Quantum Entanglement in Electron Optics*, Springer Series on Atomic, Optical, and Plasma Physics 67, DOI 10.1007/978-3-642-24070-6,

in place of (3.4c), represents the uncorrelated (arbitrarily polarized γ_r + unpolarized $\mathfrak{T}$)-system.

On using in (B.1a) and (B.1b) the density operator (3.3c) for a photon in a state $|1m_r\rangle$ of definite polarization, one reproduces the respective expressions (3.4b) and (3.4c) for the density operator ρ_i. However, for an electromagnetic wave with arbitrary polarization, the photon density matrix ρ_r is well represented in terms of the Stokes' parameters (I_r, η_1, η_2, η_3) by [61, 173, 355]

$$\begin{array}{ccc} |+1\rangle & |0\rangle & |-1\rangle \end{array}$$

$$\rho_r = \frac{1}{2} I_r \begin{pmatrix} 1+\eta_2 & 0 & -\eta_3 + i\eta_1 \\ 0 & 0 & 0 \\ -\eta_3 - i\eta_1 & 0 & 1-\eta_2 \end{pmatrix} \tag{B.2}$$

In this equation, ρ_r has been calculated in the ($|1+1\rangle$, $|10\rangle$, $|1\text{-}1\rangle$) basis. Further in (B.2), I_r is the total intensity of the arbitrarily polarized radiation incident on the target $\mathfrak{T}$ along the OZ-axis of our SF. If $I_r(\beta_r)$ is the intensity of this same beam transmitted by a Nicol prism oriented at an angle β_r with respect to the OX-axis of the SF, the remaining three Stokes' parameters are given by [61, 173, 355]

$$\eta_1 = \frac{1}{I_r}\Big[I_r(\pi/4) - I_r(3\pi/4)\Big], \tag{B.3a}$$

$$\eta_2 = \frac{1}{I_r}\Big[I_r(+) - I_r(-)\Big], \tag{B.3b}$$

$$\eta_3 = \frac{1}{I_r}\Big[I_r(0) - I_r(\pi/2)\Big]. \tag{B.3c}$$

In (B.3), $I_r(+)$ and $I_r(-)$ are the intensities of the electromagnetic wave (of total intensity I_r) transmitted along the OZ-axis by the polarization filters, which fully transmit only photons of positive and negative helicities, respectively. Hence, for an incident electromagnetic wave

$$\left.\begin{array}{rl} \text{LP along the } OX\text{-axis:} & \eta_1,\ \eta_2 = 0, \eta_3 = +1; \\ \text{LP along the } OY\text{-axis:} & \eta_1,\ \eta_2 = 0, \eta_3 = -1; \\ \text{RCP (i.e., positive helicity):} & \eta_1 = 0,\ \eta_2 = +1,\ \eta_3 = 0; \\ \text{LCP (i.e., negative helicity):} & \eta_1 = 0,\ \eta_2 = -1,\ \eta_3 = 0. \end{array}\right\} \tag{B.4}$$

The density operator for a photon beam can be expanded in terms of the tensor operators $T(1)_{KQ}$ as follows [61, 68, 173, 355]

$$\rho_r = \sum_{KQ} \left\langle T(1)^{\dagger}_{kQ} \right\rangle T(1)_{KQ} \tag{B.5a}$$

with state multipoles (or statistical tensors) given by

$$\left\langle T(1)^{\dagger}_{kQ}\right\rangle = \sqrt{2K+1}\sum_{m_r\, m_r'}(-1)^{1-m_r}$$

$$\begin{pmatrix} 1 & 1 & K \\ m_r & -m_r' & -Q \end{pmatrix}\langle 1m_r|\rho_r|1m_r'\rangle. \tag{B.5b}$$

The state multipoles for a beam of electromagnetic radiation in a pure state of polarization are readily obtained [172] by substituting the corresponding density matrix (3.3c) in (B.5b). However, in order to calculate state multipoles for an UP electromagnetic wave, one needs to use [172]

$$\rho_r = \frac{1}{2}\Big(|1+1\rangle\langle 1+1| + |1-1\rangle\langle 1-1|\Big), \tag{B.6}$$

in place of (3.3c), in (B.5b). Table B.1 [172] contains state multipoles for an LP photon beam with its electric field vector, or for RCP, LCP, or UP radiation propagating, along the OZ-axis of the SF. These results are same as those given in [172, 173].

Table B.2 [172], on the other hand, contains state multipoles for an arbitrarily polarized electromagnetic wave propagating along the OZ-axis. Entries in this table are identical to those calculated by Kleinman et al. [355].

Furthermore [172], state multipoles (or statistical tensors) given in Table B.1 for RCP, LCP, or UP γ_r propagating along the OZ-axis can also be gotten by substituting the appropriate values (B.4) of the Stokes' parameters in Table B.2. This procedure, however, cannot give us state multipoles in Table B.1 for radiation LP along the OZ-axis (i.e., for $m_r = 0$). For, both Stokes' parameters in (B.4) and

Table B.1 State multipoles for an electromagnetic wave in a pure polarization state.

Polarization of the electromagnetic wave	Monopole moment ($K = 0$)	Orientation vector[a] ($K = 1$)	Alignment tensor[b] ($K = 2$)
LP ($m_r = 0$)[c]	$\left\langle T(1)^{\dagger}_{00}\right\rangle = \frac{1}{\sqrt{3}}$	$\left\langle T(1)^{\dagger}_{10}\right\rangle = 0$	$\left\langle T(1)^{\dagger}_{20}\right\rangle = -\sqrt{\frac{2}{3}}$
RCP ($m_r = +1$)[c]	$\left\langle T(1)^{\dagger}_{00}\right\rangle = \frac{1}{\sqrt{3}}$	$\left\langle T(1)^{\dagger}_{10}\right\rangle = \frac{1}{\sqrt{2}}$	$\left\langle T(1)^{\dagger}_{20}\right\rangle = \frac{1}{\sqrt{6}}$
LCP ($m_r = -1$)[c]	$\left\langle T(1)^{\dagger}_{00}\right\rangle = \frac{1}{\sqrt{3}}$	$\left\langle T(1)^{\dagger}_{10}\right\rangle = -\frac{1}{\sqrt{2}}$	$\left\langle T(1)^{\dagger}_{20}\right\rangle = \frac{1}{\sqrt{6}}$
UP[d]	$\left\langle T(1)^{\dagger}_{00}\right\rangle = \frac{1}{\sqrt{3}}$	$\left\langle T(1)^{\dagger}_{10}\right\rangle = 0$	$\left\langle T(1)^{\dagger}_{20}\right\rangle = \frac{1}{\sqrt{6}}$

[a] Its each of the remaining two components $\left[ie, \left\langle T(1)^{\dagger}_{1\,\pm 1}\right\rangle\right]$, not given in this table, are zero

[b] Its each of the remaining four components $\left[ie, \left\langle T(1)^{\dagger}_{2\,\pm 1}\right\rangle, \left\langle T(1)^{\dagger}_{2\,\pm 2}\right\rangle\right]$, not given in this table, are zero

[c] The density operator is given by (3.3c)

[d] The density operator is given by (B.6)

Table B.2 State multipoles for an electromagnetic wave of arbitrary polarization.

Monopole moment ($K=0$)	Orientation vector ($K=1$)	Alignment tensor ($K=2$)
$\left\langle T(1)^{\dagger}_{00}\right\rangle = \frac{1}{\sqrt{3}} I_r$	$\left\langle T(1)^{\dagger}_{10}\right\rangle = \frac{1}{\sqrt{2}} I_r \eta_2$	$\left\langle T(1)^{\dagger}_{20}\right\rangle = \frac{1}{\sqrt{6}} I_r$
	$\left\langle T(1)^{\dagger}_{1\pm1}\right\rangle = 0$	$\left\langle T(1)^{\dagger}_{2\pm1}\right\rangle = 0$
		$\left\langle T(1)^{\dagger}_{2\pm2}\right\rangle = \frac{1}{2}\Big(- \eta_3 \pm i\eta_1\Big) I_r$

state multipoles in the Table B.2 are for electromagnetic wave LP along the OX- or OY-axis.

The inverse of (B.5b) is

$$\langle 1m_r|\rho_r|1m_r'\rangle = (-1)^{1-m_r} \sum_{KQ} \sqrt{2K+1} \times \begin{pmatrix} 1 & 1 & K \\ m_r & -m_r' & -Q \end{pmatrix} \left\langle T(1)^{\dagger}_{kQ}\right\rangle. \tag{B.7}$$

On substituting (B.7) for the matrix elements of the photon density operator ρ_r, one readily expresses the density operator ρ_i in (B.1) in terms of the state multipoles of the arbitrarily polarized ionizing photon γ_r. If one is interested in studying entanglement between the particles produced in any of the processes (1.1)–(1.8), wherein an inner-shell vacancy is created due to the absorption of an arbitrarily polarized γ_r, the consequent expression for ρ_i should be used in deriving the corresponding density matrix. However, this change will not at all affect the Coulombic entanglement discussed in this monograph, The entanglement generated in the absence of SDIs is always independent of the polarization of the ionizing radiation. Use of [(B.1), (B.7)] should, on the other hand, certainly have an effect on the fine-structure entanglement because it very much involves polarization of the γ_r used in (1.1)–(1.8) on account of the fact that the presence of the SDIs couples the non-local correlation with both the kinematics and dynamics of these processes.

Appendix C
Decoherence and Dissipation of Quantum Entanglement

This appendix contains a very brief, elementary and qualitative description of decoherence and dissipation of entanglement in a state of a system due to its interaction with ambient external environment.

The first time a student comes across the word coherence in physics is, probably, in the context of electromagnetic waves when phenomena like interference (in, e.g., young's double slit experiment, etc.) and/or Fraunhofer and Fresnel diffractions are introduced to him/her in optics [356]. Two electromagnetic waves are said to be coherent if the phase difference between them does not change with time; otherwise, they are considered to be incoherent waves. Electromagnetic radiations produced in the spontaneous decay of excited states of atoms or molecules in the solid, liquid, or gas phase of matter are known to be incoherent; whereas, laser or maser generated in induced (i.e., stimulated) decay of an excited state are the well known examples of coherent electromagnetic waves. Another popular example is a coherent state in quantum optics [25,27].

Superpositions of two or more electromagnetic waves, or of states in quantum mechanics, can also be of two kinds: Namely (see, for example, [76, 356]), coherent or incoherent superpositions. A coherent superposition consists of addition of complex amplitudes with appropriate multiplying constants which may even be complex. In an incoherent superposition, on the other hand, intensities (i.e., square of the modulus of amplitudes) are added with appropriate weights. Coherent superposition generates interference of amplitudes which gives rise, for example, to a stable pattern of maxima and minima in the above-mentioned phenomena of interference and diffraction in optics, provided [76,356] superposed waves are also coherent. Such interference terms can, of course, never be present in an incoherent superposition.

Both, time-dependent and time-independent Schrödinger equations in quantum mechanics are linear as well as homogeneous (see, for example, [58,59]). These two properties mean that a linear combination (i.e., a superposition) of their solutions

N. Chandra and R. Ghosh, *Quantum Entanglement in Electron Optics*, Springer Series on Atomic, Optical, and Plasma Physics 67, DOI 10.1007/978-3-642-24070-6,

is also a solution of the respective Schrödinger equation. Of the two kinds of superpositions mentioned in the preceding paragraph, separable states like (2.27), (2.39a), etc. represent an incoherent superposition of the product states. Similarly, (A.4) too is an example of an incoherent superposition. On the other hand, the simplest possible coherent superposition in quantum mechanics is, probably, the one-particle state (2.1). This state gives rise to terms whose coefficients are $C_0C_1^*$ and $C_0^*C_1$, in addition to those multiplied by $|C_0|^2$ and $|C_1|^2$, when probability for finding a result on measuring an observable of the system represented by the state (2.1) is to be calculated. The former two (i.e., $C_0C_1^*$ and $C_0^*C_1$) are known as interference terms representing local coherence inherently present in any linear superposition [e.g., like (2.1)] of one (or more)-particle states. The incoherent state (2.27), on the other hand, is a statistical mixture, which can never give rise to any interference terms similar to those which may be generated by a coherent superposition.

An entangled state is also a coherent superposition of the product states of more than one particle. For example, Bell states (2.17) are four different coherent superpositions of the products ($|00\rangle$, $|01\rangle$, $|10\rangle$, $|11\rangle$) of the states of two qubits (e.g., photons, two-level atoms, spin-$\frac{1}{2}$ particles). Similar is the case with the bipartite states [e.g., (4.13), (5.14), (6.10b), etc.] discussed in various chapters of this monograph. In an entangled state, correlation is present among its constituent particles [i.e., between two qubits in (2.17)] as well. These particles may or may not be interacting among themselves and may be located at places far away from each other. Such a non-local correlation is also called distributed coherence. Thus, both local and non-local coherences are present in an entangled state. The former exists for each particle constituting the system; while, the non-separability of the state of this system is due to the non-local correlation/distributed coherence. It has already been mentioned at several places in this monograph that non-diagonal elements of a density matrix represents coherence effects. Thus, the non-diagonal elements in the matrix representation of an entangled state contain both local and non-local coherence. It is, nevertheless, possible that local coherence may be present also in a separable state. (For a more detailed discussion of local and non-local coherence in a bipartite entangled state of two qubits, see, for example, [53].)

Hence, in a density matrix form [e.g., (A.30), *etc*] of a linear superposition [e.g., (2.17), *etc*], the non-diagonal terms arise due to coherence (or, interference); whereas diagonal terms represent probabilities for the outcome of a measurement of an observable performed on a coherent superposition of states. Sometimes, depending upon the observable being measured, the diagonal elements are called populations as well. It is obvious from the discussion given herein that coherence in states like (2.1), (2.17), *etc*, is a consequence of the phases of various terms present in a superposition; whereas these phases play no role whatsoever in determining the populations represented by these states.

Coherence is considered to be a major resource in quantum mechanics. Chapter 2 explains, in brief, the importance of entangled state, i.e., the availability of non-local coherence, in quantum information. However, due to the presence of noisy backgrounds and interaction with the ambient external environment, an entangled state always tends to lose its both local and non-local coherences. Such loss of coherence for reasons external to a system is known as decoherence (see, for example, [357–361]). Thus, decoherence affects the phase relationship among the terms present in a coherent superposition of states. This, in other words, means that value of the exponent ϕ may be affected by decoherence of a single qubit state (2.2b) if it is subjected to a noisy environment. Hence, decoherence is the decay of the off-diagonal matrix elements of the density matrix of a system due to its coupling to an ambient external environment. There have recently been several studies of decoherence, including dynamics of dis-entanglement (see, for examples, reviews [160, 348, 351]).

On the other hand, changes in the diagonal elements (i.e., populations) in a density matrix representation of a coherent superposition of a state of a system due to its interaction with the external ambient environment is known as dissipation. Thus, in dissipation, magnitudes—rather than phases—of the various terms present in a coherent superposition change if a system represented by this state is experiencing a noisy external background. References [180–182], *etc*, are some of the studies of both decoherence and dissipation in quantum information.

Appendix D
A Computer Programme for Calculating 3-j, 6-j, and 9-j Symbols in Powers of Prime Numbers

It is obvious from discussions given in Chaps. 3–10 of this monograph that calculations of the density matrices needed for studying entanglement properties of various bipartite states of flying particles generated in the processes (1.1)–(1.8) capable of taking place in an atom or a linear molecule require a large number of 3-j, 6-j, and/or 9-j symbols [65]. Most of the books on quantum mechanics (e.g., [58, 59]), angular momentum (e.g., [65–68, 187]), atomic physics (e.g., [10, 60, 184]), nuclear physics (e.g., [64]), for example, contain tables of 3-j, 6-j, and/or 9-j symbols in one form or the other—specifically, for smaller values of the angular momentum quantum numbers. Among these, and several others, the excellent book by Rotenberg et al. [187] is exclusively dedicated to the tables of 3-j and 6-j symbols. In addition, a number of different computer programmes for calculating all or some of these n-j (i.e., 3-j, 6-j, and/or 9-j) symbols are also readily available in the literature (see, for example, [68]). One may, then, wonder as to what is the need for writing a new computer code for calculating these n-j symbols and then including it in the present monograph?

First, the book cited in [187] was published more than fifty years ago. It has been out of print since long and, consequently, it is presently available only at few places (e.g., libraries, universities, etc.) in the world. Moreover, Rotenberg et al. [187] have given tables only for 3-j and 6-j symbols, but none for 9-j symbols. The expressions for the density matrices derived herein and elsewhere (e.g., [56, 57]) contain, on the other hand, also the 9-j symbols, in addition to the 3-j and 6-j symbols. To the best of our knowledge, there are no computer codes for calculating an n-j symbol (or, square of it, for that matter) in powers of prime numbers. Almost all of the available programmes compute values of these symbols in decimal forms which results in the considerable loss of accuracy. Moreover, in an analytical calculation, it is far more convenient to use a value of an n-j symbol in terms of prime numbers than in a decimal form. It is for these and other such

N. Chandra and R. Ghosh, *Quantum Entanglement in Electron Optics*, Springer Series on Atomic, Optical, and Plasma Physics 67, DOI 10.1007/978-3-642-24070-6,

reasons that computer programmes which calculate square of a n-j symbol in terms of powers of prime numbers have been written in Chandra's [362] group.

Although this [362] computer programme can not be claimed to be either elegant or optimum, its accuracy has, however, been checked by various methods. For example, results obtained from this programme for a 3-j symbol and a 6-j symbol agree with those given in [67,187].[1] A 9-j symbol, on the other hand was calculated using two independent expressions (D.4) and (D.5) given on pages 275 and 276 herein. These two methods gave identical results which, in addition, agreed also with their values in [67]. This programme for n-j symbols was subsequently used for calculating density matrices needed for studying both Coulombic entanglement and fine-structure entanglement.

D.1 Methods of Calculation

A 3-j symbol has been computed after suitably modifying the expression originally derived by Racah [363] for Clebsch–Gordon coefficients and written in a symmetrical form by Edmonds (see (3.6.11) and (3.7.3) in [65]). Accordingly, one can write [65, 187]

$$\begin{pmatrix} A & B & C \\ a & b & c \end{pmatrix} = (-1)^{A-B-c}\,(2C+1)^{-\frac{1}{2}}\Big(Aa\;Bb|AB\,C-c\Big) \tag{D.1a}$$

$$\begin{aligned} &= (-1)^{A-B-c}\,\Delta\big(A,B,C\big)\,\delta_{0,a+b+c} \\ &\times \sqrt{(A+a)!(A-a)!(B+b)!(B-b)!(C+c)!(C-c)!} \\ &\times \Sigma_{k\geq 0}\,(-1)^k\Big[k!(A+B-C-k)!(A-a-k)!(B+b-k)! \\ &\times (C-B+a+k)!(C-A-b+k)!\Big]^{-1}. \end{aligned} \tag{D.1b}$$

Here,

$$\Delta\big(A,B,C\big) = \left[\frac{(A+B-C)!(A-B+C)!(-A+B+C)!}{(A+B+C+1)!}\right]^{\frac{1}{2}}. \tag{D.2}$$

In (D.1b), one needs to sum over all those positive integral values of k for which arguments of none of the factorials functions present in the denominator of this

[1] All the tables given by Rotenberg et al. [187] contain values of the squares of 3-j and of 6-j symbols in prime numbers; whereas, those given in [67] have n-j symbols in both prime numbers and in decimals forms.

expression becomes negative [see, for example, footnote (3) on page 277 in the present Appendix D].

A 6-j symbol, used in this monograph, is related to Racah's W-coefficient [363] by [67, 187]

$$\begin{Bmatrix} A & B & C \\ D & E & F \end{Bmatrix} = (-1)^{A+B+D+E}\, W\big(ABED; CF\big). \tag{D.3a}$$

This can finally be written as [65, 67, 68]

$$\begin{aligned} \begin{Bmatrix} A & B & C \\ D & E & F \end{Bmatrix} &= \Delta(ABC)\,\Delta(AEF)\,\Delta(DBF)\,\Delta(DEC) \\ &\times \Sigma_{k\geq 0}\,(-1)^k\,(k+1)!\Big[(k-A-B-C)!(k-A-E-F)!(k-D-B-F)! \\ &\times (k-D-E-C)!(A+B+D+E-k)!(B+C+E+F-k)! \\ &\times (C+A+F+D-k)!\Big]^{-1}. \end{aligned} \tag{D.3b}$$

Similar to (D.1b), the upper limit for the positive integer k in the sum (D.3b) is determined so that none of the factorials in this expression is negative.

There are various equivalent expressions (see, for example, [64, 65, 67, 68, 364]) available in the literature for computing a 9-j symbol. It can be written as a sum of the product of six 3-j symbols, sum of a triple product of 6-j symbols, or in the form of an algebraic expression. The two expressions used herein are

$$\begin{aligned} \begin{Bmatrix} A & B & C \\ D & E & F \\ G & H & J \end{Bmatrix} &= \Sigma_{k\geq 0}\,(-1)^{2k}\,(2k+1)\begin{Bmatrix} A & D & G \\ H & J & k \end{Bmatrix} \\ &\quad \begin{Bmatrix} B & E & H \\ D & k & F \end{Bmatrix}\begin{Bmatrix} C & F & J \\ k & A & B \end{Bmatrix} \end{aligned} \tag{D.4}$$

in terms of 6-j symbols {e.g., (6.4.3) in [65]} and the following algebraic relation

$$\begin{aligned} \begin{Bmatrix} A & B & C \\ D & E & F \\ G & H & J \end{Bmatrix} &= (-1)^{C+F-J}\,\frac{\Delta'(DAG)\,\Delta'(BEH)\,\Delta'(JGH)}{\Delta'(DEF)\,\Delta'(BAC)\,\Delta'(JCF)} \\ &\times \Sigma_{p\geq 0}\,\Sigma_{q\geq 0}\,\Sigma_{r\geq 0}\,\frac{(-1)^{p+q+r}}{p!\,q!\,r!}\quad \frac{(2F-p)!\,(2A-r)!}{(2H+1+q)!\,(A+D+G+1-r)!} \\ &\times \frac{(D+E-F+p)!\,(C+J-F+p)!}{(E+F-D-p)!\,(C+F-J-p)!} \end{aligned}$$

$$\times \frac{(E+H-B+q)!\,(G+H-J+q)!}{(B+E-H-q)!\,(G+J-H-q)!}$$

$$\times \frac{(B+C-A+r)!}{(A+D-G-r)!\,(A+C-B-r)!}$$

$$\times \frac{(A+D+J-H-q-r)!}{(D+H-B-F+p+q)!\,(B-F-A+J+p+r)!} \tag{D.5a}$$

with

$$\Delta'(A,B,C) = \left[\frac{(A+B-C)!(A-B+C)!(A+B+C+1)!}{(-A+B+C)!}\right]^{\frac{1}{2}}. \tag{D.5b}$$

originally obtained by Jucys and Bandzaitis [364] and reproduced in [(A-6)–(A-8)] in [68]. In the three sums present in (D.5a), only those positive integral values of each of p, q, and r are to be considered for which argument of none of the factorials present therein is negative.

D.2 A Short Description of the Programme

The programme `threenj` consists of a master driver which reads the input data and directs the flow of the whole programme through its eight subroutines.[2] This driver programme also writes the results calculated. Details of the working of the programme are described in its comment cards. In the following, we give, in alphabetical order, a brief description of each of the 10 parts of the present programme `threenj`. A complete listing of this programme is available from the Springer's website <http://extra.springer.com>.

D.2.1 Subroutine `delta`

This subroutine is called in each of the three subroutines `ninej_JB` (Sect. D.2.5), `sixj` (Sect. D.2.9), and `threej` (Sect. D.2.10). It is needed for calculating

---

[2]The driver of `threenj` calls the subroutines of this programme in the following sequence: (*i*) `primnum`, (*ii*) `gamafun`, (*iii*) `threej`, (*iv*) `printr`, (*v*) `sixj`, (*vi*) `printr`, (*vii*) `nine_JB`, (*viii*) `printr`, (*ix*) `nine_6j`, (*x*) `printr`.
The `logical function`, in addition to the subroutine `printr`, is used by each of `threej`, `sixj`, `nine_JB`, and `nine_6j`.

$\Delta\big(A\ B\ C\big)$ [in `threej` and `sixj`] or $\Delta'\big(A\ B\ C\big)$ [in `ninej_JB`], defined in (D.2) or (D.5b), respectively. The value of the input integer "iselect" decides as to which of the two deltas is to be calculated:

If "iselect" = 1, $\Delta\big(A\ B\ C\big)$ of (D.2) is computed;
if "iselect" = 2, $\Delta'\big(A\ B\ C\big)$ of (D.5b) will be calculated.

Delta, calculated in the subroutine `delta`, is printed if "iprint(4)" is a non-zero integer.

D.2.2 *The Driver*

It reads the input data, controls the flow of the programme, and prints results. The input data is described in the Sect. D.3 (page 280). On the other hand, the output of a test run is available from the web page <http://extra.springer.com> which contains also a complete listing of the present programme `threenj`.

D.2.3 *Subroutine* `gamafun`

This is a subroutine for calculating Gamma functions in terms of prime numbers and their powers.[3] Gamma functions are calculated for all the integers in the range from n = 1 to prime(nprim). The calculated Gamma functions are stored in the 2-dimensional integer matrix "gamma.[4]" This matrix is passed to other parts of the programme `threenj` through a common block.

Matrix "gamma" is printed if "iprint(2)" is a non-zero integer.

---

[3] Note [259] that $\Gamma(n) = (n-1)!$; $0! = 1$; $n! = \pm\infty$ when n is a *negative* integer.

[4] For example, for "nprim" = 10, this subroutine will calculate first ten prime numbers (excluding 1), namely: 2, 3, 5, 7, 11, 13, 17, 19, 23, 29. Then, the Γ functions are calculated for the first thirty integers, *ie*, 1, 2, 3, 4, …, 29, 30. Thus, elements gama(1, i) = 0 (with i = 1–10) contain the value of $\Gamma(1)(= 0! = 1 = 2^0 \times 3^0 \times 5^0 \times \ldots 29^0)$. Similarly, gama (2, i) = 0 (with i = 1–10) give the value of $\Gamma(2)(= 1! = 1 = 2^0 \times 3^0 \times 5^0 \times \ldots 29^0)$; gama (3, 1) = 1, gama (3, i) = 0 (with i = 2–10) give the value of $\Gamma(3)$ $(= 2! = 2 = 2^1 \times 3^0 \times 5^0 \times \ldots 29^0)$; gama(4, 1) = 1, gama(4, 2) = 1, gama (4, i) = 0 (with i = 3–10) give the value of $\Gamma(4)$ $(= 3! = 6 = 2^1 \times 3^1 \times 5^0 \times \ldots 29^0)$. Further, gama(29, 1) = 25, gama(29, 2) = 13, gama(29, 3) = 6, gama(29, 4) = 4, gama(29, 5) = 2, gama(29, 6) = 2, gama(29, 7) = 1, gama(29, 8) = 1, gama(29, 9) = 1, gama(29, 10) = 0, give the value of $\Gamma(29)$ $(= 28! = 2^{25} \times 3^{13} \times 5^6 \times 7^4 \times 11^2 \times 13^2 \times 17^1 \times 19^1 \times 23^1 \times 29^0)$. Lastly, gama(30, 1) = 25, gama(30, 2) = 13, gama(30, 3) = 6, gama(29, 4) = 4, gama(30, 5) = 2, gama(30, 6) = 2, gama(30, 7) = 1, gama(30, 8) = 1, gama(30, 9) = 1, gama(30, 10) = 1, is the value of $\Gamma(30)$ $(= 29! = 2^{25} \times 3^{13} \times 5^6 \times 7^4 \times 11^2 \times 13^2 \times 17^1 \times 19^1 \times 23^1 \times 29^1)$.
Thus, gama(j, i) is the power of the i-th prime number present in $\Gamma(j)$.

D.2.4 Function `logical function`

This function tests whether the three angular momenta (say, j1, j2, j3) of a 3n-j symbol satisfy the well-known triangular condition (i.e., $|j1 - j2| \leq j3 \leq j1 + j2$) applicable [63–68] to their vector addition in quantum mechanics. It is called at various places in the subroutine `threenj` in order to avoid those calls of the subroutines `ninej_JB`, `ninej_6j`, `sixj`, and `threej` wherein any of the 3-j, 6-j, or 9-j symbols identically vanishes.

Value of the logical variable "ltc," calculated in this function, is printed if iprint(7) is a non-zero integer.

D.2.5 Subroutine `ninej_JB`

This is a subroutine for calculating square of a 9-j symbol using the algebraic expression (D.5). Its output is in powers of prime numbers. The input arguments are: "j1" = 2*A, "j2" = 2*B, "j12" = 2*C, "j3" = 2*D, "j4" = 2*E, "j34" = 2*F, "j13" = 2*G, "j24" = 2*H, and "jj" = 2*J, i.e., TWICE of the actual values of angular momenta. Hence, each of the nine input arguments is always a positive and real integer. The respective triads (j1, j2 j12), (j3, j4, j34), and (j13, j24, jj) of the angular momenta form the first, second, and third rows of a 9-j symbol [see, for example, left-hand sides of (D.4) and of (D.5a)]. The integer variables "sign9j," "n9j," "sum9j," and the integer array "v9j" contain this subroutine's output.

The driver, discussed in the Sect. D.2.2, will call the subroutine `ninej_JB` if input integer "i9" = 1, or $\geq$ 3.

A 9-j symbol, calculated in this subroutine is printed if "iprint(9)" is a non-zero integer.

A "*" symbol in front of a printed value of a 9-j symbol means that its negative square root is to be used.

D.2.6 Subroutine `ninej_6j`

This subroutine uses the relation (D.4) for computing square of a 9-j symbol in terms of the prime numbers. Its input arguments are exactly the same as in the case of the subroutine `ninej_JB`, discussed in the immediately preceding Sect. D.2.5.

The subroutine `ninej_6j` is called by the driver (see the Sect. D.2.2) of the present programme `threenj` if the input integer "i9" $\geq$ 2.

A 9-j symbol, calculated in the subroutine `ninej_6j`, is printed if "iprint(10)" is a non-zero integer.

A "*" symbol in front of a printed value of a 9-j symbol means that its negative square root is to be used.

Thus, if "i9" > 2, a required 9-j symbol will be calculated by using the relation (D.4) as well as the algebraic expression (D.5). This provides a very good check for each of the three subroutines `sixj` (Sect. D.2.9), `ninej_JB` (Sect. D.2.5), and `ninej_6j` (Sect. D.2.6), provided the last two subroutines compute identical values for a given 9-j symbol.

D.2.7 *Subroutine* `primnum`

This subroutine calculates first "nprim" prime numbers. These prime numbers are stored in the 1-dimensional, integer array called "prim." [See footnotes (2) and (4).] Both "nprim" and the array "prim" are passed to other parts of the programme through a common block.

Array "prim" is printed if "iprint(1)" is a non-zero integer.

D.2.8 *Subroutine* `printr`

The driver (Sect. D.2.2) as well as each of the four subroutines `ninej_JB` (Sect. D.2.5), `nine_6j` (Sect. D.2.6), `sixj` (Sect. D.2.9), and `threej` (Sect. D.2.10) calls the subroutine `printr` for printing a 3-j, 6-j, and/or 9-j symbol calculated in the present programme `threenj`.

D.2.9 *Subroutine* `sixj`

This subroutine computes the square of a 6-j symbol in terms of the powers of prime numbers. The definition used is that given in (D.3). Its input arguments ("j1" = 2*A, "j2" = 2*B, "j3" = 2*C, "l1" = 2*D, "l2" = 2*E, and "l3" = 2*F) are TWICE of the actual values of angular momenta for which a 6-j symbol is required. Hence, each of these six input arguments is always a positive and real integer. The output of `sixj` is given by the integers "sign6j," "n6j," "sum6j," and the integer array "v6j." A "*" sign means that a negative square root of a computed value is to be used.

A 6-j symbol, calculated in the subroutine `sixj`, is printed if "iprint(6)" is a non-zero integer.

D.2.10 *Subroutine* `threej`

The square of a 3-j symbol is computed in this subroutine in terms of prime numbers and their powers. It is defined in (D.1) and (D.2). The input arguments of the present subroutine are: "j1" = 2*A, "j2" = 2*B, "j3" = 2*C, "m1" = 2*a, "m2" = 2*b, and "m3" = 2*c, i.e., TWICE of the actual angular momenta and of their projections. Hence, arguments ("j1, j2, j3, m1, m2, m3") of `threej` are always real integers. Here, "j1, j2, j3" are necessarily non-negative; whereas -j1 $\leq$ m1 $\leq$ j1, -j2 $\leq$ m2 $\leq$ j2, -j3 $\leq$ m3 $\leq$ j3 such that (m1 + m2 + m3) = 0. The output of this subroutine is contained in the integer variables "sign3j, n3j, sum3j," and in the integer array "v3j." Results preceded by a '*' (for "sign3j" < 0) means that one should use negative of the square root of the computed value.

A 3-j symbol, calculated in the subroutine `threej`, is printed if "iprint(3)" is a non-zero integer.

D.3 Input

The whole of the input information needed for a successful run of `threenj` is read in the `driver` segment of this programme. Some of it (i.e., integer variable "nprim," and integer array "iprint") is passed to the other parts of the programme `threenj` through a common block; whereas the remaining in the form of the arguments of the various subroutines already described herein.

1. nprim (1 $\leq$ nprim $\leq$ 100) Described in the Sects. D.2.3 and D.2.7.
2. irpt (1 $\leq$ irpt $\leq$ 11) Controls the re-looping of the programme `threenj`:

irpt = 1, input for the statement number 10 is read;
irpt = 2-4, input for the statement number 30 (for a 3-j symbol) is read;
irpt = 5-7, input for the statement number 60 (for a 6-j symbol)is read;
irpt = 8-10, input for the statement number 90 (for a 9-j symbol)is read;
irpt = 11, programme `threenj` ends.

3. iprint(i) Controls the printing in various subroutines.
4. (j1, j2, j3, m1, m2, m3) Described in the Sect. D.2.10.
5. (j1, j2, j3, l1, l2, l3) Described in the Sect. D.2.9.
6. (j1, j2, j12, j3, j4, j34, j13, j24, jj) Described in the Sects. D.2.5 and D.2.6.
7. i9 Described in the Sects. D.2.5 and D.2.6.

D.4 Programme Listing

It is available from Springer's web page <http://extra.springer.com>.

D.5 Test Run Output

It too is available from Springer's web page <http://extra.springer.com>.

Appendix E
A Generic Computer Programme for the Coefficient A's Needed in the Calculation of the Density Matrix for Fine-Structure Entanglement in 2-DPI

This appendix is a brief write-up of the computer code `coefficient_A` used [188] for calculating the coefficient A's present in the density matrix (7.19) needed for studying fine-structure entanglement between the spins of a photoelectron e_p and Auger electron e_a generated in the 2-DPI process (1.3). This is a generic programme as it can readily be adapted for calculating density matrices needed for similar studies of entanglement between two electronic qubits, which may be generated in the 2-DPI process taking place in a linear molecular target or in 1-DPI (1.2) in an atom or a molecule. Furthermore, the following programme `coefficient_A` can readily be generalized also to the two-step process (1.4) for studying the hybrid, fine-structure entanglement between the spin of photoelectron e_p and polarization of the emitted fluorescence γ_d.

E.1 Method of Calculation

For a given set of values of the integer variables $(S_p, S_a, N_{S_p}, N_{S_a}, m_r)$ [(7.19b)], programme `coefficient_A` calculates A's in terms of the products of spherical harmonics $\left[Y_{L_p}^{M_{L_p}}(\hat{\boldsymbol{k}}_p)\right]^* \left[Y_{L_a}^{M_{L_a}}(\hat{\boldsymbol{k}}_a)\right]^*$ as well as of the photoionization and Auger decay amplitudes $P_e(J_{1^+*}\; j_p;\; J_0\, 1;\; j_t;\; \ell_p)$ $\left[P_e(J_{1^+*}\; j_p';\; J_0\, 1;\, j_t';\; \ell_p')\right]^*$ and $A_e(J_{2^+}\; j_a;\; J_{1^+*};\; \ell_a)$ $\left[A_e(J_{2^+}\; j_a';\; J_{1^+*};\; \ell_a')\right]^*$, respectively. For the spherical harmonics, the phase convention used is the one given in (2.5.29) in Edmond's book [65] or, for examples, (1)–(5) on page 59 in [365]; whereas the latter two amplitudes are defined by the respective (7.20). This programme substitutes neither the trigonometric [i.e., in terms of the spherical angles (θ_p, ϕ_p) of $\hat{\boldsymbol{k}}_p$, or (θ_a, ϕ_a) of $\hat{\boldsymbol{k}}_a$] nor numerical values of a spherical harmonics $\left[Y_{L_p}^{M_{L_p}}(\hat{\boldsymbol{k}}_p)\right]^*$ or of $\left[Y_{L_a}^{M_{L_a}}(\hat{\boldsymbol{k}}_a)\right]^*$. It, on the other hand, determines values of (L_p, M_{L_p}) and (L_a, M_{L_a}) for which the two required spherical harmonics are to be used. The advantage of doing this is

N. Chandra and R. Ghosh, *Quantum Entanglement in Electron Optics*, Springer Series on Atomic, Optical, and Plasma Physics 67, DOI 10.1007/978-3-642-24070-6,

that on substituting these A's in the expression (7.19a), one is free to compute the density matrix $\langle J_{2+};\ \mu_p\, \hat{\boldsymbol{u}}_p\, \boldsymbol{k}_p;\ \mu_a\, \hat{\boldsymbol{u}}_a\, \boldsymbol{k}_a |\, \rho^{(2\text{-DPI})}\, | J_{2+};\ \mu_p'\, \hat{\boldsymbol{u}}_p\, \boldsymbol{k}_p;\ \mu_a'\, \hat{\boldsymbol{u}}_a\, \boldsymbol{k}_a \rangle$ for any experimental geometry (i.e., directions $\hat{\boldsymbol{k}}_p$ and $\hat{\boldsymbol{k}}_a$) one desires without having to recalculate the A's for that specific 2-DPI process (1.3). Thus, need for calculating A's for each new $\hat{\boldsymbol{k}}_p$ and/or $\hat{\boldsymbol{k}}_a$ is completely the eliminated.

Similarly, it is more likely than not that more than one amplitudes $P_e(J_{1+^*}\ j_p;\ J_0\, 1;\ j_t;\ \ell_p)$ [each with a different set of values of the quantum numbers $(\ell_p,\ j_p,\ j_t)$] and $A_e(J_{2+}\ j_a;\ J_{1+^*};\ \ell_a)$ [differing in $\ell_a,\ j_a$] contribute to the photoionization (1.3a)] and Auger decay (1.3b), respectively. In view of these possibilities, although the programme `coefficient_A` does inform its user as to which of the $P_e(J_{1+^*}\ j_p;\ J_0\, 1;\ j_t;\ \ell_p)$ and $A_e(J_{2+}\ j_a;\ J_{1+^*};\ \ell_a)$ is to be used where; it, however, does not require numerical/algebraic values of either amplitudes. These values are used when A's are substituted in (7.19a) for the final calculation of the density matrix. This feature of `coefficient_A` makes it possible to compute the density matrix (7.19a) for the same participating transitions but in different atoms without requiring recalculation of A's. In other words, the A's obtained from the programme `coefficient_A` are only transition, but neither atom nor geometry, specific.

However, in spite of the above-discussed two features of the programme `coefficient_A`, one still requires to compute A's for each different polarization (specified by $m_r = -1,\ 0,\ +1$) of the incident photon γ_r used for ionization of the atomic target $\mathfrak{T}$ in the 2-DPI process (1.3).

For the calculation of the various n-j (i.e., 3-j, 6-j, and/or 9-j) symbols, present in the expression (7.19b), `coefficient_A` uses the subroutines given in the programme `threenj` (but without its driver part) described in the Appendix D on pages 273–280.

E.2 A Short Description of the Programme

The programme `coefficient_A` contains twelve subroutines, including a `block_data` and a `driver`. Out of these, nine (namely: `delta`, `gamafun`, `logical_function`, `ninej_JB`, `ninej_6j`, `primnum`, `printr`, `sixj`, `threej`) are those needed to calculate a 3n-j symbol and have already been explained in the Appendix D. In the following, therefore, the remaining three parts of the present programme are explained.

E.2.1 The Driver

It reads the input data, i.e., "nprim," "iprint," and "irpt" (each explained in the Sect. D.3 on page 280); all the ten elements of the array "iwrite" (explained in the source code of the subroutine main of the present programme). In addition, the values of the parameters $(S_p, S_a, N_{S_p}, N_{S_a}, m_r)$,[1] needed to specify an $A^{S_p\,S_a}_{N_{S_p}\,N_{S_a}}(m_r;\,\hat{\boldsymbol{k}}_p\,\hat{\boldsymbol{k}}_a)$ in (7.19), are also read in this driver part of the programme `coefficient_A`. It then calls the subroutine `main`.

E.2.2 `block data`

This part of the present programme stores the variables corresponding to the total angular momenta $(J_0,\ J_{1^{+*}},\ J_{2^+})$ of $(\mathfrak{T},\ \mathfrak{T}^{1+^*},\ \mathfrak{T}^{2+})$ as well as values $(\ell_p,\ \ell_a)$ of the orbital angular momenta of the partial waves of (photoelectron e_p, Auger electron e_a) participating in the 2-DPI process (1.3).

In this `block_data`, (nl1v, nl2v) are the number of the values of (ℓ_p, ℓ_a), which need to be taken in to account for a given 2-DPI process (1.3). These values of (ℓ_p, ℓ_a) are stored in the integer arrays (l1v, l2v).

In the given version of the programme `coefficient_A`, each of the arrays (l1v, l2v) has been declared to have dimension 05. This can, however, readily be changed according to one's requirements. If either or both of (nl1v, nl2v) have values less than the declared dimensions of (l1v, l2v), then the remaining elements of these two arrays are filled with some un-physical value (e.g., say, -1), which neither of the partial waves (ℓ_p, ℓ_a) can ever have.

E.2.3 Subroutine `main`

This subroutine is the main part of the present programme `coefficient_A`. It performs the following tasks:

1. Determines the possible values of all the variables [other than those (i.e., S_p, S_a, N_{S_p}, N_{S_a}, m_r) read in the driver (Sect. E.2.1) or stored (i.e., J_0, $J_{1^{+*}}$, J_{2^+}, ℓ_p, ℓ_a in the `block data` (Sect. E.2.2)).
2. Calculates all the 3n-j symbols contributing to (7.19b).
3. Multiplies the product of these 3n-j symbols with appropriate spherical harmonics $\left[Y^{M_{L_p}}_{L_p}(\hat{\boldsymbol{k}}_p)\right]^*$, $\left[Y^{M_{L_a}}_{L_a}(\hat{\boldsymbol{k}}_a)\right]^*$ [after determining the proper values of $(L_p,\ M_{L_p})$ and of $(L_a,\ M_{L_a})$] and by the photoionization amplitudes

[1] Recall that each of S_p, S_a, N_{S_p}, N_{S_a}, and m_r is always an integer. Their values are: $m_r = 0$ [for LP ionizing radiation γ_r], or $+1$ (-1) [for CP ionizing radiation γ_r with negative (positive) helicity] in (7.19). $N_{S_p} = 0$ (for $S_p = 0$); -1, 0, $+1$ (for $S_p = +1$). $N_{S_a} = 0$ (for $S_a = 0$); -1, 0, $+1$ (for $S_a = +1$). Programme `coefficient_A` does not need any information about either the unit vectors $(\hat{\boldsymbol{k}}_p, \hat{\boldsymbol{k}}_a)$, or amplitudes (P_e, A_e).

$\left[P_e(J_{1+*}\ j_p;\ J_0\,1;\ j_t;\ \ell_p),\ \left[P_e(J_{1+*}\ j_p';\ J_0\,1;\ j_t';\ \ell_p')\right]^*\right]$ as well as Auger decay amplitudes $\left[A_e(J_{2+}\ j_a;\ J_{1+*};\ \ell_a),\ \left[A_e(J_{2+}\ j_a';\ J_{1+*};\ \ell_a')\right]^*\right]$ belonging to the appropriate values of $[(j_p\ j_t\ \ell_p),\ (j_p'\ j_t'\ \ell_p')]$ and $[(j_a\ \ell_a),\ (j_p'\ \ell_a')]$, respectively.

4. These products are then added to obtain the final expression for a desired $A^{S_p\,S_a}_{N_{S_p}\,N_{S_a}}(m_r;\ \hat{\boldsymbol{k}}_p\,\hat{\boldsymbol{k}}_a)$ according to the definitions given in (7.19) and (7.20).
5. Complete calculation in this programme is performed in integer algebra.

E.3 Input

The integer variables ("nprim," and "iprint") are read in the beginning of the driver part (i.e., Sect. E.2.1) of the programme `coefficient_A`. Integer variables $[S_p,\ S_a,\ N_{S_p},\ N_{S_a},\ m_r]$ (see footnote (1) on page 283), "irpt," and all the ten elements of the array "iwrite" are supplied in the second read statement of the driver. A value of the integer variable "irpt" decides the flow of the programme `coefficient_A` in its driver part.

E.4 Programme Listing

Folder *programme_coefficient_A*, available at the web page <http://extra.springer.com> contains a complete listing of the present programme coefficient_A.

E.5 Test Run Output

Folder *programme_coefficient_A*, available at the web page <http://extra.springer.com> contains output of a test run of coefficient_A. The test run given therein corresponds to the 2-DPI transition (7.21).

References

1. A. Einstein, Annalen der Physik (Leipzig) **17**, 132 (1905)
2. A. Einstein, Dissertation toward the Degree of Doctor of Philosophy granted by the Faculty of Philosophy (Department of Mathematics & Natural Sciences), of the University of Zurich, 1905. Published in Annalen der Physik (Leipzig) **19**, 289 (1906) with the addendum in Annalen der Physik (Leipzig) **19**, 305 (1906)
3. A. Einstein, Annalen der Physik (Leipzig) **17**, 549 (1905)
4. A. Einstein, Annalen der Physik (Leipzig) **17**, 891 (1905)
5. A. Einstein, Annalen der Physik (Leipzig) **18**, 639 (1905)
6. H.R. Hertz, Annalen der Physik (Berlin) **267**, 983 (1887)
7. P. Lenard, Annalen der Physik (Leipzig) **8**, 149 (1902)
8. A.H. Compton, Phys. Rev. **21**, 483 (1923)
9. M. Planck, Verh. Dtsch. Phys. Ges. **2**, 237 (1900)
10. B.H. Bransden, C.J. Joachain, *Physics of Atoms and Molecules*, 2nd edn. (Benjamin Cummings, New York, 2003)
11. A. Einstein, B. Podolsky, N. Rosen, Phys. Rev. **47**, 777 (1935)
12. J.S. Bell, Physics (Long Island City, N.Y.) **1**, 195 (1965), reprinted in *Speakable and non-Speakable in Quantum Mechanics*, 2nd ed. (Cambridge University Press, Cambridge, England, 2004), pp. 14; in *Foundations of Quantum Mechanics*, ed. B. d'Espagnat (Academic Press, New York, 1971), pp. 171
13. A.K. Ekert, Phys. Rev. Lett. **67**, 661 (1991)
14. C.H. Bennett, G. Brassard, C. Crépeau, R. Jozsa, A. Peres, W.K. Wootters Phys. Rev. Lett. **70**, 1895 (1993)
15. E. Ekert, R. Jozsa, Revs. Mod. Phys. **68**, 733 (1996)
16. A. Steane, Reps. Prog. Phys. **61**, 117 (1998)
17. M.A. Nielsen, I.L. Chuang, *Quantum Computation and Quantum Information*, 10 Anv. ed. (Cambridge University Press, Cambridge, England, 2011)
18. B. Schumacher, Phys. Rev. A **51**, 2738 (1995)
19. J.G. Rarity, P.R. Tapster, Phys. Rev. Lett. **64**, 2495 (1990)
20. P.R. Tapster, J.G. Rarity, P.C.M. Owen, Phys. Rev. Lett. **73**, 1923 (1994)
21. M. Bourennane, M. Eibl, C. Kurtsiefer, S. Gaertner, H. Weinfurter, O. Gühne, P. Hyllus, D. Bruβ, M. Lewenstein, A. Sanpera, Phys. Rev. Lett. **92**, 087902 (2004)
22. M. Barbieri, F. De Martini, G. Di Nepi, P. Mataloni, G.M. D'Ariano, C. Macchiavello, Phys. Rev. Lett. **91**, 227901 (2003)
23. M. Barbieri, F. De Martini, G. Di Nepi, P. Mataloni, Phys. Rev. Lett. **92**, 177901 (2003)
24. O. Gühne, G. Tóth, Phys. Rep. **474**, 1 (2009)
25. L. Mandel, E. Wolf, *Optical Coherence and Quantum Optics*, (Cambridge University Press, New York, 1995)

N. Chandra and R. Ghosh, *Quantum Entanglement in Electron Optics*, Springer Series on Atomic, Optical, and Plasma Physics 67, DOI 10.1007/978-3-642-24070-6,

26. Z.Y. Ou, L. Mandel, Phys. Rev. Lett. **61**, 50 (1988)
27. M.O. Scully, M.S. Zubairy, *Quantum Optics* (Cambridge University Press, New York, 1997)
28. A. Aspect, J. Dalibard, G. Roger, Phys. Rev. Lett. **49**, 1804 (1982)
29. A. Aspect, P. Grangier, G. Roger, Phys. Rev. Lett. **49**, 91 (1982)
30. T. Haji-Hassan, A. J. Duncan, W. Perrie, H. Kleinpoppen, E. Merzbacher, Phys. Rev. Lett. **62**, 237 (1989)
31. V.B. Braginsky, F.Y. Khalili, *Quantum Measurement* (Cambridge University Press, Cambridge, England, 1992)
32. C.M. Caves, K.S. Thorne, R.W.P. Drever, V.D. Sandberg, M. Zimmermann, Revs. Mod. Phys. **52**, 341 (1980)
33. D. Loss, D.P. DiVicenzo, Phys. Rev. A **57**, 120 (1998)
34. D. Bouwmeester, A.K. Ekert, A. Zeilinger (eds.), *The Physics of quantum Information* (Springer, Berlin, 2000)
35. D.P. DiVicenzo, Science **270**, 255 (1995)
36. S. Das Sarma, J. Fabian, X. Hu, I. Žutić, Solid State Commun. **119**, 207 (2001)
37. I. Žutić, J. Fabian, S. Das Sarma, Revs. Mod. Phys. **76**, 323 (2004)
38. C.H. Bennett, D.P. DiVincenzo, Nature **404**, 247 (2000)
39. P. Recher, E.V. Sukhorukov, D. Loss, Phys. Rev. B **63**, 165314 (2001)
40. P. Recher, D. Loss, Phys. Rev. B **65**, 165327 (2002)
41. See, for example, D.S. Saraga, G. Burkard, J.C. Egues, H.-A. Engel, P. Recher, D. Loss, Turk. J. Phys. **27**, 427 (2003) and references therein
42. D.S. Saraga, D. Loss, Phys. Rev. Lett. **90**, 166803 (2003)
43. G. Alber, T. Berth, M. Horodecki, P. Horodecki, R. Horodecki, M. Rötteler, H. Weinfurter, R. Werner, A. Zeilinger (eds.), *Quantum Information: An Introduction to Basic Theoretical Concepts and Experiments* (Springer, Berlin, 2001)
44. M. Murao, D. Jonathan, M.B. Plenio, V. Vederal, Phys. Rev. A **59**, 156 (1999)
45. R. Cleve, H. Buhrman, Phys. Rev. A **56**, 1201 (1997)
46. P.J. dos Reis, S.S. Sharma, Phys. Rev. A **79**, 012326 (2009)
47. W. Son, M.S. Kim, J. Lee, D. Ahn, J. Mod. Opt. **49**, 1739 (2002)
48. H.-Fu Wang, X.-Q. Shao, Y.-F. Zhao, S. Zhang, K.-H. Yeon, J. Phys. B **42**, 175506 (2009)
49. C. Cabrillo, J.I. Cirac, P. García-Fernández, P. Zoller, Phys. Rev. A **59**, 1025 (1999)
50. Z.J. Deng, M. Feng, K.L. Gao, Phys. Rev. A **75**, 024302 (2007)
51. XuBo Zou, K. Pahlke, W. Mathis, Phys. Rev. A **68**, 024302 (2003)
52. XuBo Zou, K. Pahlke, and W. Mathis, Phys. Rev. A **69**, 052314 (2004)
53. T. Yu, J.H. Eberly, Phys. Rev. Lett. **93**, 140404 (2004)
54. M.P. Almeida, F. de Melo, M. Hor-Meyll, A. Salles, S.P. Walborn, P.H.S. Ribeiro, L. Davidovich, Science **316**, 579 (2007)
55. O.J. Farías, C.L. Latune, S.P. Walborn, L. Davidovich, P.H.S. Ribeiro, Science **324**, 1414 (2009)
56. S. Parida, N. Chandra, R. Ghosh, Eup. Phys. J. D **65**, 303 (2011)
57. S. Parida, N. Chandra, Phys. Lett. A **373**, 1852 (2009); Phys. Rev. A **79**, 062501 (2009)
58. C. Cohen-Tannoudji, B. Diu, F. Laloë, *Quantum Mechanics*, Vol. I & II (John Wiley, New York, 1978)
59. L.I. Schiff, *Quantum Mechanics*, 3rd rev. ed. (McGraw-Hill, New York, 1968)
60. M. Weissbluth, *Atoms and Molecules* (Academic, New York, 1978)
61. K. Blum, *Density Matrix Theory and Applications*, 3rd edn. (Springer, Berlin, 2012)
62. U. Fano, Revs. Mod. Phys. **29**, 74 (1957)
63. D.M. Brink, G.R. Satchler, *Angular Momentum*, 3rd ed. (Oxford University Press, Oxford, 1994)
64. A. de-Shalit, I. Talmi, *Nuclear Shell Theory* (Dover, New York, 2004)
65. A.R. Edmonds, *Angular Momentum in Quantum Mechanics* (Princeton University Press, Princeton, 1996)
66. M.R. Rose, *Elementary theory of Angular Momentum* (Dover, 2011)

67. D.A. Varshalovich, A.N. Moskalev, V.K. Khersonskii, *Quantum Theory of Angular Momentum* (World Scientific, Singapur, 1988)
68. R.N. Zare, *Angular Momentum* (Wiley-Interscience, 1988)
69. M. Mizushima, *Theory of Rotating Diatomic Molecules* (Wiley, 1975)
70. M. Tinkham, *Group Theory and Quantum Mechanics* (Dover, 2003)
71. T.A. Carlson, *Photoelectron and Auger Spectroscopy* (Plenum, New York, 1975)
72. H.A. Bethe, E.E. Salpeter, *Quantum Mechanics of One- and Two-Electron Atoms* (Springer, Berlin, 1957)
73. P. Auger, Commn. Royal Acad. Sci. Paris **178**, 929 (1924)
74. P. Auger, Commn. Royal Acad. Sci. Paris **178**, 1535 (1924)
75. J. Xie, R.N. Zare, J. Chem. Phys. **93**, 3033 (1990)
76. M. Born, E. Wolf, *Principles of Optics* (Cambridge University Press, Cambridge, UK, 1997)
77. C. Soanes (ed.), *The Compact Oxford Reference Dictionary* (Oxford University Press, Oxford, 2001)
78. R. Landauer, IBM J. Res. Dev. **5**, 183 (1961)
79. R. Landauer, Phys. Today **44**, 23 (May 1991)
80. R. Landauer, Phys. Lett. A **217**, 188 (1996)
81. I.A. Khan, J.C. Howell, Phys. Rev. A **73**, 031801(R) (2006)
82. P.G. Kwiat, A.M. Steinberg, R.Y. Chiao, Phys. Rev. A **47**, R2472 (1993)
83. J.C. Howell, R.S. Bennink, S.J. Bentley, R.W. Boyd, Phys. Rev. Lett. **92**, 210403 (2004)
84. A. Mair, A. Vaziri, G. Weihs, A. Zeilinger, Nature **412**, 313 (2001)
85. A. Vaziri, G. Weihs, A. Zeilinger, Phys. Rev. Lett. **89**, 240401 (2002)
86. L.A. Allen, S.M. Barnett, M.J. Padgett, *Optical Angular Momentum* (Institute of Physics Publishing, Bristol, UK, 2003)
87. W.K. Wootters, W.H. Zurek, Nature **299**, 802 (1982)
88. E. Schrödinger, Die Naturwissenschaften **23**, 807, 823, 844 (1935); English translations in Proc. Camb. Phil. Soc. **31**, 555 (1935); *ibid* **32**, 446 (1936); Proc. Am. Phil. Soc. **124**, 323 (1980)
89. D. Bohm, *Quantum Theory* (Prentice-Hall, NJ, 1951)
90. M. Genovese, Phys. Rep. **413**, 319 (2005)
91. B.M. Terhal, M.W. Wolf, A.C. Doherty, Phys. Today **56**, 46 (April 2003)
92. J.F. Clauser, A. Shimony, Reps. Prog. Phys. **41**, 1881 (1978)
93. A. Peres, Am. J. Phys. **46**, 745 (1978)
94. M.A. Rowe, D. Kielpinski, V. Meyer, C.A. Sackett, W.M. Itano, C. Monroe, D.J. Wineland, Nature **409**, 791 (2001)
95. A. Zeilinger, Revs. Mod. Phys. **71**, S288 (1999)
96. A. Aspect, Nature **398**, 189 (1999)
97. G. Weihs, T. Jennewein, C. Simon, H. Weinfurter, A. Zeilinger, Phys. Rev. Lett. **81**, 5039 (1998)
98. J.F. Clauser, M.A. Horne, A. Shimony, R.A. Holt, Phys. Rev. Lett. **23**, 880 (1969)
99. J. Preskill, *Quantum Information and Computation* (Lecture Notes for ph219/cs219, California Institute of Technology, California, 2 November 2001), Chapter 4, pp.16
100. A. Bohm, *Quantum Mechanics: Foundations and Applications*, 2nd ed. (Springer, New York, 1986)
101. M Freyberger, P.K. Arvind, M.A. Horne, A. Shimony, Phys. Rev. A **53**, 1232 (1996)
102. E.S. Fry, T. Walther, in *Quantum [Un]speakables*, eds. R.A. Bertlmann, A. Zeilinger, (Springer, New York, 2002) pp. 103
103. L. Amico, R. Fazio, A. Osterloh, V. Vedral, Revs. Mod. Phys. **80**, 517 (2008)
104. J. Hald, J.I. Sørensen, C. Schori, E.S. Polzik, Phys. Rev. Lett. **83**, 1319 (1999)
105. O. Mandel, M. Greiner, A. Widera, T. Rom, T. Hänsch, I. Bloch, Nature **425**, 937 (2003)
106. S. Ghosh, T.F. Rosenbaum, G. Aeppli, S.N. Coopersmith Nature **425**, 48 (2003)
107. V. Vedral, Nature **425**, 28 (2003)
108. C.H. Bennett, Phys. Scr. **T76**, 210 (1998)
109. C.H. Bennett, G. Brassard, N.D. Mermin, Phys. Rev. Lett. **68**, 557 (1992)

110. G. Benenti, G. Casati, G. Strini, *Principles of Quantum Computation and Information*, Volume I: Basic Concepts (World Scientific, Singapore, 2005)
111. C.H. Bennett, S. Wiesner, Phys. Rev. Lett. **69**, 2881 (1992)
112. D. Deutsch, Proc. Roy. Soc. A **425**, 73 (1989)
113. P.W. Shor, in *Proc. of the 35th Annual Symposium on the Foundations of Computer Science*, ed. S. Goldwasser (IEEE Computer Society Press, Los Alamos, California, 1994) p. 124
114. R. Jozsa, N. Linden, Proc. Roy. Soc. A **459**, 2011 (2003)
115. R. Raussendorf, H. Briegel, Phys. Rev. Lett. **86**, 5188 (2001)
116. R. Raussendorf, D.E. Browne, H. Briegel, Phys. Rev. A **68**, 022312 (2003)
117. J. Eisert, M. Wilkens, M. Lewenstein, Phys. Rev. Lett. **83**, 3077 (1999)
118. C. Brukner, M. Žukowski, J.-W. Pan, A. Zeilinger, Phys. Rev. Lett. **92**, 127901 (2004)
119. S. Huelga, C. Macchiavello, T. Pellizzari, A. Ekert, M.B. Plenio, J. Cirac, Phys. Rev. Lett. **79**, 3865 (1997)
120. A.N. Boto, P. Kok, D.S. Abrams, S.L. Braunstein, C.P. Williams, , J.P. Dowling, Phys. Rev. Lett. **85**, 2733 (2000)
121. R.F. Werner, Phys. Rev. A **40**, 4277 (1988)
122. M. Seevinck, Jos Uffink, Phys. Rev. A **78**, 032101 (2008)
123. L.P. Hughston, R. Jozsa, W.K. Wootters, Phys. Lett. A **183**, 14 (1993)
124. A. Peres, Phys. Rev. Lett. **77**, 1413 (1996)
125. M. Horodecki, P. Horodecki, R. Horodecki, Phys. Lett. A **223**, 1 (1996)
126. R. Horodecki, P. Horodecki, M. Horodecki, K. Horodecki, Revs. Mod. Phys. **81**, 865 (2009)
127. L.M. Ioannou, Quant. Inf. Comput. **7**, 335 (2007), quant-ph/0603199v7
128. N. Gisin, Phys. Lett. A **210**, 151 (1996)
129. S. Popescu, Phys. Rev. Lett. **72**, 797 (1994)
130. S. Popescu, Phys. Rev. Lett. **74**, 2619 (1995)
131. R. Horodecki, P. Horodecki, M. Horodecki, Phys. Lett. A **200**, 340 (1995)
132. M. Horodecki, P. Horodecki, R. Horodecki, in *Quantum Information: An Introduction to Basic Theoretical Concepts and Experiments*, eds. G. Alber, T. Berth, M. Horodecki, P. Horodecki, R. Horodecki, M. Rötteler, H. Weinfurter, R. Werner, A. Zeilinger (Springer, Berlin, 2001)
133. R.F. Werner, M.M. Wolf, quant-ph/0107093V2
134. A. Ekert, P.L. Knight, Am. J. Phys. **63** (1995) 415
135. B.M. Terhal, *LANL e-print quant-ph*/0101032, 2001
136. W. Dür, J.I. Cirac, R. Tarrach, Phys. Rev. Lett. **83**, 3562 (1999)
137. W. Dür, J.I. Cirac, Phys. Rev. A **61**, 042314 (2000)
138. P. Horodecki, Phys. Lett. A **232**, 333 (1997)
139. B. Kraus, J.I. Cirac, S. Karnas, M. Lewenstein, Phys. Rev. A **61**, 062302 (2000)
140. P. Horodecki, M. Lewenstein, G. Vidal, I. Cirac, Phys. Rev. A **62**, 03231 (2000)
141. A. Sanpera, R. Terrach, G. Vidal, Phys. Rev. A **58**, 826 (1998)
142. T.-C. Wei, K. Nemoto, P.M. Goldbart, P.G. Kwiat, W.J. Munro, F. Verstraete, Phys. Rev. A **67**, 022110 (2003)
143. A. Sanpera, R. Terrach, G.Vidal, *LANL e-print quant-ph*/9707041, 1997
144. M.M. Wolf, *Partial Transposition in Quantum Information Theory*, (Ph.D. thesis. Technical University of Braunschweig, Germany, 2003)
145. E. H. Lloyd, *The Mathematical Gazette*, **37**, 29 (1953)
146. M. Horodecki, P. Horodecki, Phys. Rev. A **59**, 4206 (1999)
147. J. Eisert, M.B. Plenio, J. Mod. Opt. **46**, 145 (1999)
148. M. Horodecki, Quant. Inf. Comput. **1**, 3 (2001)
149. W.K. Wootters, Quant. Inf. Comput. **1**, 27 (2001)
150. M.B. Plenio, V. Vedral, Contemp. Phys. **39**, 431 (1998)
151. M.B. Plenio, S. Virmani, Quant. Inf. Comput. **7**, 1 (2007)
152. M. Horodecki, P. Horodecki, R. Horodecki, Phys. Rev. Lett. **84**, 2014 (2000)
153. V. Vedral, M.B. Plenio, M.A. Rippin, P.L. Knight, Phys. Rev. Lett. **78**, 2275 (1997)
154. V. Vedral, M.B. Plenio, Phys. Rev. A **57**, 1619 (1998)

155. J. von Neumann, *Mathematical Foundations of Quantum Mechanics*, (Princeton University Press, Princeton, 1996)
156. M.A. Nielson, Phys. Rev. Lett. **83**, 436 (1999)
157. K. Zyczkowski, P. Horodecki, A. Sanpera, M. Lewenstein, Phys. Rev. A **58**, 883 (1998)
158. M. Reed, B. Simon, *Methods in Modern Mathematical Physics: Functional Analysis*, Volume 1 (Academic New York, 1981).
159. G. Vidal, F. Werner, Phys. Rev. A **65**, 032314 (2002)
160. F. Mintert, A.R.R. Carvalho, M. Kus, A. Buchleitner, Phys. Rep. **415**, 207 (2005)
161. C.H. Bennett, D.P. DiVincenzo, J. Smolin, W.K. Wooters, Phys. Rev. A **54**, 3824 (1996)
162. S. Hill, W.K. Wootters, Phys. Rev. Lett. **78**, 5022 (1997)
163. W.K. Wootters, Phys. Rev. Lett. **80**, 2245 (1998)
164. P. Rungta, V. Bůzek, C.M. Caves, M. Hillery, C.-J. Milburn, Phys. Rev. A **64**, 042315 (2001)
165. F. Mintert, M. Kuś, A. Buchleitner, Phys. Rev. Lett. **95**, 260502 (2005)
166. S.P. Walborn, P.H.S. Ribeiro, L. Davidovich, F. Mintert, A. Buchleitner, Nature **440**, 1022 (2006)
167. S.P. Walborn, P.H.S. Ribeiro, L. Davidovich, F. Mintert, A. Buchleitner, Phys. Rev. A **75**, 032338 (2007)
168. F. Mintert, A. Buchleitner, Phys. Rev. Lett. **98**, 140505 (2007)
169. L Aolita, A. Buchleitner, F. Mintert, Phys. Rev. A **78**, 022308 (2008)
170. C.-J. Zhang, Y.-X. Gong, Y.-S. Zhang, G.-C. Guo, Phys. Rev. A **78**, 042308 (2008)
171. Y.-F. Huang, X.-L. Niu, Y.-X. Gong, J. Li, L. Peng, C.-J. Zhang, Y.-S. Zhang, G.-C. Guo, Phys. Rev. A **79**, 052338 (2009)
172. N. Chandra, in *Frontiers in Atomic. Molecular, and Optical Physics*, eds. S.S. Bhattacharyya, S.C. Mukherjee, Special publication for the 75th year of the Indian Journal of Physics, **3**, 279 (2003)
173. V.L. Jacobs, J. Phys. B **5**, 2257 (1972)
174. N. Chandra, Chem. Phys. **108**, 301 (1986)
175. P. Auger, Commn. Royal Acad. Sci. Paris **182**, 776 (1926)
176. P. Auger, Commn. Royal Acad. Sci. Paris **177**, 169 (1923)
177. T. Aberg, G. Howat, Handbuch der Physik **31**, 469 (1982)
178. V.B. Berestetski, E.M. Lifshitz, L.P. Pitaevskii, *Relatistic quantum Theory* (Pergmon Press, N.Y.), Part 1, pp. 136
179. G. Breit, H.A. Bethe, Phys. Rev. **93**, 888 (1954)
180. M. Dubé, P.C.E. Stamp, Int. J. Mod. Phys. B **12**, 1191 (1998)
181. M. Governale, M. Grifoni, G. Schön, Chem. Phys. **268**, 273 (2001)
182. M. Thorwart, P. Hänggi, Phys. Rev. A **65**, 012309 (2002)
183. B. Lohmann, *Angle and Spin Resolved Auger Emission: Theory and Applications to Atoms and Molelcues* (Springer Series on Atomic, Optical, and Plasma Physics, Springer-Verlag, Berlin, Vol. **46**, 2009)
184. E.U. Condon, R. Shortley, *The Theory of Atomic Spectra*, (Cambridge at the University Press, England, 1970)
185. J.M. Brown, A.C. Carrington, *Rotational Spectroscopy of Diatomic Mollecules* (Cambridge University Press, Cambridge, UK, 2003)
186. I.I. Sobelman, *Atomic Spectra and Radiative Transitions*, 2nd ed. (Springer, Germany, 1992)
187. M. Rotenberg, R. Bivis, N. Metropolis, J.K. Wooten, Jr., *The 3-j and 6-j Symbols* (The Technology Press, MIT, Massachusetts, 1959)
188. N. Chandra, R. Ghosh, Phys. Rev. A **74**, 052329 (2006)
189. U. Fano, Phys. Rev. **178**, 131 (1969); Addendum: Phys. Rev. **184**, 250 (1969)
190. N. Chandra, Phys. Rev. A **42**, 4050 (1990)
191. P.G. Burke, N. Chandra, F.A. Gianturco, J. Phys. B **5**, 2212 (1972)
192. N. Chandra, J. Phys. B **20**, 3405, 3417 (1987)
193. N. Chandra, Phys. Rev. A **36**, 3163 (1987)
194. N. Chandra, Chem. Phys. Lett. **145**, 173 (1988)
195. N. Chandra, J. Chem. Phys. **89**, 5987 (1988)

196. N. Chandra, Phys. Rev. A **39**, R2256 (1989)
197. N. Chandra, Phys. Rev. A **40**, 752, 768 (1989)
198. N. Chandra, J. Chem. Phys. **92**, 5814 (1990)
199. N. Chandra, Phys. Rev. A **43**, 304 (1991)
200. N. Chandra, M. Chakraborty, J. Chem. Phys. **95**, 6382 (1991)
201. N. Chandra, J. Phys. Chem. A **96**, 7207 (1992)
202. K.L. Reid, I. Powis, J. Chem. Phys. **100**, 1066 (1994)
203. N. Chandra, M. Chakraborty, Z. Phys. D **41**, 43 (1997)
204. N. Chandra, M. Chakraborty, Eur. Phys. J. D **2**, 253 (1998)
205. V.V. Kuznetsov, N.A. Cherepkov, G.H. Fecher, G. Schönhense, J. Chem. Phys. **110**, 9997 (1999)
206. V.V. Kuznetsov, N.A. Cherepkov, G.H. Fecher, G. Schönhense, J. Chem. Phys. **117**, 7180 (2002)
207. N. Chandra, J. Phys. B **37**, 3865 (2004)
208. D. Toffoli, P. Decleva, J. Chem. Phys. **128**, 234101 (2008)
209. M. Yamazaki, J.-i. Adachi, Y. Kimura, M. Stener, P. Decleva, A. Yagishita, J. Chem. Phys. **133**, 164301 (2010)
210. N. Chandra, M. Chakraborty, J. Chem. Phys. **97**, 236 (1992)
211. N. Chandra, M. Chakraborty, J. Chem. Phys. **99**, 7314 (1993)
212. F. Hund, Handbuch der Physik **24**, 561 (1933)
213. N. Chandra, R. Ghosh, Phys. Rev. A **69**, 012315 (2004)
214. N. Chandra, S. Sen, J. Chem. Phys. **98**, 5242 (1993)
215. R. Ghosh, Ph.D. Thesis (Indian Institute of Technology, Kharagpur, India, 2008), unpublished
216. N. Chandra, R. Ghosh, Rad. Phys. Chem. **75**, 1808 (2006)
217. R. Ghosh, N. Chandra, S. Parida, Eur. Phys. J. Special Topics **169**, 117 (2009)
218. N. Chandra, S. Sen, J. Chem. Phys. **102**, 8359 (1995)
219. M. Horodecki, R. Horodecki, Quant. Inf. Comput. **1**, 45 (2001)
220. S. Parida, N. Chandra, Phys. Rev. A **86**, 062302 (2012)
221. A.F. Starace, Handbuch der Physik **31**, 1 (1982)
222. Y.S. Kim, Y.J. Kim, R.H. Pratt, Phys. Scr. **T110**, 79 (2004)
223. P. Horodecki, J. Smolin, B. Terhal, A. Tnhaplyial, quant-ph/99101222
224. D. Bruβ, A. Peres, Phys. Rev. A **61**, 030301 (2000)
225. H. Bennett, G. Brassard, S. Popescu, B. Schumacher, J.A. Smolin, W.K. Wootters, Phys. Rev. Lett. **76**, 722 (1996), *ibid* **78**, 2031 (1997)
226. S. Bandyopadhyay, Phys. Rev. A **62**, 032308 (2000)
227. C.H. Bennett, H.J. Bernstein, S. Popescu, B. Schumacher, Phys. Rev. A **53**, 2046 (1996)
228. S. Bose, V. Vedral, P.L. Knight, Phys. Rev. A **60**, 194 (1999)
229. H.-K. Lo, S. Popescu, e-print quant-ph/9707038
230. M. Murao, M.B. Plenio, S. Popescu, V. Vederal, P.L. Knight Phys. Rev. A **57**, 4075 (1998)
231. D. Deutsch, A. Ekert, R. Jozsa, C. Macchiavello, S. Popescu, A. Sanpera, Phys. Rev. Lett. **77**, 2818 (1996)
232. S.J. van Enk, J.I. Cirac, P. Zoller, Phys. Rev. Lett. **78**, 4293 (1997)
233. S.J. van Enk, J.I. Cirac, P. Zoller, Science **279**, 205 (1998)
234. D. Jonathan, M.B. Plenio, Phys. Rev. Lett. **83**, 1455 (1999)
235. G. Vidal, Phys. Rev. Lett. **83**, 1046 (1999)
236. M. Horodecki, P. Horodecki, R. Horodecki, Phys. Rev. Lett. **78**, 574 (1997)
237. P. Horodecki,521 M. Horodecki, R. Horodecki, Phys. Rev. Lett. **82**, 1056 (1999)
238. C.H. Bennett, D.P. DiVincenzo, T. Mor, P.W. Shor, J.A. Smolin, B.M. Terhal, Phys. Rev. Lett. **82**, 5385 (1999)
239. L. Clarisse, Phys. Lett. A **359**, 603 (2006)
240. D. Chruściński, J. Jurkowski, , A. Kossakowski, Phys. Rev. A **77**, 022113 (2008)
241. D. Chruściński, A. Kossakowski, Phys. Rev. A **74**, 022308 (2006)
242. D. Chruściński, A. Kossakowski, Phys. Rev. A **76**, 032308 (2007)
243. J.M. Leinaas, J. Myrheim, P.Ø. Sollid, Phys. Rev. A **81**, 062329 (2010)
244. J.M. Leinaas, J. Myrheim, P.Ø. Sollid, Phys. Rev. A **81**, 062330 (2010)

245. J.M. Leinaas, J. Myrheim, E. Ovrum, Phys. Rev. A **76**, 034304 (2007)
246. M. Piani, Phys. Rev. A **73**, 012345 (2006)
247. J.S. Briggs, V. Schmidt, J. Phys. B **33**, R1 (2000)
248. J.H. McGuire, Adv. At. Mol. Phys. **29**, 217 (1992)
249. J.H. McGuire, N. Berrah, R.J. Bartlett, J.A.R. Samson, J.A. Tanis, C.J. Cocke, A.S. Schlachter, J. Phys. B **28**, 913 (1995)
250. V. Schmidt, in *X-Ray and Inner-Shell Processes*, edited by T.A. Carlson, M.O. Krause, S.T. Manson, AIP Conf. Proc. No. 215 (AIP, New York, 1990), p. 559
251. V. Schmidt, *Electron Spectroscopy of Atoms Using Synchrotron Radiation* (Cambridge University Press, Cambridge, 1997)
252. N. Chandra, M. Chakraborty, J. Phys. B **35**, 2219 (2002)
253. T. Pattard, T. Schneider, J.M. Rost, J. Phys. B **36**, L189 (2003)
254. T. Pattard, J. Burgdörfer, Phys. Rev. A **64**, 042720 (2001)
255. J.A.R. Samson, Phys. Rev. Lett. **65**, 2861 (1990)
256. J. Colgan, M.S. Pindzola, J. Phys. B **37**, 1153 (2004)
257. Y. Hikosaka, P. Lablanquie, F. Penent, P. Selles, T. Kaneyasu, E. Shigemasa, J.H.D. Eland, K. Ito, Phys. Rev. A **80**, 031404(R) (2009)
258. N. Chandra, Phys. Rev. A **56**, 1879 (1997)
259. G. Arfken, *Mathematical Methods for Physicists: A Comprehensive Guide*, 7th edn. (Academic, New York, 2012)
260. R. Grobe, K. Rzazewski, J.H. Eberly, J. Phys. B **27**, L503 (1994)
261. P. Auger, J. Phys. Radium **6**, 205 (1925)
262. N. Chandra, R. Ghosh, Phys. Rev. A **70**, 060306(R) (2004)
263. N. Chandra, R. Ghosh, Quant. Inf. Comput. **9**, 36 (2006)
264. T. Radtke, S. Fritzsche, A. Surzhykov, Phys. Lett. A **347**, 73 (2005)
265. T. Radtke, S. Fritzsche, A. Surzhykov, Phys. Rev. A **74**, 032709 (2006)
266. N.A. Cherepkov, Adv. At. Mol. Phys. **19**, 395 (1983)
267. V.V. Balashov, A.N. Grum-Grzhimailo, N.M. Kabachnik, *Polarization and Correlation Phenomena in Atomic Collisions: A Practical Theory Course* (Springer, Germany, 2010)
268. B. Kämmerling, V. Schmidt, Phys. Rev. Lett. **67**, 1848 (1991)
269. B. Kämmerling V. Schmidt, J. Phys. B **26**, 1141 (1993)
270. B. Kämmerling, V. Schmidt, J. Phys. B **25**, 3621 (1992)
271. B. Schmidtke, M. Drescher, N.A Cherepkov, U. Heinzman, J. Phys. B **33**, 2451 (2000)
272. W.R. Johnson, K.T. Cheng, Phys. Rev. Lett. **69**, 1144 (1992)
273. W.R. Johnson, K.T. Cheng, Phys. Rev. A **20**, 978 (1979)
274. W.R. Johnson, C.D. Lin, Phys. Rev. A **20**, 964 (1979)
275. S. Wolfram, *The Mathematica Book*, 4th edn. (Cambridge University Press, Cambridge, 1999)
276. J.M. Raimond, M. Brune, S. Haroche, Revs. Mod. Phys. **73**, 565 (2001)
277. B.B. Blinov, D.L. Moehring, L.-M. Duan, C. Monroe, Nature **428**, 153 (2004)
278. D.N. Matsukevich, T. Chaneliére, M. Bhattacharya, S.-Y. Lan, S.D. Jenkins, T.A.B. Kennedy, A. Kuzmich, Phys. Rev. Lett. **95**, 040405 (2005)
279. D.L. Moehring, M.J. Madsen, B.B. Blinov, C. Monroe, Phys. Rev. Lett. **93**, 090410 (2004)
280. J. Volz, M. Weber, D. Schlenk, W. Rosenfeld, J. Vrana, K. Saucke, C. Kurtsiefer, H. Weinfurter, Phys. Rev. Lett. **96**, 030404 (2006)
281. K.W. Chan, C.K. Law, J.H. Eberly, Phys. Rev. Lett. **88**, 100402 (2002)
282. K.W. Chan, C.K. Law, J.H. Eberly, Phys. Rev. A **68**, 022110 (2003)
283. R. Guo, H. Guo, Phys. Rev. A **73**, 012103 (2006)
284. C. Flindt, A.S. Sørensen, M.D. Lukin, J.M. Taylor, Phys. Rev. Lett. **98**, 240501 (2007)
285. J. Kessler, *Polarized Electrons* (Springer, New York, 1985)
286. S. Devons, J.B. Goldfarb, Handbuch der Physik **42**, 362 (1957)
287. H. Frauenfelder, H. Steffen, in *Alpha-, Beta-, and Gamma-Ray Spectroscopy*, vol. II., edited by K. Siegbhan (North-Holland, Amsterdam, 1965), p. 997
288. W.T. Silfvast, IEEE J. Selected Topics in Quantum Electronics **6**, 1400 (2000)
289. R.B. Cairns, H. Harrison, R.I. Schoen, J. Chem. Phys. **51**, 5440 (1969)

290. H. Harrison, J. Chem. Phys. **52**, 901 (1970)
291. G.V. Marr, J.M. Austin, Proc. Roy. Soc. A **310**, 137 (1969)
292. G. Schonhense, J. Phys. B **14**, L187 (1981)
293. S.P. Shannon, K. Codling, J. Phys. B **11**, 1193 (1978)
294. S.L. Carter, H.P. Kelly, J. Phys. B **11**, 2467 (1978)
295. E.J. McGuire, Phys. Rev. **175**, 20 (1968)
296. C.E. Theodosiou, J. Phys. B **12**, L673 (1979)
297. B.R. Tambe, W. Ong, S.T. Manson, Phys. Rev. A **23**, 799 (1981)
298. C.E. Theodosiou, A.F. Starace, B.R. Tambe, S.T. Manson, Phys. Rev. A **24**, 301 (1981)
299. T.E.H. Walker, J.T. Waber, J. Phys. B **7**, 674 (1974)
300. S. Flügge, W. Mehlhorn, V. Schmidt, Phys. Rev. Lett. **29**, 7 (1972)
301. E.G. Berezhko, N.M. Kabachnik, V.S. Rostovsky, J. Phys. B **11**, 1749 (1978)
302. K. Blum, Kleinpoppen, Phys. Rep. **52**, 203 (1979)
303. U. Fano, J.H. Macek, Revs. Mod. Phys. **45**, 553 (1973)
304. C.H. Greene, R.N. Zare, Phys. Rev. A **25**, 2031 (1982)
305. H. Klar, J. Phys. B **12**, L409 (1979)
306. H. Klar, J. Phys. B **13**, 2037 (1980)
307. H. Klar, J. Phys. B **15**, 4535 (1982)
308. C.D. Caldwell, R.N. Zare, Phys. Rev. A **16**, 255 (1977)
309. W. Mauser, W. Mehlhorn, Extended Abstract of the VI International Conference on Vacuum Ultraviolet Radiation Physics, Charlottesville, Virginia, 1980 (unpublished), II-VII
310. N. Chandra, S. Sen, Europ. Phys. J. D **6**, 457 (1999)
311. S. Sen, N. Chandra, Phys. Rev. A **62**, 052702 (2000)
312. G. Dujardin, M.J. Besnard, L. Hellner, Y. Malinovitch, Phys. Rev. A **35**, 5012 (1987)
313. R. Dörner, H. Bräuning, O. Jagutzki1, V. Mergell, M. Achler, R. Moshammer, J. M. Feagin, T. Osipov, A. Bräuning-Demian, L. Spielberger, J. H. McGuire, M. H. Prior, N. Berrah, J.D. Bozek, C. L. Cocke, H. Schmidt-Böcking, Phys. Rev. Lett. **81**, 5776 (1998)
314. R.I. Hall, G. Dawber, A. McConkey, M.A. MacDonald, G.C. King, Phys. Rev. Lett. **68**, 2751 (1992)
315. H. Kossmann, O. Schwarzkopf, B. Kämmerling, V. Schmidt, Phys. Rev. Lett. **63**, 2040 (1989)
316. T. Masuoka, E. Nakamura, Phys. Rev. A **48**, 4379 (1993)
317. F. Penent, R. I. Hall, R. Panajotović, J. H. D. Eland, G. Chaplier, P. Lablanquie, Phys. Rev. Lett. **81**, 3619 (1998)
318. H. Le Rouzo, J. Phys. B **19**, L677 (1986)
319. H. Le Rouzo, Phys. Rev. A **37**, 1512 (1988)
320. H.R. Sadeghpour, A. Dalgarno, Phys. Rev. A **47**, R2458 (1993)
321. N. Chandra, Chem. Phys. Lett. **237**, 545 (1995)
322. I. Powis, T. Baer, C.Y. Ng (eds.), *High Resolution Laser Photoionization and Photoelectron Studies* (Wiley, New York, 1995)
323. D.Dill, J.R. Swanson, S. Wallace, J.L. Dehmer, Phys. Rev. Lett. **45**, 1393 (1980)
324. O. Hemmers, F. Heiser, J. Eiben, R. Wehlitz, U. Becker, Phys. Rev. Lett. **71**, 987 (1993)
325. K. *Zä*hringer, H.-D. Meyer, L.S. Cederbaum, Phys. Rev. A **46**, 5643 (1992)
326. K. Bonhoff, S. Nahrup, B. Lohmann, K. Blum, J. Chem. Phys. **104**, 7921 (1996)
327. J. Lehmann, K. Blum, J. Phys. B **30**, 633 (1997)
328. N. Chandra, S. Sen, Phys. Rev. A **48**, 2084 (1993)
329. N. Chandra, S. Sen, Phys. Rev. A **52**, 2820 (1995)
330. G. Öhrwall, P. Baltzer, J. Bozek, Phys. Rev. Lett. **81**, 546 (1998)
331. G. Öhrwall, P. Baltzer, Phys. Rev. A **58**, 1960 (1998)
332. G. Öhrwall, P. Baltzer, J. Bozek, Phys. Rev. A **59**, 1903 (1999)
333. N.A. Cherepkoc, V.V. Kuznetsov, V.A. Verbitskii, J. Phys. B **28**, 1221 (1995)
334. A.N. Grum-Grzhimailo, M. Meyer, Eur. Phys. J. Spec. Top., **169**, 43 (2009)
335. K.J. Kollath, J. Phys. B **13**, 2901 (1980)
336. H. Klar, H. Kleinpoppen, J. Phys. B **15**, 933 (1982)
337. A.N. Grum-Grzhimailo, D. Cubaynes, E. Heinecke, P. Hoffmann, P. Zimmermann, M. Meyer, J. Phys. B **43**, 201002 (2010)

338. A.N. Grum-Grzhimailo, E.V. Gryzlova, D. Cubaynes, E. Heinecke, M. Yalcinkaya, P. Zimmermann, M. Meyer, J. Phys. B **42**, 171002 (2009)
339. M. Meyer, D. Cubaynes, F.J. Wuilleumier, E. Heinecke, T. Richter, P. Zimmermann, S.I. Strakhova, A.N. Grum-Grzhimailo, J. Phys. B **39**, L153 (2006)
340. B. Ritchie, Phys. Rev. A **12**, 567 (1975)
341. N.A. Cherepkoc, G. Schonhense, Eurphys. Lett. **24**, 79 (1993)
342. L. Baungarten, C.M. Schneider, H. Petersen, F. Schäfers, J. Kirschner, Phys. Rev. Lett. **65**, 492 (1990)
343. N.A. Cherepkov, V.V. Kuznetsov, J. Phys. B **25**, L445 (1992)
344. Ch. Roth, H.B. Rose, F.U. Hillebrecht, E. Kisker, Solid State Commun. **86**, 647 (1993)
345. D.M. Greenberger, M. Horne, A. Zeilinger, in *Bell's Theorem, Quantum Theory, and Conceptions of the Universe*, edited by M. Kafatos (Springer, Germany, 2010), p. 69
346. D.M. Greenberger, M.A. Horne, A. Shimony, A. Zeilinger, Am. J. Phys. **58**, 1131 (1990)
347. W. Dür, G. Vidal, J.I. Cirac, Phys. Rev. A **62**, 062314 (2000)
348. J.H. Eberly, T. Yu, Science **316**, 555 (2007)
349. J-H Huang, L-G Wang, S-Y Zhu, Phys. Rev. A **81**, 064304 (2010)
350. T. Yu, J.H. Eberly, Optics Commun. **264**, 393 (2006)
351. T. Yu, J.H. Eberly, Science **323**, 598 (2009)
352. T. Yu, J.H. Eberly, Optics Commun. **283**, 676 (2010)
353. J.-S. Xu, C.-F. Li, M. Gong, X.-B. Zou, C.-H. Shi, G. Chen, G.-C. Guo, Phys. Rev. Lett. **104**, 100502 (2010)
354. J. von Neumann, Göttinger Nachrichten **1**, 245 (1927)
355. U. Kleinman, B. Lohmann, K. Blum, J. Phys. B **32**, 309 (1999)
356. A. Ghatak, *Optics* (Tata McGraw-Hill, New Delhi, 2005)
357. C. Anastopoulos, B.L. Hu, Phys. Rev. A **62**, 033821 (2000)
358. D. Braun, F. Haake, W.T. Strunz, Phys. Rev. Lett. **86**, 2913 (2001)
359. E. Joos, H.D. Zeh, Z. Phys. B 9extbf59, 223 (1985)
360. W.H. Zurek, Phys. Today **44**, 36 (October 1991)
361. W.H. Zurek, Prog. Theor. Phys. **89**, 281 (1993)
362. N. Chandra (unpublished)
363. G. Racah, Phys. Rev. **438**, 62 (1942)
364. A.P. Jucys, A.A.Bandzaitis, *Theory of Angular Momentum in Quantum Mechanics* (Mokslas, Vilnuis, 1977)
365. S.L. Altmann, P. Herzig, *Point-Group Theory Tables*, (Clarendon Press, Oxford, 1994)

Index

N. Chandra and R. Ghosh, *Quantum Entanglement in Electron Optics*, Springer Series on Atomic, Optical, and Plasma Physics 67, DOI 10.1007/978-3-642-24070-6,

MIX
Papier aus verantwortungsvollen Quellen
Paper from responsible sources
FSC® C105338

If you have any concerns about our products,
you can contact us on
ProductSafety@springernature.com

In case Publisher is established outside the EU,
the EU authorized representative is:
Springer Nature Customer Service Center GmbH
Europaplatz 3, 69115 Heidelberg, Germany

Printed by Libri Plureos GmbH
in Hamburg, Germany